Health and Quality Control in Food Industry

Edited by **Lisa Jordan**

New York

Published by Callisto Reference,
106 Park Avenue, Suite 200,
New York, NY 10016, USA
www.callistoreference.com

Health and Quality Control in Food Industry
Edited by Lisa Jordan

International Standard Book Number: 978-1-63239-417-0 (Hardback)

Printed in the United States of America.

Contents

Preface

The main aim of this book is to educate learners and enhance their research focus by presenting diverse topics covering this vast field. This is an advanced book which compiles significant studies by distinguished experts in the area of analysis. This book addresses successive solutions to the challenges arising in the area of application, along with it; the book provides scope for future developments.

Substantial information regarding food, quality control and health is provided in this book. Food is an essential necessity for the survival of humankind and this book presents an extensive analysis of food industry that runs the supply of food, its current trends and the way forward to ensure food security and health. The global food industry serves a population of seven billion people and has the biggest market. This book highlights the issues of world hunger and discusses how food shortages endanger water and energy supply. It is organized under two sections: health aspects and quality control. Food producers, industry experts, corrosion practitioners, academicians and designers of food processing equipment have made valuable contributions to this book based on their extensive knowledge and experience. They present different perspectives and approaches in the diverse aspects of food science and technology.

It was a great honour to edit this book, though there were challenges, as it involved a lot of communication and networking between me and the editorial team. However, the end result was this all-inclusive book covering diverse themes in the field.

Finally, it is important to acknowledge the efforts of the contributors for their excellent chapters, through which a wide variety of issues have been addressed. I would also like to thank my colleagues for their valuable feedback during the making of this book.

Editor

Part 1

Health Aspects

Yeast, the Man's Best Friend

Joana Tulha, Joana Carvalho, Rui Armada,
Fábio Faria-Oliveira, Cândida Lucas, Célia Pais,
Judite Almeida and Célia Ferreira
University of Minho/CBMA (Centre of Molecular and Environmental Biology)
Portugal

1. Introduction

In most cultures, bread making depends on a fermentation step. The flour leavening ability was, at first, most probably dependent on spontaneous fermentation. It became a controlled process by the maintenance of fresh innocula from one preparation to the next and this kind of environmental constraints eventually generated a particular type of yeast and bacteria biodiversity, adapted to ferment a certain type of flour mixture, conferring specific organoleptic characteristics to the dough. Nowadays, although some types of bread are still prepared using dough carried over from previous makings as a starter, the baking industry generally uses commercially available strains of *Saccharomyces cerevisiae* for bread making. Although the flour types, geographical origin and mixtures introduce organoleptic diversity in bread, the leavening is a crucial step in order to achieve the traditional specific flavours and textures of each population and region. The trend of globalization of baker's yeast market decreased worldwide bread diversity and the cultural values associated, simultaneously increasing the dependence of local producers on world-scale yeast producers. Sustainability demands assessing yeast biodiversity and devising simple and cheap methods for maintaining dough and multiply yeast. The following sub-chapters address these possibilities.

1.1 Biodiversity at the bakery

The better example of traditional practises still available is the use of sourdough, an extremely diverse fermented product widely used for the production of bread and sweet leavened baked goods. The production of sourdough bread can be traced back to ancient times (Rothe et al., 1973). The products are characterized by their unique flavour, enhanced shelf life and nutritional value, and favourable technological properties (Hämmes & Gänzle, 1998; Salovaara, 1998). Traditionally, sourdoughs have been used to produce many types of bread with rye, maize or wheat flours. This variety underlies the generalized utilization of the term sourdough as a synonym of leavening bread found in the literature. Although the primary purpose of the sourdough is leavening by the yeasts, a simultaneous souring action also takes place due to the activity of the lactic acid bacteria (LAB) present, resulting in bread with a good grain, an elastic crumb and, usually, the characteristic sensory quality of sourdough bread (Gobbetti et al., 1994a).

1.2 Sourdoughs: A yeast and bacteria synergistic ecosystem

Three types of sourdough have been defined based on common principles used in artisanal and industrial processes (Bocker et al., 1995). One type is produced with traditional techniques and is characterized by continuous propagation to keep the microorganisms in an active state, as indicated by high leavening ability. Examples of baked goods so obtained are San Francisco sourdough, French bread, the Italian panettone, and three-stage sourdough rye bread. The industrialization of the baking process for rye bread led to the development of another type of sourdoughs, which serve mainly as dough acidifiers. These sourdoughs are fermented for long periods (up to 5 days) at temperatures of 30° C, and high dough yields permit pumping of the dough. The microorganisms are commonly in the late stationary phase and therefore exhibit restricted metabolic activity only. Another type of sourdoughs is dried doughs, which are used as acidifier supplements and aroma carriers. These two last types of dough require the addition of baker's yeast for leavening (Meroth et al., 2003a, 2003b). Although the fermentation process runs under nonaseptic conditions, microbial associations present may last for years, as shown for certain industrial sourdough processes (de Vuyst & Vancanneyt, 2007).

Hundreds of different types of traditional sourdough breads exist in Europe, in particular in Italy. They differ in the type of flour, other ingredients, and the applied technology and fermentation process. Because of their artisan and region-dependent handling, sourdoughs are an important source of diverse LAB species and strains that are metabolically active or can be reactivated upon addition of flour and water. Some of these strains play a crucial role during the sourdough fermentation process and are or can be used as sourdough starters. Sourdough is a unique food ecosystem in that it (i) selects for LAB strains that are adapted to their environment, and (ii) harbours LAB species specific for sourdough (Dal Bello et al., 2005; de Vuyst & Neysens, 2005; Gobbetti et al., 2005).

Sourdoughs are stable ecosystems obtained using daily propagations of LAB and yeasts. Dough acidification is due mostly to homofermentative and heterofermentative LAB species, mainly belonging to the *Lactobacillus* genus (Hämmes & Vogel, 1995), while yeasts are primarily responsible for the leavening action through the production of carbon dioxide and consequent increase in dough volume.

The microorganisms present in sourdough usually originate from flour, dough ingredients or the environment. The variety and number of species in the dough are influenced by several endogenous and exogenous factors, such as type of flour, temperature and time of fermentation, redox potential, and length of time that the "starter dough" has been maintained (Hämmes & Gänzle, 1998). In fact, strong effects are exerted by process parameters such as dough yield, amount and composition of the starter, number of propagation steps, and fermentation time. The impact of these parameters during continuous propagation of sourdough causes the selection of a characteristic microflora consisting of LAB and usually yeasts (Gobbetti et al., 1994b).

Extensive research efforts have been directed towards the study of the species diversity and identification of lactic acid bacteria involved in sourdough fermentation processes. Lactobacilli, obligatory homofermentative and facultative or obligatory heterofermentative, are the typical sourdough LAB. These usually belong to the genus *Lactobacillus*, but occasionally *Leuconostoc* spp., *Weissella* spp., *Pediococcus* spp., and *Enterococcus* spp. have been found. In general, heterofermentative *Lactobacillus* species dominate the sourdough microbiota (de Vuyst & Neysens, 2005). *Lactobacillus sanfranciscensis* (Trüper & De' Clari, 1997), *Lb. plantarum* and *Lb. brevis* are the most frequently isolated lactobacilli.

Recent biodiversity studies of particular sourdough ecosystems throughout Europe resulted in the description of new LAB species. During the last years, several new LAB species have been isolated from traditional sourdoughs that were continuously propagated by back-slopping (repeated cyclic re-inoculation) at ambient temperature: *Lb. mindensis, Lb. spicheri, Lb. rossiae, Lb. zymae, Lb. acidifarinae, Lb. hammesii,* and *Lb. nantensis.* Some of these species have been described on one single isolate only. The distribution of the taxa of LAB is highly variable from one sourdough ecosystem to another. Therefore, it is difficult to define correlations between population composition and both the type of sourdough or the geographic location (Gobbetti, 1998).

Adaptations of certain LAB to a sourdough environment include (i) a unique central metabolism and/or transport of specific carbohydrates such as maltose and fructose, being maltose the most abundant fermentable carbohydrate and fructose being an important alternative electron acceptor; (ii) an activated proteolytic activity and/or arginine deiminase pathway; (iii) particular stress responses; and (iv) production of antimicrobial compounds. For instance, dough acidification is a prerequisite for rye baking to inhibit the flour α-amylase (de Vuyst & Vancanneyt, 2007).

Although LAB initially isolated from sourdough are not necessarily unique for sourdough ecosystems, some correlations can be seen between specific LAB species and the type of sourdough, and sometimes the origin of the sourdough. In practice, sourdoughs are either continuously propagated by using a piece of dough from the preceding fermentation process, or produced by using once a week a commercial starter followed by back-slopping for several days. Therefore, large differences can often be seen in species composition within and among sourdough types (de Vuyst &Vancanneyt, 2007).

This is well illustrated in a more recent study of spontaneously fermented wheat sourdoughs from two regions of Greece, where a total of 136 lactic acid bacteria strains were isolated. *Lactobacillus sanfranciscensis* were dominant in the sourdoughs from Thessaly and *Lb. plantarum* sub spp. *plantarum* in the sourdoughs from Peloponnesus. The latter was accompanied by *Pediococcus pentosaceus* as secondary microbiota. In this case, none of the lactic acid bacteria strains isolated produced antimicrobial compounds (Paramithiotis et al., 2010).

1.3 The particular roles of yeasts in sourdoughs

Several of the most important functions in bread making are fulfilled by yeast. They contribute to leavening and produce metabolites such as alcohols, esters, and carbonyl compounds, which are involved in the development of the characteristic bread flavour (Corsetti et al., 1998; Damiani et al., 1996; B. Hansen & Å. Hansen, 1994; Martinez-Anaya et al., 1990a, 1990b). Furthermore, the enzymatic activities of yeasts by enzymes such as proteases, lecithinases, lipases α-glucosidase, β-fructosidase, and invertase have an influence on the dough stickiness and rheology as well as on the flavour, crust colour, crumb texture, and firmness of the bread (Antuna & Martinez-Anaya, 1993; Collar et al., 1998; Meroth et al., 2002).

S. cerevisiae is the species most frequently found in sourdoughs but several other yeast species may be present in these ecosystems. In early studies the amount of *S. cerevisiae* may have been overestimated due to the lack of reliable systems for identifying and classifying yeasts from this habitat (Vogel, 1997). To study the sourdough yeast microbiota traditional cultivation methods in combination with phenotypic (physiological and biochemical) and/or genotypic (randomly amplified polymorphic DNA [RAPD]-PCR and restriction fragment length polymorphism [RFLP] analysis) identification methods have commonly been used (Corsetti et al., 2001; Galli et al., 1988; Mäntynen et al., 1999; Paramithiotis et al., 2000; Rocha & Malcata, 1999). These studies focused mainly on the characterization of ripe doughs and revealed the presence of 23 yeast species belonging especially to the genera *Saccharomyces* and *Candida* (Brandt, 2001; Ottogalli et al., 1996; Rossi, 1996). In particular *S. exiguus* (imperfect state *Torulopsisholmii* or *Candida holmii*, physiologically similar to *C. milleri*), and *C. krusei*, *Pichia norvegensis* and *P. anomala* are yeasts associated with LAB in sourdoughs. The LAB/yeast ratio in sourdoughs is generally 100:1 (Gobbetti et al., 1994a). Like other fermented foods produced by mixed microflora, the organoleptic, health and nutritional properties of baked sourdough goods depend on the cooperative activity of LAB and yeasts (Gobbetti, 1998). No data are available on the competitiveness of yeasts; thus, the effects of ecological factors and process conditions on the development of yeast biota during sourdough fermentation processes are virtually unknown (Meroth et al., 2003a). In a simulation of the complex natural sourdough ecosystem, the competition for substrates was studied in model co-cultures showing that yeasts only partially compete with the LAB for the nitrogen sources present and synthesize and excrete essential and stimulatory amino acids which enhance the cell yield of the LAB (Gobetti et al., 1994a). These findings contribute to the interpretation of some of the complex interactions which occur during sourdough leavening and which are difficult to understand because of the extensive proteolytic activity that occurs in sourdough (Spicher & Nierle, 1984).

Several recent studies have given emphasis to the yeast microbiota associated with spontaneous sourdough fermentations (Paramithiotis et al., 2010; Valmorri et al., 2010; Vrancken et al., 2010). A total of 167 yeast and 136 lactic acid bacteria strains were isolated from spontaneously fermented wheat sourdoughs from two regions of Greece, namely Thessaly and Peloponnesus. Identification of the isolates exhibited dominance of *Torulaspora delbrueckii* with sporadic presence of *S. cerevisiae* (Paramithiotis et al., 2010). In another study, conducted in 20 sourdoughs collected from central Italy, PCR-RFLP analysis identified 85% of the isolates as *S. cerevisiae*, with the other dominant species being *C. milleri* (11%), *C. krusei* (2.5%), and *T. delbrueckii* (1%). RAPD-PCR analysis performed with primers M13 and LA1, highlighted intraspecific polymorphism among the *S. cerevisiae* strains. The diversity of the sourdoughs from the Abruzzo region is reflected in the chemical composition, yeast species, and strain polymorphism (Valmorri et al., 2010). In contrast with the Greek study, the high presence of *S. cerevisiae* had already been reported in Italian sourdoughs by other authors (Corsetti et al., 2001; Gobbetti et al., 1994b; Iacumin et al., 2009; Pulvirenti et al., 2004; Succi et al., 2003). The current general opinion is that cross contamination from bakery equipment and working environment by baker's yeast is commonly associated with the presence of *S. cerevisiae* in sourdoughs. The other species detected (*Candida milleri*, *C. krusei* and *Torulaspora delbrueckii*) are typically associated with sourdoughs (Corsetti et al., 2001; Garofalo et al., 2008; Gobbetti et al., 1994b; Halm et al., 1993; Iacumin et al., 2009; Obiri-Danso, 1994; Ottogalli et al., 1996; Pulvirenti et al., 2004; Rossi, 1996; Succi et al., 2003; Sugihara et al., 1971; Vernocchi et al., 2004a, 2004b).

In rural areas in the north of Portugal a corn and rye bread is still prepared using a piece of dough usually kept in cool places, covered with a layer of salt. Prior to bread making this piece of dough is mixed with fresh flour and water and, when fully developed, serves as the inoculum for the bread dough. This starter dough is a natural biological system characterized by the presence of yeast and lactic acid bacteria living in complex associations in a system somewhat similar to that existing in sourdough. In a survey carried in 33 dough samples from farms mainly located in the north of Portugal, 73 yeast isolates were obtained belonging to eight different species. The predominant species was *S. cerevisiae* but other yeasts also occurred frequently, among which *Issachenkia orientalis, Pichia membranaefaciens* and *Torulaspora delbrueckii* were the most abundant, being present in about 40% of the doughs examined. Only six of the doughs contained a single yeast species. Associations of two species were found in 48% of the bread doughs, 30% presented three different species and the remainder consisted of a mixture of four yeast species. Associations of *S. cerevisiae* and *T. delbrueckii, I. orientalis* and/ or *P. membranaefaciens* were the most frequent. All mixed populations included at least one fermentative species with the exception of the association between *P. anomala, P. membranaefaciens*and *I. orientalis,* which was found in one of the doughs (Almeida & Pais, 1996a). Apparently this dough is somehow similar to the San Francisco sour dough in which maltose-negative *S. exiguus* is predominantly found and the fermentation may be carried out by lactic acid bacteria (Sugihara et al., 1971). In another Portuguese study in which, besides sourdough, maize and rye flour were examined the most frequently isolated yeasts were *S. cerevisiae* and *C. pelliculosa* (Rocha & Malcata, 1999).

In conclusion, yeasts and lactic acid bacteria (LAB) are often encountered together in the fermentation of wheat and rye sourdough breads. To optimize control of the fermentation, there has been an increased interest in understanding the interactions that occur between the LAB and yeasts in the complex biological ecosystem of sourdough.

2. Sustainability: The old made new

Sustainability aims are all about using simple ideas, mixing with old procedures and new materials, adding inventive solutions, generating innovation. In the baking market, in particular in the baking industry, there is considerable space for improvement. The present procedure of bread making in developed countries consists of using block or granular baker's yeast identically produced all around the world. Consequently, the above mentioned old procedure of using the previous leaven to the next leavening step was lost as baking became progressively an industrialized process. Producing and conserving large amounts of yeast requires energy wasting biotechnological plants and expensive technical support, favouring the standardization and centralization of production. Additionally, the conservation processes involving freezing temperatures compromise *S. cerevisiae* viability but also its desirable leavening ability and organoleptic properties.

2.1 Frozen yeast and frozen dough
2.1.1 Yeast response to cold
All living organisms, from prokaryotes to plants and higher eukaryotes are exposed to environmental changes. Cellular organisms require specific conditions for optimal growth and function. Growth is considered optimal when it allows fast multiplication of the cells, and the preservation of a favourable cell/organism internal composition, *i.e.,* homeostasis.

Therefore, any circumstance that provokes unbalance in a previous homeostatic condition may generally be considered stressful, as is the case of sudden changes in the external environment. These generally cause disturbances in the metabolism/regulation of the cells, tissues or organs, eventually disrupting their functions and preventing growth. Cellular organisms have to face this constant challenge and, therefore, rapidly adapt to the surroundings, adjusting their internal milieu to operate under the new situation. For this purpose, uncountable strategies have been developed to sustain the homeostasis. Whereas, multicellular organisms can make use of specialized organs and tissues to provide a relatively stable and homogenous internal environment, unicellular organisms have built up independent mechanisms in order to adjust to drastic environmental changes. Several approaches have been described for the most diverse microbes, from bacteria to fungi, involving responses at the level of gene expression as well as metabolism adaptation by faster processes like protein processing, targeting and inactivation, or iRNA interference (K.R. Hansen et al., 2005), just to mention the more general processes.

Yeasts, in particular, in their natural habitat can be found living in numerous, miscellaneous and changeable environments, since they can live as saprophytes on, either plants, or animals. As examples we can name fruits and flowers, humans, animals, etc. Likewise, in their substrates, yeasts are also exposed to highly variable milieus. On such diverse ambiences, it can be expected that yeasts regularly withstand fluctuations in the types and quantities of available nutrients, acidity and osmolarity, as well as temperature of their environment. In fact, the most limiting factors cells have to cope are the low water activity (a_w), i.e. availability of water, and temperature. Being yeast unicellular organisms, cell wall and plasma membrane are the first barriers to defeat environment and its alterations. Both changes on the water content and temperature lead to physical and functional modifications on plasma membrane, altering its permeability that are on the basis of cell lyses and ultimately cell death (D'Amico et al., 2006; Simonin et al., 2007).

Actually, the variations on the permeability of plasma membrane, attributed to transitions of the phospholipid phase in the membranes (Laroche & Gervais, 2003), are associated to loss of viability during dehydration/rehydration stress (Laroche & Gervais, 2003; Simonin et al., 2007). In a physical perspective, membrane phospholipid bilayers, under an optimum temperature level and favourable availability of water, are supposedly in a fluid lamellar liquid-crystalline phase. When temperature levels drop or under any other cause of dehydration, such organization suffers alterations as the hydrophilic polar head groups of phospholipids compulsorily gather. This phenomenon leads to the loss liquid-crystalline regular phase and conversion into a gel phase and consequent reduction of membrane fluidity (D'Amico et al., 2006; Simonin et al., 2007; Aguilera et al., 2007).

Still, a decline in temperature has other effects besides the reduction in membrane fluidity. Aside with the alterations on plasma membrane permeability (primarily but on the other physiological membranes as well) and hence changes on the transport of nutrients and waste products occurs the formation of intracellular ice crystals, which damage all cellular organelles and importantly reduces the a_w, under near-freeze temperatures. Furthermore, it has been evoked that temperature downshifts can cause profound alterations on protein biosynthesis, alterations in molecular topology or modifications in enzyme kinetics (Aguilera et al., 2007). Other crucial biological activities involving nucleic acids, such as DNA replication, transcription and translation can also suffer from exposure to low temperatures. This happens through the formation and stabilization of RNA and RNA secondary or super-coiled structures (D'Amico et al., 2006; Simonin et al., 2007; Aguilera et al., 2007). In turn, the stabilization of secondary structures of RNAs takes place, for instance, at the level of the inhibition of the expression of several genes that would be unfavourable for cell growth at low temperatures (Phadtare & Severinov, 2010). The latter occurs since the transcription of the mentioned genes is impaired, as well as the RNA degradation becomes ineffective (Phadtare & Severinov, 2010).

In yeast, as in most organisms, the adaptive response to temperature downshifts, commonly referred to as the cold-shock response comprises orchestrated adjustments on the lipid composition of membranes and on the transcriptional and translational machinery, including protein folding. These adjustments are mostly elicited by a drastic variation in the gene expression program (Aguilera et al., 2007; Simonin et al., 2007). Still, some authors name cold-shock response to temperature falls in the region of 18- 10°C and near-freezing response to downshifts below 10°C. In fact, yeast cells appear to initiate quite different responses to one or another situation, which can be rationalized since yeast can actively grow at 10–18 °C, but growth tends to stop at lower temperatures (Al-Fageeh & Smales, 2006). To withdraw misinterpretations we will focus mainly in low/near-freezing temperatures, which is also a cold response. Some works on genome-wide expression analysis have explored the genetic response of *S. cerevisiae* to temperature downshifts (L. Zhang et al., 2001; Rodrigues-Vargas et al., 2002; Sahara et al., 2002; Murata et al., 2006). In *S. cerevisiae* exposed to low temperature, 4°C, together with the enhanced expression of the general stress response genes, other groups of genes were induced as well. These include genes involved in trehalose and glycogen synthesis (*TPS1, GDB1, GAC1*, etc.), which may suggest that biosynthesis and accumulation of those reserve carbohydrates are necessary for cold tolerance and energy preservation. Genes implicated on phospholipids biosynthesis (*INO1, OPI3*, etc.), seripauperin proteins (*PAU1, PAU2, PAU4, PAU5, PAU6* and *PAU7*) and cold shock proteins (*TIP1, TIR1*, etc.) displayed as well increased expression, which is consistent with membrane maintenance and increased permeability of the cell wall. Conversely, the observed induction of Heat Shock genes (*HSP12, HSP104, SSA4*, etc.) can possibly be linked with the demand of enzyme activity revitalization, and the induction of glutathione related genes (*TTR1, GTT1, GPX1*, etc.) required for the detoxification of active oxygen species. On the other hand, it is also described the down-regulation of some genes, like the ones associated with protein synthesis (*RPL3, RPS3*, etc.), reflecting the reduction of cell growth, which in turn may be a sign of a preparation for the following adjustment to the novel conditions (Fig. 1). A rationalization of all the data from genome-wide expression analysis and also from the numerous works on yeast cold response developed on the last years, led to the idea that there are two separated responses to temperature downshifts (Aguilera et al., 2007; Al-Fageeh & Smales, 2006). One is a general response, which involves certain clusters of genes. These include members of the *DAN/TIR* family encoding putative cell-wall mannoproteins, temperature shock inducible genes (*TIR1/SRP1, TIR2* and *TIR4*) and seripauperins family, which have some phospholipids interacting activity. The other is a time dependent separated response, meaning that the transcriptional profile changes are divided in a time succession (Aguilera et al., 2007; Al-Fageeh & Smales, 2006). For instance, within the first two hours would be observed an over-expression of genes involved in phospholipid synthesis (like *INO1, OPI3*, etc.), in fatty-acid desaturation (*OLE1*), genes related to transcription, including RNA helicases, polymerase subunits and processing proteins, and also some ribosomal protein genes.

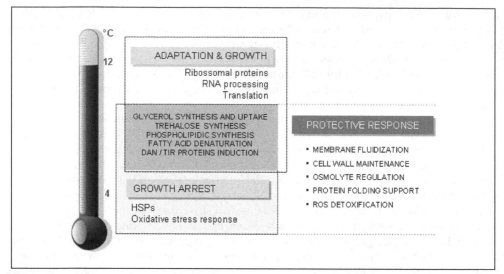

Fig. 1. *S. cerevisiae* major response to a temperature downshift (Adapted from Aguilera et al., 2007).

Whereas, in a second stage the latter genes (transcription related ones) are silenced and is promoted the induction of another set of genes, such as some of the heat shock protein (*HSP*) genes, also of genes associated with the accumulation of glycogen (*GLG1*, *GSY1*, *GLC3*, *GAC1*, *GPH1* and *GDB1*) and trehalose (*TPS1*, *TPS2* and *TSL1*), of genes in charge of the detoxification of reactive oxygen species (ROS) and defence against oxidative stress (including catalase, *CTT1*; glutaredoxin, *TTR1*; thioredoxin, *PRX1* and glutathione transferase, *GTT2*) (Aguilera et al., 2007; Al-Fageeh & Smales, 2006).

2.1.2 Improving baker's yeast frozen dough performance

Preservation by low temperatures is widely accepted as a suitable method for long-term storage of various types of cells. Specially, freezing has become an important mean of preservation and storage of strains used for many types of industrial and food processing, such as those used in the production of wine, cheese and bread. Bread, in particular, is a central dietary product in most countries of the world, and presently frozen dough technology is extensively used in the baking industry. Yet, the loss of leavening ability, and organoleptic properties, but mainly the loss of viability of the yeasts after thawing the frozen dough is a problem that persists nowadays.

In-depth knowledge concerning yeast genetics, physiology, and biochemistry as well as engineering and fermentation technologies has accumulated over the time, and naturally, there have been several attempts to improve freeze-thaw stress tolerance in *S. cerevisiae*. A recent work described that genes associated with the homeostasis of metal ions were upregulated after freezing/thawing process and that mutants in some of these genes, as *MAC1* and *CTR1* (involved in copper homeostasis), exhibited freeze-thaw sensitivity (Takahashi et al., 2009). Furthermore, the researchers showed that cell viability after freezing/thawing process was considerably improved by supplementing the broth with copper ions. Those results suggest that insufficiency of copper ion homeostasis may be one of the causes of freeze-thaw injury; yet, these ions toxicity does not allow their easy incorporation in food products. A very promising study reported an improved freeze-resistant industrial strain, in which the aquaporin was overexpressed (Tanghe et al., 2002). Nonetheless, this enhancement was not attained in larger dough preparations (under industrial conditions), wherein freezing rate is not that rapid (Tanghe et al., 2004). Another recent approach addressed the impact of unsaturated fatty acids on freeze-thaw tolerance by assaying the overexpression of two different desaturases (*FAD2-1* and *FAD2-3*) from sunflower in *S. cerevisiae*. This resulted into increased membrane fluidity and freezing tolerance (Rodriguez-Vargas et al., 2007). Also the heterologous expression of antifreeze proteins (antifreeze peptide GS-5 from the polar fish grubby sculpin (*Myxocephalusaenaeus*)) was tested in an industrial yeast strain, leading to both improved viability and enhanced gas production in the frozen dough (Panadero et al., 2005). A very current study confirmed the role of hydrophilins in yeast dehydration stress tolerance yeast cells, since overexpression of *YJL144W* and *YMR175W* (*SIP18*) become yeast more tolerant to desiccation and to freezing (Dang & Hincha, 2011). An alternative work, showed improved freezing resistance by expressing of *AZI1* (Azelaic acid induced 1) from *Arabidopsis thaliana* in *S. cerevisiae* (Xu et al., 2011). Other approaches devoid of genetic engineering were also taken. Cells were cultured in diverse conditions, including media with high concentration of trehalose or glycerol (Hirasawa et al., 2001; Izawa et al., 2004a); with poly-γ-glutamate (Yokoigawa et al., 2006), and with soy peptides (Izawa et al., 2007) acquiring improved tolerance to freeze-thaw stress and also retaining high leavening ability.

The benefits of cryoprotectants, substances that promote the excretion of water, thus decreasing the formation of ice crystals that happens during the freezing process, were also addressed. These include Me_2SO (Momose et al., 2010); proline (Terao et al., 2003; Kaino et al., 2008) and charged aminoacids as arginine and glutamate (Shima et al., 2003); trehalose (Kandror et al., 2004) as well as glycerol (Izawa et al., 2004a, 2004b; Tulha et al., 2010). A comparative analysis of yeast transcriptional responses to Me_2SO and trehalose revealed that exposure to cryoprotectants prior to freezing not only reduce the freeze-thaw damage, but also provide various process to the recovery from freeze-thaw injury (Momose et al., 2010). Yet, the use of Me_2SO in food preparation is not possible due to its toxicity. Intracellular proline accumulation was found to enhance freeze-thaw tolerance, thus several engineering strains emerged, overexpressing glutamyl metabolic related enzymes *PRO1* and *PRO2* or specific alleles (Terao et al., 2003), and self-cloned strains in which *PRO1* specific alleles combined with disruption of proline oxidase *PUT1* (Kaino et al., 2008). Moreover, it was shown that an arginase mutant (disrupted on *CAR1* gene) accumulates high levels of arginine and/or glutamate (depending on the cultivation conditions), with increased viability and leavening ability during the freeze-thaw process (Shima et al., 2003).

Trehalose and glycerol are not only cryoprotectants but also confer resistance to osmotic stress. A correlation between the intracellular trehalose content and freeze-thaw stress tolerance in *S cerevisiae* was described (Kandror et al., 2004). The same correlation has been made for glycerol (Izawa et al., 2004a, 2004b; Tulha et al., 2010). Furthermore, it has been reported that, beyond the cryoprotection, an increased level of intracellular glycerol has several benefits for the shelf life of wet yeast products and for the leavening activity (Myers et al., 1998; Hirasawa & Yokoigawa 2001; Izawa et al., 2004a) and no effect on final bread quality in terms of flavour, colour, and texture (Myers et al., 1998).

2.1.3 Role of glycerol for the baker's yeast frozen dough

S. cerevisiae accumulates intracellular glycerol as an osmolyte under osmotic stress but also under temperature (high and low) stress through the high osmolarity glycerol signaling pathway (HOG pathway) (Siderius et al., 2000; Hayashi & Maeda, 2006; Ferreira & Lucas, 2007; Tulha et al., 2010). Moreover, it was reported that a pre-treatment of yeast cells with osmotic stress was an effective way to acquire freeze tolerance, probably due to the intracellular glycerol accumulation attained. Some engineering approaches were performed in order to increase the intracellular glycerol accumulation in baker's yeast. For instance, Izawa and co-authors (Izawa et al., 2004a) showed that the quadruple mutant on the glycerol dehydrogenase genes ($ara1\Delta gcy1\Delta gre3\Delta ypr1\Delta$), responsible for the alternative pathway of glycerol dissimilation (Fig. 2) has an increased level of intracellular glycerol with concomitant freeze-thaw stress resistance. Similarly, the overexpression of the isogenes *GPD1* and *GPD2* that encode for glycerol-3-phosphate dehydrogenase (Fig. 2) (Ansell et al., 1997) also lead to an increase in intracellular glycerol levels (Michnick et al., 1997; Remize et al., 1999) and probably improved freeze-thaw tolerance. One of the most promising genetic modifications was the deletion of *FPS1* encoding the yeast glycerol channel. Fps1p channel opens/closes, regulating extrusion and retention of massive amounts of glycerol in response to osmotic hyper- or hipo-osmotic shock (Luyten et al., 1995; Tamás et al., 1999). The engineered cells deleted on *FPS1* showed an increased intracellular glycerol accumulation accompanied by higher survival after 7 days at -20° C (Izawa et al., 2004b). Yet, the dynamics of the channel under this type of stress remains unexplored. The mentioned study was considered quite innovative, it was even suggested the possibility that the *fps1Δ* mutant strain could be applicable to frozen dough technology. This because the *fps1Δ* mutant strain displayed the higher intracellular glycerol content attained so far, and (similarly to the previous engineered strains) avoided the exogenous supply of glycerol into the culture medium, which was at the time too expensive for using at an industrial scale. Our group has recently described a simple recipe with high biotechnological potential (Tulha et al., 2010), which also avoids the use of transgenic strains. We found that yeast cells grown on glycerol based medium and subjected to freeze-thaw stress displayed an extremely high expression of the glycerol/H$^+$ symporter, Stl1p (Ferreira et al., 2005), also visible at activity level. This permease plays an important role on the fast accumulation of glycerol; under those conditions, the strains accumulated more than 400 mM glycerol (whereas the mutant *stl1Δ* presented less than 1 mM) and survived 25-50% more. Therefore, any *S. cerevisiae* strain already in use can become more resistant to cold/freeze-thaw stress just by simply adding glycerol (presently a cheap substrate) to the broth. Moreover, as mentioned above glycerol also improves the leavening activity and has no effect on final bread quality (Izawa et al., 2004a; Myers et al., 1998).

3. Low-cost yeasts, a new possibility

The industrial production of baker's yeast is carried out in large fermentors with working volumes up to 200.000 l, using cane or sugar beet molasses as carbon source. These are rich but expensive substrates. Quite the opposite, glycerol, once a high value product, is fast becoming a waste product due to worldwide large surplus from biofuels industry, with disposal costs associated (Yazdani & Gonzales, 2007). Glycerol represents approximately 10% of the fatty acid/biodiesel conversion yield. Due to its chemical versatility, glycerol has countless applications, yet, new applications have to be found to cope with the amounts presently produced. This underlies the global interest for glycerol, which became an attractive cheap substrate for microbial fermentation processes (Chatzifragkou et al., 2011).

3.1 Metabolism of glycerol in yeasts

A significant number of bacteria are able to grow anaerobically on glycerol (da Silva et al., 2009). In the case of yeasts, most of the known species can grow on glycerol (Barnett et al., 2000) but this is achieved under aerobic conditions. *S. cerevisiae* is a poor glycerol consumer, presenting only residual growth on synthetic mineral medium with glycerol as sole carbon and energy source. In order to obtain significant growth on this medium a starter of 0.2% (w/v) glucose is needed (Sutherland et al., 1997). Yet, glycerol is a very important metabolite in yeasts, including *S. cerevisiae*. Importantly, its pathway is central for bulk cell redox balance, because it couples the cytosolic potential with mitochondria's. Furthermore, glycerol is the only osmolyte known to yeasts, in which accumulation, cells depend for survival under high sugar, high salt (Hohmann, 2009), high and low temperature (Siderius et al., 2000), anaerobiosis and oxidative stress (Påhlman et al., 2001).

Recently, it was suggested that *S. cerevisiae* glycerol poor consumption yields could be due to a limited availability of energy for gluconeogenesis, and biomass synthesis (X. Zhang et al., 2010). Nevertheless, the weak growth performances have long been attributed to a redox unbalance caused by the intersection of glycerol pathway with glycolysis at the level of glycerol-P shuttle (Fig. 2) (Larsson et al., 1998). Fermenting cultures of *S. cerevisiae* produce glycerol to reoxidize the excess NADH generated during biosynthesis of aminoacids and organic acids, since mitochondrial activity is limited by oxygen availability, and ethanol production is a redox neutral process (van Dijken & Scheffers, 1986) (Fig. 2). This is the reason why glycerol is a major by-product in ethanol and wine production processes. Consistently, the mutant defective in the above mentioned isogenes encoding the glycerol 3-P dehydrogenases *(Δgpd1Δgpd2)* is not able to grow anaerobically (Ansell et al., 1997; Påhlman et al., 2001). This ability was partially restored supplementing the medium with acetic acid as electron acceptor (Guadalupe Medina et al., 2010).

S. cerevisiae takes up glycerol through the two transport systems above mentioned, the Fps1 channel and the Stl1 glycerol/H^+ symporter (Ferreira et al., 2005). Fps1p is expressed constitutively (Luyten et al., 1995; Tamás et al., 1999), while *STL1* is complexly regulated by a number of conditions (Ferreira et al., 2005; Rep et al., 2000). It is derepressed by starvation, and inducible by transition from fermentative to respiratory metabolism, as happens during diauxic shift at the end of exponential growth on rich carbon sources. Additionally, it is also the most expressed gene under hyper-osmotic stress (Rep et al., 2000), highly expressed at high temperature, overcoming glucose repression (Ferreira & Lucas, 2007) and under low-near-freeze temperatures (Tulha et al., 2010). In yeasts glycerol can be consumed through two alternative pathways (Fig. 2), the most important of which involving the glycerol 3-P shuttle above mentioned, directing glycerol to dihydroxyacetone-P through respiration and mitochondria. According to very disperse literature, other yeasts, better glycerol consumers than *S. cerevisiae*, appear to have equivalent pathways, though they should differ substantially in the underlying regulation to justify the better performance. At the level of transcription, significant ability to consume glycerol depends on the constitutive expression of active transport (Lages et al., 1999). Possibly, unlike in *S. cerevisiae* where *GUT1* is under glucose repression (Rønnow & Kielland-Brandt, 1993; Grauslund et al., 1999), glycerol consumption enzymes could be identically regulated. This should be in accordance with the yeasts respiratory/fermentative ability. Related or not, *S. cerevisiae* respiratory chain differs from a series of other yeasts classified as respiratory, which are resistant to cyanide (CRR - Cyanide resistant respiration) (Veiga et al., 2003). Cyanide acts at the level of Cytochrome Oxidase complexes. CRR owes its resistance to an alternative oxidase (AOX) that short-circuits the main respiratory chain, driving electrons directly from ubiquinone to oxygen, by-passing complex III and IV. Although exhaustive data are not available, CRR appears to occur quite frequently in yeasts that are Crabtree[1] negative or simply incapable of aerobic fermentation (Veiga et al., 2003), all of which are good glycerol consumers (Lages et al., 1999; Barnett et al., 2000). Interestingly, CRR may not be constant, occurring only under specific physiological conditions like diauxic shift, in *P. membranifaciens* and *Y. lipolytica*, or early exponential phase, in *D. hanseni* (Veiga et al., 2003). In *S. cerevisiae*, both conditions highly and transiently induce the glycerol transporter *STL1* expression (Ferreira et al., 2005; Rep et al., 2000, Lucas C. unpublished results).

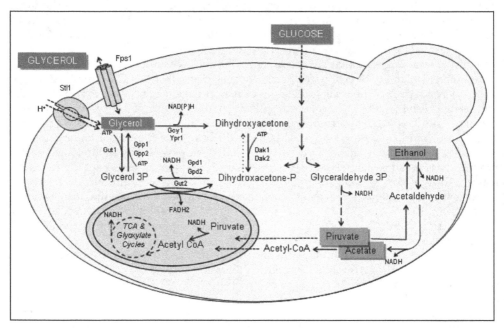

Fig. 2. Glycerol transport and metabolism in *S. cerevisiae* and coupling to main metabolic pathways.

Baker's yeast is a Crabtree[1] positive yeast. Fermentation begins instantly when a glucose pulse is added to glucose-limited, aerobically grown cells. Crabtree effect has been seldom addressed in the last two decades, although it is still a recognized important variable in industrial processes (Ochoa-Estopier et al., 2011). The molecular regulation and main players of this process remain obscure. A relation of Crabtree effect with respiration was discarded (van Urk et al., 1990). Instead, the piruvate decarboxilase levels were found to be 6 times higher in the Crabtree positive yeasts *S. cerevisiae*, *T. glabrata* (today *C. glabrata*) and *S. pombe*. This presented an increased glucose consumption rate that the authors attributed

---

[1] Crabtree effect is the phenomenon whereby *S. cerevisiae* produces ethanol aerobically in the presence of high external glucose concentrations. Instead, Crabtree negative yeasts instead produce biomass via TCA. In *S. cerevisiae*, high concentrations of glucose accelerate glycolysis, producing appreciable amounts of ATP through substrate-level phosphorylation. This reduces the need of oxidative phosphorylation done by the TCA cycle via the electron transport chain, inhibits respiration and ATP synthesis, and therefore decreases oxygen consumption.

to glucose uptake (van Urk et al., 1990), that did not correspond to equivalent growth improvement, but instead to ethanol production through fermentation. Concurrently, growth on glycerol is supposedly entirely oxidative (Gancedo et al., 1968; Flores et al., 2000), which underlies the good and bad performance of respectively Crabtree negative and positive yeasts. In order to turn glycerol broths commercially attractive for *S. cerevisiae*-based biotechnology, in particular baker's yeast cultivation, several approaches were assayed.

One of the most straightforward strategies is metabolic engineering, obtained through genetic manipulation (Randez-Gil et al., 1999). However, this needs precise knowledge on the strain/species genome, available molecular tools (mutants and vectors to the least), and deep knowledge of the metabolic process involved, which are not always available. Additionally, cellular processes are hardly under the control of a single gene and simply regulated. Because of this, available molecular and informatics tools are combined for engineering industrial strains of interest (Patnaik, 2008). In the particular case of baker's yeast, the industrial strains are mostly aneuploids and homothallic, impairing easy genetic improvement (Randez-Gil et al., 1999). In view of the Crabtree effect regulation complexity, and these genetic characteristics, the improvement of baker's yeast glycerol consumption can hardly be possible by genetic engineering. All this, and the general skepticism of consumers towards the use of genetically modified organisms in the food industry, led to the search of alternative strategies for the baking industry.

3.2 Improbable hybrids

The traditional way of producing new strains is by the generation of hybrids through mating. This approach allows the indirect *in vivo* genetic recombination and the propagation of phenotypes of interest. It can be achieved through intra- or inter-specific hybridization. The most resourceful way is the intra-specific recombination of strains with desirable phenotypes. To achieve this, it is necessary to induce sporulation of the target diploid strains, usually by nitrogen starvation. The haploid ascospores are then isolated and their mating type determined, followed by the mating of ascospores from opposite mating type, and the formation of a new heterozygous diploid. Several wine and baker's yeast strains available commercially are the result of such hybridization (Higgins et al., 2001; Pretorius & Bauer, 2002; Marullo et al., 2006).

These strategies demand for a deep knowledge of the phenotypes and the underlying metabolic and molecular processes. As an example, Higgins and collaborators (Higgins et al., 2001) generated a *S. cerevisiae* strain able to combine efficient maltose metabolism, indispensable for fermentative ability of unsugared dough's, with hyperosmotic resistance for optimization of growth on sugared dough's. Loading *S. cerevisiae* with glycerol has been shown to improve the fermentation of sweet doughs (Myers et al., 1998), therefore the selection for osmotolerant phenotype. On the other hand, unlagged growth on maltose is due to the constitutive derepression of maltase and maltose permease (Higgins et al., 1999), as well as of invertase (Myers et al., 1997), but this was previously reported to negatively influence the leavening of sweet doughs (Oda et al., 1990). This difficulty was overcome by the use of massive random mating upon sporulation enrichment, yielding approximately 10% of interesting isolates for further detailed screening (Higgins et al., 2001).

In the particular case of baker's yeast, the sporulation ability of industrial strains is extremely reduced and most strains are homothallic yielding random-mating spores. This is due to their frequent aneuploidy and the consequent heterogenous coupling of their chromosomes during meiosis. This raises the need of using *assexual* approaches, as spheroplast fusion or cell-spore mating (Sauer, 2001), as well as other mass mating strategies that may circumvent isolated spores inability to mate (Higghins et al., 2001). In spheroplast fusion, after appropriate cell wall digestion, it is possible to force the fusion of cells with different levels of ploidy. These are though in many cases phenotypically and reproductively unstable non-resilient multinuclear cells unfit for industry.

3.3 Evolutionary engineering

The alternative solution to extensive and expensive genetic manipulation is evolutionary engineering (Chatterjee & Yuan; 2006, Fong, 2010). This strategy allows the improvement of complex phenotypes of interest, for example stress resistance combined with carbon source utilization. The methodology is based in the combination of confined environmental selection and natural variability. It was first used in a work of Butler and co-workers (Butler et al., 1996), who selected different genetic strains of *Streptomyces griseus* under selective conditions. Evolutionary engineering aims the creation of an improved strain based in selection of behavioral differences between individual cells within a population. For this reason, the generation of genetic variability is vital to this approach, accelerating the adaptive confined evolution based on spontaneous mutations which demands extremely prolonged cultivation under selective conditions (Aguillera et al., 2010; Faria-Oliveira F., Ferreira C. & Lucas C. unpublished results).

One of the simplest ways of generating variability within a population is the introduction of random genetic mutations. Within a population, there is naturally occurring mutagenesis, either through local changes in the genome or larger modifications like DNA rearrangements and horizontal transfers (Sauer, 2001). Nevertheless, spontaneous mutations occur at very low rate, mainly due to the DNA proof-reading mechanism of the organisms and high fidelity of the DNA polymerases. However, it is known that under adverse conditions the mutation rate is enhanced. This feature is crucial to increase the genetic variability within the population to a level propitious to adaptation to challenging environmental constraints. This selection through survival is the basic principle behind the evolutionary engineering.

Several methodologies are available for the generation of variability, namely physical or chemical mutagenesis, sporulation followed by mating, spheroplast fusion, whole genome shuffling, and so on (Fong, 2010; Petri & Schmidt-Dannert, 2004). Mutagenesis is the most common practice, being technically simple and applicable to most organisms (Fong, 2010). The most common mutagens are either chemicals, like ethyl methane sulfonate (EMS), ethidium bromide (EB), or radiation, namely ultraviolet (UV). These mutagens are rather unspecific, and for this reason are widely used (Sauer, 2001). The main drawback of such approaches is the low rate of useful mutations, and the high rate of lethal and neutral mutations. Most chemicals, like EB, introduce preferentially alterations to the DNA like nucleotide exchanges or frame shifts, but other like EMS can induce deletions (Nair & Zhao, 2010). These are responsible for important DNA rearrangements and severe phenotypic alterations. Yet, some chemicals have affinity for certain genome sub-regions, and its utilization in sequential rounds of mutation/selection can be rather reductive. Physical mutagens, namely UV radiation and X-rays, are more prone to chromosomal structural changes and nucleotide frame shifts.

The simplest method to ameliorate a baker's yeast strain relies, as mentioned above, on spontaneous mutations and prolonged cultivations. This strategy has the advantage of doing without the manipulation of dangerous chemicals or radiation, and the disadvantage of the long time needed to obtain results. These batch or fed-batch cultivations have to cover more than 100 generations (Aguilera et al., 2010; Merico et al., 2011; Ochoa-Estopier et al., 2011; Faria-Oliveira F., Ferreira C. & Lucas C. unpublished results) and can last for several months depending on the severity of the environmental constraints. This procedure was applied with success to transform *S. cerevisiae* into a good lactose consumer (Guimarães et al., 2008), to improve freeze and salt tolerance (Aguilera et al., 2010), and turning baker's yeast into an efficient consumer of glycerol as sole carbon and energy source to industrially acceptable biomass yields (Faria-Oliveira F., Ferreira C. & Lucas C. unpublished results). This was obtained through the use of a simple evolution strategy consisting of sequential aerobic batch cultivations on synthetic glycerol-based media for 150 generations, followed by several cycles of cultivation on rich media for ensuring phenotypic/mutation stability. The resulting strains were able to grow up to 4 g biomass dry weight in 2% (w/v) raw biodiesel centrifuged glycerol[2] which corresponds to a biomass yield of 0,4 g.g^{-1} glycerol. Although yields were not significantly different from cultures obtained on reagent grade glycerol, specific growth rates were 3 times higher in raw glycerol (μ_g 0.12 h^{-1}) and lag phase was reduced to a minimum of 2 h. Merico and collaborators (Merico et al., 2011) describe the selection and characterization of an identically evolved *S. cerevisiae* strain which, additionally, also exhibited a high resistance to freeze and thaw stress after prolonged storage at -20° C. Screening capacity can nowadays be expanded by high throughput techniques, but more importantly, as exemplified above, there has to be prior extensive knowledge to be able to design clever phenotype selection platforms.

Alternatively, yeast species other than *S. cerevisiae*, displaying good characteristics for the baking industry can be used. For example, the above mentioned *T. delbrueckii* strains isolated from corn and rye traditional bread doughs display dough-raising capacities and yield production similar to the ones found in commercial baker's yeasts (Almeida & Pais, 1996a), and maintain approximately the same leavening ability during storage of frozen doughs for 30 days, showing a very high tolerance to freezing (Almeida & Pais, 1996b). Furthermore, in one of these strains no loss of cell viability was observed after 120 days of freezing at -20°C, whereas a loss of 80% was observed in a commercial baker's yeast after 15 days (Alves-Araújo et al., 2004). These characteristics make them candidates of great potential value to the baking industry, mainly to be used in frozen dough products.

4. Conclusion

Nowadays, the man made baker's yeast strains, as well as their associated technological process particularities are fast disappearing due to globalization of yeast and dough industry. Nevertheless, sustainability demands contradict globalization trends, asking for solutions empowering local populations with tools that decrease their dependence on global markets. Yeasts, as always, have a role in this desired change of paradigm.

---

[2] Raw glycerol was diluted with water (1:3) and the pH adjusted to 4.5 with HCl. This was centrifuged at 5000 rpm for 15 min at 4°C. Fat separates from glycerol forming an upper layer which is sucked with a vacuum pump to liberate the cleaner glycerol fraction. The pH adjustment increases the separation efficiency.

Going after the regional tastes for unique types of bread and other bakery products could improve the revenue for the local economical players. This can be achieved through the reintroduction of lost biodiversity in the leavening processes. The industrial production of such yeasts and bacteria can be done using unconventional substrates like biodiesel-derived glycerol, since most bacteria and yeast can consume this substrate naturally, in opposition to the traditional baker's yeast strains. Yet still, if necessary, baker's yeast can be improved for producing interesting biomass yields at the expense of glycerol by clever and simple accelerated and confined evolution strategies. Finally, the glycerol-based broths can by themselves improve the shelf-life span of doughs and leavens.

Biotechnological processes need improvement in order to meet sustainability objectives. No simple unique solution exists. Instead, sustainability can be achieved by increasing diversity of processes, tools and products, for which clever simplicity-generating solutions can be devised.

5. Acknowledgment

Authors would like to acknowledge Hugh S. Johnson for the several critical readings of the manuscript regarding proper English usage. Fábio Faria-Oliveira is supported by a PhD grant from FCT-SFRH/BD/45368/2008. This work was financed by FEDER through COMPETE Programme (Programa Operacional Factores de Competitividade) and national funds from FCT (Fundação para a Ciência e a Tecnologia) project PEst-C/BIA/UI4050/2011.

6. References

Aguilera, J., Andreu, P., Randez-Gil, F. & Prieto, J.A. (2010). Adaptive evolution of baker's yeast in a dough-like environment enhances freeze and salinity tolerance. *Microb Biotechnol*, Vol.3, No.2, pp. 210-221, ISSN 1751-7915

Aguilera, J., Randez-Gil, F. & Prieto, J.A. (2007). Cold response in *Saccharomyces cerevisiae:* new functions for old mechanisms. *FEMS Microbiology Reviews*, Vol.31, No.3, pp. 327-341, ISSN 1574-6976

Al-Fageeh, M.B. & Smales, C.M. (2006). Control and regulation of the cellular responses to cold shock: the responses in yeast and mammalian systems. *Biochem J*, Vol.397, No.2, pp. 247-259

Almeida, M.J. & Pais, C. (1996b). Leavening ability and freeze tolerance of yeasts isolated from traditional corn and rye bread doughs. *Appl Environ Microbiol*, Vol.62, No.12, pp. 4401-4404, ISSN 0099-2240

Almeida, M.J. & Pais, C.S. (1996a). Characterization of the yeast population from traditional corn and rye bread doughs. *Letters in Applied Microbiology*, Vol.23, No.3, pp. 154-158, ISSN 1472-765X

Alves-Araújo, C., Almeida, M.J., Sousa, M.J. & Leão, C. (2004). Freeze tolerance of the yeast *Torulaspora delbrueckii:* cellular and biochemical basis. *FEMS Microbiol Lett*, Vol.240, No.1, pp. 7-14, ISSN 0378-1097

Ansell, R., Granath, K., Hohmann, S., Thevelein, J.M. & Adler, L. (1997). The two isoenzymes for yeast NAD+-dependent glycerol 3-phosphate dehydrogenase encoded by *GPD1* and *GPD2* have distinct roles in osmoadaptation and redox regulation. *EMBO J*, Vol.16, No.9, pp. 2179-2187, ISSN 0261-4189

Antuna, B. & Martinez-Anaya, M.A. (1993). Sugar uptake and involved enzymatic activities by yeasts and lactic acid bacteria: their relationship with bread making quality. *Int J Food Microbiol*, Vol.18, No.3, pp. 191-200, ISSN 0168-1605

Barnett, J., Payne R.W. & Yarrow D. (Ed(s).) (2000). *Yeasts: Characteristics and identification* (3rd), Cambridge University Press, ISBN 978-052-1573-96-2, Cambridge

Bocker, G., Stolz, P. & Hammes, P. (1995). Neue Erkenntnisse zum Okosystem Sauerteig und zur Physiologie der sauerteigtypischen Stamme *Lactobacillus sanfrancisco* und *Lactobacillus pontis*. *Getreide Mehl Brot*, Vol.49, pp. 370–374

Brandt, M.J. (2001). Mikrobiologische Wechselwirkungen von technologischer Bedeutung. Ph.D. thesis. University of Hohenheim, Stuttgart, Germany.

Butler, P.R., Brown, M. & Oliver, S.G. (1996). Improvement of antibiotic titers from *Streptomyces* bacteria by interactive continuous selection. *Biotechnol Bioeng*, Vol.49, No.2, pp. 185-196, ISSN 0006-3592

Chatterjee, R. & Yuan, L. (2006). Directed evolution of metabolic pathways. *Trends Biotechnol*, Vol.24, No.1, pp. 28-38, ISSN 0167-7799

Chatzifragkou, A., Makri, A., Belka, A., Bellou, S., Mavrou, M., Mastoridou, M., Mystrioti, P., Onjaro, G., Aggelis, G. & Papanikolaou, S. (2011). Biotechnological conversions of biodiesel derived waste glycerol by yeast and fungal species. *Energy*, Vol.36, No.2, pp. 1097-1108, ISSN 0360-5442

Collar, C., Andreu, P. & Martínez-Anaya, M.A. (1998). Interactive effects of flour, starter and enzyme on bread dough machinability. *Zeitschrift für Lebensmitteluntersuchung und - Forschung A*, Vol.207, No.2, pp. 133-139, ISSN 1431-4630

Corsetti, A., Gobbetti, M., Balestrieri, F., Paoletti, F., Russi, L. & Rossi, J. (1998). Sourdough Lactic Acid Bacteria Effects on Bread Firmness and Stalin. *Journal of Food Science*, Vol.63, No.2, pp. 347-351, ISSN 1750-3841

Corsetti, A., Lavermicocca, P., Morea, M., Baruzzi, F., Tosti, N. & Gobbetti, M. (2001). Phenotypic and molecular identification and clustering of lactic acid bacteria and yeasts from wheat (species *Triticum durum* and *Triticum aestivum*) sourdoughs of Southern Italy. *Int J Food Microbiol*, Vol.64, No.1-2, pp. 95-104, ISSN 0168-1605

da Silva, G.P., Mack, M. & Contiero, J. (2009) Glycerol: A promising and abundant carbon source for industrial microbiology. *Biotechnology Advances*, Vol.27, No.1, pp. 30-39, ISSN 0734-9750

Dal Bello, F., Walter, J., Roos, S., Jonsson, H. & Hertel, C. (2005). Inducible Gene Expression in *Lactobacillus reuteri* LTH5531 during Type II Sourdough Fermentation. *Appl. Environ. Microbiol.*, Vol.71, No.10, pp. 5873-5878

Damiani, P., Gobbetti, M., Cossignani, L., Corsetti, A., Simonetti, M.S. & Rossi, J. (1996). The sourdough microflora. Characterization of hetero- and homofermentative Lactic Acid Bacteria, yeasts and their interactions on the basis of the volatile compounds produced. *Lebensmittel-Wissenschaft und-Technologie*, Vol.29, No.1-2, pp. 63-70, ISSN 0023-6438

D'Amico, S., Collins, T., Marx, J.C., Feller, G. & Gerday, C. (2006). Psychrophilic microorganisms: challenges for life. *EMBO Rep*, Vol.7, No.4, pp. 385-389, ISSN 1469-221X

Dang, N.X. & Hincha, D.K. (2011). Identification of two hydrophilins that contribute to the desiccation and freezing tolerance of yeast (*Saccharomyces cerevisiae*) cells. *Cryobiology*, Vol.62, No.3, pp. 188-193, ISSN 0011-2240 de Vuyst, L. & Neysens, P. (2005). The sourdough microflora: biodiversity and metabolic interactions. *Trends in Food Science & Technology*, Vol.16, No.1-3, pp. 43-56, ISSN 0924-2244

de Vuyst, L. & Vancanneyt, M. (2007). Biodiversity and identification of sourdough lactic acid bacteria. *Food Microbiology*, Vol.24, No.2, pp. 120-127, ISSN 0740-0020

Ferreira, C. & Lucas, C. (2007). Glucose repression over *Saccharomyces cerevisiae* glycerol/H+ symporter gene *STL1* is overcome by high temperature. *FEBS Lett*, Vol.581, No.9, pp. 1923-1927, ISSN 0014-5793

Ferreira, C., van Voorst, F., Martins, A., Neves, L., Oliveira, R., Kielland-Brandt, M.C., Lucas, C. & Brandt, A. (2005). A member of the sugar transporter family, Stl1p is the glycerol/H+ symporter in *Saccharomyces cerevisiae*. *Mol Biol Cell*, Vol.16, No.4, pp. 2068-2076, ISSN 1059-1524

Flores, C.L., Rodriguez, C., Petit, T. & Gancedo, C. (2000). Carbohydrate and energy-yielding metabolism in non-conventional yeasts. *FEMS Microbiol Rev*, Vol.24, No.4, pp. 507-529, ISSN 0168-6445

Fong, S.S. (2010). Evolutionary engineering of industrially important microbial phenotypes, In: *The metabolic pathway engineering handbook: tools and applications*, C.D. Smolke, pp. 1/1-1/13, CRC Press, ISBN 978-142-0077-65-0, New York

Galli, A., Franzetti, L. & Fortina. M.G. (1988). Isolation and identification of sour dough microflora. *Microbiol. Aliment. Nutr.*, Vol.6, pp. 345-351

Gancedo, C., Gancedo, J.M. & Sols, A. (1968). Glycerol metabolism in yeasts. Pathways of utilization and production. *Eur J Biochem*, Vol. 5, No.2, pp. 165-172, ISSN 0014-2956

Garofalo, C., Silvestri, G., Aquilanti, L. & Clementi, F. (2008). PCR-DGGE analysis of lactic acid bacteria and yeast dynamics during the production processes of three varieties of Panettone. *J Appl Microbiol*, Vol.105, No.1, pp. 243-254, ISSN 1365-2672

Gobbetti, M. (1998). The sourdough microflora: Interactions of lactic acid bacteria and yeasts. *Trends in Food Science & Technology*, Vol.9, No.7, pp. 267-274, ISSN 0924-2244

Gobbetti, M., Corsetti, A. & Rossi, J. (1994a). The sourdough microflora. Interactions between lactic acid bacteria and yeasts: metabolism of amino acids. *World Journal of Microbiology and Biotechnology*, Vol.10, pp. 275-279

Gobbetti, M., Corsetti, A., Rossi, J., La Rosa, F. & de Vincenzi, S. (1994b). Identification and clustering of lactic acid bacteria and yeasts from wheat sourdoughs of central Italy. *Italian J Food Sci*, Vol.6, pp. 85-94

Gobbetti, M., De Angelis, M., Corsetti, A. & Di Camargo, R. (2005). Biochemistry and physiology of sourdough lactic acid bacteria. *Trends in Food Science & Technology*, Vol.16, pp. 57-69

Grauslund, M., Lopes, J.M. & Rønnow, B. (1999). Expression of *GUT1*, which encodes glycerol kinase in *Saccharomyces cerevisiae*, is controlled by the positive regulators Adr1p, Ino2p and Ino4p and the negative regulator Opi1p in a carbon source-dependent fashion. *Nucleic Acids Res*, Vol.27, No.22, pp. 4391-4398, ISSN 1362-4962

Guadalupe Medina, V., Almering, M.J., van Maris, A.J. & Pronk, J.T. (2010). Elimination of glycerol production in anaerobic cultures of *Saccharomyces cerevisiae* engineered for use of acetic acid as electron acceptor. *Appl Environ Microbiol*, Vol.76, No.1, pp. 190-195, ISSN 0099-2240

Guimarães, P.M., Francois, J., Parrou, J.L., Teixeira, J.A. & Domingues, L. (2008). Adaptive evolution of a lactose-consuming *Saccharomyces cerevisiae* recombinant. *Appl Environ Microbiol*, Vol.74, No.6, pp. 1748-1756, ISSN 1098-5336

Halm, M., Lillie, A., Sorensen, A.K. & Jakobsen, M. (1993). Microbiological and aromatic characteristics of fermented maize doughs for kenkey production in Ghana. *Int J Food Microbiol*, Vol.19, No.2, pp. 135-43, ISSN 0168-1605

Hämmes, W.P. & Gänzle, M.G. (1998). Sourdough breads and related products, In: *Microbiology of Fermented Foods*, B.J.B. Wood, W.H. Holzapfel, pp. 199-216, Blackie Academic and Professional, ISBN 978-075-1402-16-2, London

Hämmes, W.P. & Vogel, R.F. (1995). The genus *Lactobacillus*, In: *The genera of lactic acid bacteria*, B.J.B. Wood, W.H. Holzapfel, pp. 19-54, Blackie Academic and Professional, ISBN 0-7514-0215X, London

Hansen, B. & Hansen, Å. (1994). Volatile compounds in wheat sourdoughs produced by lactic acid bacteria and sourdough yeasts. *Zeitschrift für Lebensmitteluntersuchung und - Forschung A*, Vol.198, No.3, pp. 202-209, ISSN 1431-4630

Hansen, K.R., Burns, G., Mata, J., Volpe, T.A., Martienssen, R.A., Bahler, J. & Thon, G. (2005). Global effects on gene expression in fission yeast by silencing and RNA interference machineries. *Mol Cell Biol*, Vol.25, No.2, pp. 590-601, ISSN 0270-7306

Hayashi, M. & Maeda, T. (2006). Activation of the HOG Pathway upon cold stress in *Saccharomyces cerevisiae*. *Journal of Biochemistry*, Vol.139, No.4, pp. 797-803

Higgins, V.J., Bell, P.J., Dawes, I.W. & Attfield, P.V. (2001). Generation of a novel *Saccharomyces cerevisiae* strain that exhibits strong maltose utilization and hyperosmotic resistance using nonrecombinant techniques. *Appl Environ Microbiol*, Vol.67, No.9, pp. 4346-4348

Higgins, V.J., Braidwood, M., Bissinger, P., Dawes, I.W. & Attfield, P.V. (1999). Leu343Phe substitution in the Malx3 protein of *Saccharomyces cerevisiae* increases the constitutivity and glucose insensitivity of *MAL* gene expression. *Curr Genet*, Vol.35, No.5, pp. 491-498, ISSN 0172-8083

Hirasawa, R. & Yokoigawa, K. (2001). Leavening ability of baker's yeast exposed to hyperosmotic media. *FEMS Microbiol Lett*, Vol.194, No.2, pp. 159-162, ISSN 0378-1097

Hirasawa, R., Yokoigawa, K., Isobe, Y. & Kawai, H. (2001). Improving the freeze tolerance of bakers' yeast by loading with trehalose. *Biosci Biotechnol Biochem*, Vol.65, No.3, pp. 522-526, ISSN 0916-8451

Hohmann, S. (2009). Control of high osmolarity signalling in the yeast *Saccharomyces cerevisiae*. *FEBS Letters*, Vol.583, No.24, pp. 4025-4029, ISSN 0014-5793

Iacumin, L., Cecchini, F., Manzano, M., Osualdini, M., Boscolo, D., Orlic, S. & Comi, G. (2009). Description of the microflora of sourdoughs by culture-dependent and culture-independent methods. *Food Microbiol*, Vol.26, No.2, pp. 128-135, ISSN 1095-9998

Izawa, S., Ikeda, K., Maeta, K. & Inoue, Y. (2004b). Deficiency in the glycerol channel Fps1p confers increased freeze tolerance to yeast cells: application of the *fps1Δ* mutant to frozen dough technology. *Appl Microbiol Biotechnol*, Vol.66, No.3, pp. 303-305, ISSN 0175-7598

Izawa, S., Ikeda, K., Takahashi, N. & Inoue, Y. (2007). Improvement of tolerance to freeze-thaw stress of baker's yeast by cultivation with soy peptides. *Appl Microbiol Biotechnol*, Vol.75, No.3, pp. 533-537, ISSN 0175-7598

Izawa, S., Sato, M., Yokoigawa, K. & Inoue, Y. (2004a). Intracellular glycerol influences resistance to freeze stress in *Saccharomyces cerevisiae*: analysis of a quadruple mutant in glycerol dehydrogenase genes and glycerol-enriched cells. *Appl Microbiol Biotechnol*, Vol.66, No.1, pp. 108-114, ISSN 0175-7598

Kaino, T., Tateiwa, T., Mizukami-Murata, S., Shima, J. & Takagi, H. (2008). Self-cloning baker's yeasts that accumulate proline enhance freeze tolerance in doughs. *Appl Environ Microbiol*, Vol.74, No.18, pp. 5845-5849, ISSN 1098-5336

Kandror, O., Bretschneider, N., Kreydin, E., Cavalieri, D. & Goldberg, A.L. (2004). Yeast adapt to near-freezing temperatures by *STRE*/Msn2,4-dependent induction of trehalose synthesis and certain molecular chaperones. *Molecular Cell*, Vol.13, No.6, pp. 771-781, ISSN 1097-2765

Lages, F., Silva-Graça, M. & Lucas, C. (1999). Active glycerol uptake is a mechanism underlying halotolerance in yeasts: a study of 42 species. *Microbiology*, Vol.145, No.9, pp. 2577-2585, ISSN 1350-0872

Laroche, C. & Gervais, P. (2003). Achievement of rapid osmotic dehydration at specific temperatures could maintain high *Saccharomyces cerevisiae* viability. *Appl Microbiol Biotechnol*, Vol.60, No.6, pp. 743-747, ISSN 0175-7598

Larsson, C., Pahlman, I.L., Ansell, R., Rigoulet, M., Adler, L. & Gustafsson, L. (1998). The importance of the glycerol 3-phosphate shuttle during aerobic growth of *Saccharomyces cerevisiae*. *Yeast*, Vol.14, No.4, pp. 347-357, ISSN 0749-503X

Luyten, K., Albertyn, J., Skibbe, W.F., Prior, B.A., Ramos, J., Thevelein, J.M. & Hohmann, S. (1995). Fps1, a yeast member of the MIP family of channel proteins, is a facilitator for glycerol uptake and efflux and is inactive under osmotic stress. *EMBO J*, Vol.14, No.7, pp. 1360-1371, ISSN 0261-4189

Mäntynen, V.H., Korhola, M., Gudmundsson, H., Turakainen, H., Alfredsson, G.A., Salovaara, H. & Lindstrom, K. (1999). A polyphasic study on the taxonomic position of industrial sour dough yeasts. *Syst Appl Microbiol*, Vol.22, No.1, pp. 87-96, ISSN 0723-2020

Martinez-Anaya, M.A., Pitarch, B., Bayarri, P. & Barber, C.B. (1990a). Microflora of the sourdoughs of wheat flour bread. Interactions between yeasts and lactic acid bacteria in wheat doughs and their effects on bread quality. *Cereal chem*, Vol.67, pp. 85-91

Martinez-Anaya, M.A., Torner, J.M. & de Barber, C.B. (1990b). Microflora of the sour dough of wheat flour bread. *Zeitschrift für Lebensmitteluntersuchung und -Forschung A*, Vol.190, No.2, pp. 126-131, ISSN 1431-4630

Marullo, P., Bely, M., Masneuf-Pomarede, I., Pons, M., Aigle, M. & Dubourdieu, D. (2006). Breeding strategies for combining fermentative qualities and reducing off-flavor production in a wine yeast model. *FEMS Yeast Res*, Vol.6, No.2, pp. 268-279, ISSN 1567-1356

Merico, A., Ragni, E., Galafassi, S., Popolo, L. & Compagno, C. (2011). Generation of an evolved *Saccharomyces cerevisiae* strain with a high freeze tolerance and an improved ability to grow on glycerol. *J Ind Microbiol Biotechnol*, Vol.38, No.8, pp. 1037-1044, ISSN 1476-5535

Meroth, C.B., Brandt, M.J. & Hamme, W.P. (2002). Die hefe macht's. *Brot Backwaren*, Vol.4, pp. 33-34

Meroth, C.B., Hammes, W.P. & Hertel, C. (2003a). Identification and population dynamics of yeasts in sourdough fermentation processes by PCR-denaturing gradient gel electrophoresis. *Appl Environ Microbiol*, Vol.69, No.12, pp. 7453-7461

Meroth, C.B., Walter, J., Hertel, C., Brandt, M.J. & Hammes, W.P. (2003b). Monitoring the bacterial population dynamics in sourdough fermentation processes by using PCR-denaturing gradient gel electrophoresis. *Appl Environ Microbiol*, Vol.69, No.1, pp. 475-482

Michnick, S., Roustan, J.L., Remize, F., Barre, P. & Dequin, S. (1997). Modulation of glycerol and ethanol yields during alcoholic fermentation in *Saccharomyces cerevisiae* strains overexpressed or disrupted for *GPD1* encoding glycerol 3-phosphate dehydrogenase. *Yeast*, Vol.13, No.9, pp. 783-793, ISSN 0749-503X

Momose, Y., Matsumoto, R., Maruyama, A. & Yamaoka, M. (2010). Comparative analysis of transcriptional responses to the cryoprotectants, dimethyl sulfoxide and trehalose, which confer tolerance to freeze-thaw stress in *Saccharomyces cerevisiae*. *Cryobiology*, Vol.60, No.3, pp. 245-261, ISSN 1090-2392

Murata, Y., Homma, T., Kitagawa, E., Momose, Y., Sato, M.S., Odani, M., Shimizu, H., Hasegawa-Mizusawa, M., Matsumoto, R., Mizukami, S., Fujita, K., Parveen, M., Komatsu, Y. & Iwahashi, H. (2006). Genome-wide expression analysis of yeast response during exposure to 4 degrees C. *Extremophiles*, Vol.10, No.2, pp. 117-128, ISSN 1431-0651

Myers, D.K., Joseph, V.M., Pehm, S., Galvagno, M. & Attfield, P.V. (1998). Loading of *Saccharomyces cerevisiae* with glycerol leads to enhanced fermentation in sweet bread doughs. *Food Microbiology*, Vol.15, No.1, pp. 51-58, ISSN 0740-0020

Myers, D.K., Lawlor, D.T. & Attfield, P.V. (1997). Influence of invertase activity and glycerol synthesis and retention on fermentation of media with a high sugar concentration by *Saccharomyces cerevisiae*. *Appl Environ Microbiol*, Vol.63, No.1, pp. 145-150

Nair, N.U. & Zhao, H. (2010). The metabolic pathway engineering handbook: tools and applications, In: *The metabolic pathway engineering handbook: tools and applications*, C.D. Smolke, pp. 2/1-2/37, CRC Press, ISBN 978-142-0077-65-0, New York

Obiri-Danso, K. (1994). Microbiological studies on corn dough ferment. *Cereal chem*, Vol.7, pp. 186-188

Ochoa-Estopier, A., Lesage, J., Gorret, N. & Guillouet, S.E. (2011). Kinetic analysis of a *Saccharomyces cerevisiae* strain adapted for improved growth on glycerol: Implications for the development of yeast bioprocesses on glycerol. *Bioresour Technol*, Vol.102, No.2, pp. 1521-1527, ISSN 1873-2976

Oda, Y. & Ouchi, K. (1990). Effect of invertase activity on the leavening ability of yeast in sweet dough. *Food Microbiology*, Vol.7, No.3, pp. 241-248, ISSN 0740-0020

Ottogalli, G., Galli, A. & Foschino, R. (1996). Italian bakery products obtained with sour dough: characterization of the tipical microflora. *Adv Food Sci*, Vol.18, pp. 131-144

Påhlman, I.L., Gustafsson, L., Rigoulet, M. & Larsson, C. (2001). Cytosolic redox metabolism in aerobic chemostat cultures of *Saccharomyces cerevisiae*. *Yeast*, Vol.18, No.7, pp. 611-620, ISSN 0749-503X

Panadero, J., Randez-Gil, F. & Prieto, J.A. (2005). Validation of a flour-free model dough system for throughput studies of baker's yeast. *Appl Environ Microbiol*, Vol.71, No.3, pp. 1142-1147, ISSN 0099-2240

Paramithiotis, S., Muller, M.R.A., Ehrmann, M.A., Tsakalidou, E., Seiler, H., Vogel, R. & Kalantzopoulos, G. (2000). Polyphasic identification of wild yeast strains isolated from Greek sourdoughs. *Syst Appl Microbiol*, Vol.23, No.1, pp. 156-164, ISSN 0723-2020

Paramithiotis, S., Tsiasiotou, S. & Drosinos, E. (2010). Comparative study of spontaneously fermented sourdoughs originating from two regions of Greece: Peloponnesus and Thessaly. *European Food Research and Technology*, Vol.231, No.6, pp. 883-890, ISSN 1438-2377

Patnaik, R. (2008). Engineering complex phenotypes in industrial strains. *Biotechnol Prog*, Vol.24, No.1, pp. 38-47, ISSN 8756-7938

Petri, R. & Schmidt-Dannert, C. (2004). Dealing with complexity: evolutionary engineering and genome shuffling. *Curr Opin Biotechnol*, Vol.15, No.4, pp. 298-304, ISSN 0958-1669

Phadtare, S. & Severinov, K. (2010). RNA remodeling and gene regulation by cold shock proteins. *RNA Biol*, Vol.7, No.6, pp. 788-795, ISSN 1555-8584

Pretorius, I.S. & Bauer, F.F. (2002). Meeting the consumer challenge through genetically customized wine-yeast strains. *Trends Biotechnol*, Vol.20, No.10, pp. 426-432, ISSN 0167-7799

Pulvirenti, A., Solieri, L., Gullo, M., de Vero, L. & Giudici, P. (2004). Occurrence and dominance of yeast species in sourdough. *Lett Appl Microbiol*, Vol.38, No.2, pp. 113-117, ISSN 0266-8254

Randez-Gil, F., Sanz, P. & Prieto, J.A. (1999). Engineering baker's yeast: room for improvement. *Trends Biotechnol*, Vol.17, No.6, pp. 237-244, ISSN 0167-7799

Remize, F., Roustan, J.L., Sablayrolles, J.M., Barre, P. & Dequin, S. (1999). Glycerol overproduction by engineered *Saccharomyces cerevisiae* wine yeast strains leads to substantial changes in by-product formation and to a stimulation of fermentation rate in stationary phase. *Appl Environ Microbiol*, Vol.65, No.1, pp. 143-149, ISSN 1098-5336

Rep, M., Krantz, M., Thevelein, J.M. & Hohmann, S. (2000). The transcriptional response of *Saccharomyces cerevisiae* to osmotic shock. *Journal of Biological Chemistry*, Vol.275, No.12, pp. 8290-8300

Rocha, J.M. & Malcata, F.X. (1999). On the microbiological profile of traditional Portuguese sourdough. *J Food Prot*, Vol.62, No.12, pp. 1416-1429, ISSN 0362-028X

Rodriguez-Vargas, S., Estruch, F. & Randez-Gil, F. (2002). Gene expression analysis of cold and freeze stress in Baker's yeast. *Appl Environ Microbiol*, Vol.68, No.6, pp. 3024-3030

Rodriguez-Vargas, S., Sanchez-Garcia, A., Martinez-Rivas, J.M., Prieto, J.A. & Randez-Gil, F. (2007). Fluidization of membrane lipids enhances the tolerance of *Saccharomyces cerevisiae* to freezing and salt stress. *Appl Environ Microbiol*, Vol.73, No.1, pp. 110-116

Rønnow, B. & Kielland-Brandt, M.C. (1993). *GUT2*, a gene for mitochondrial glycerol 3-phosphate dehydrogenase of *Saccharomyces cerevisiae*. *Yeast*, Vol.9, No.10, pp. 1121-1130, ISSN 0749-503X

Rossi, J. (1996). The yeasts in sourdough. *Adv Food Sci*, Vol.18, pp. 201-211

Rothe, M., Schneeweiss, R. & Ehrlich. R. (1973). Zur historischen entwicklung von getreideverarbeitung und getredeverzehr. *Ernahrungsfirschung*, Vol.18, pp. 249-283

Sahara, T., Goda, T. & Ohgiya, S. (2002). Comprehensive expression analysis of time-dependent genetic responses in yeast cells to low temperature. *J Biol Chem*, Vol.277, No.51, pp. 50015-50021, ISSN 0021-9258

Salovaara, H. (1998). Lactic acid bacteria in cereal based products, In: *Lactic Acid Bacteria - Technology and Health Effects*, Salminen, S., von Wright, A., pp. 115-338, Marcel Dekker, New York

Sauer, U. (2001). Evolutionary engineering of industrially important microbial phenotypes. *Adv Biochem Eng Biotechnol*, Vol.73, pp. 129-169, ISSN 0724-6145

Shima, J., Sakata-Tsuda, Y., Suzuki, Y., Nakajima, R., Watanabe, H., Kawamoto, S. & Takano, H. (2003). Disruption of the *CAR1* gene encoding arginase enhances freeze tolerance of the commercial baker's yeast *Saccharomyces cerevisiae*. *Appl Environ Microbiol*, Vol.69, No.1, pp. 715-718

Siderius, M., van Wuytswinkel, O., Reijenga, K.A., Kelders, M. & Mager, W.H. (2000). The control of intracellular glycerol in *Saccharomyces cerevisiae* influences osmotic stress response and resistance to increased temperature. *Mol Microbiol*, Vol.36, No.6, pp. 1381-1390, ISSN 0950-382X

Simonin, H., Beney, L. & Gervais, P. (2007). Sequence of occurring damages in yeast plasma membrane during dehydration and rehydration: mechanisms of cell death. *Biochim Biophys Acta*, Vol.1768, No.6, pp. 1600-1610, ISSN 0006-3002

Spicher, G. & Nierle, W. (1984). The microflora of sourdough. The influence of yeast on the proteolysis during sourdough fermentation. *Zeitschrifl flit Lebensmittel-Untersuchung und-Forschung*, Vol.179, pp. 109-112

Succi, M., Reale, A., Andrighetto, C., Lombardi, A., Sorrentino, E. & Coppola, R. (2003). Presence of yeasts in southern Italian sourdoughs from *Triticum aestivum* flour. *FEMS Microbiol Lett*, Vol.225, No.1, pp. 143-148, ISSN 0378-1097

Sugihara, T.F., Kline, L. & Miller, M.W. (1971). Microorganisms of the San Francisco sour dough bread process. Yeasts responsible for the leavening action. *Appl Microbiol*, Vol.21, No.3, pp. 456-458, ISSN 0003-6919

Sutherland, F., Lages, F., Lucas, C., Luyten, K., Albertyn, J., Hohmann, S., Prior, B. & Kilian, S. (1997). Characteristics of Fps1-dependent and -independent glycerol transport in *Saccharomyces cerevisiae*. *J. Bacteriol.*, Vol.179, No.24, pp. 7790-7795

Takahashi, S., Ando, A., Takagi, H. & Shima, J. (2009). Insufficiency of copper ion homeostasis causes freeze-thaw injury of yeast cells as revealed by indirect gene expression analysis. *Appl Environ Microbiol*, Vol.75, No.21, pp. 6706-6711, ISSN 1098-5336

Tamás, M.J., Luyten, K., Sutherland, F.C., Hernandez, A., Albertyn, J., Valadi, H., Li, H., Prior, B.A., Kilian, S.G., Ramos, J., Gustafsson, L., Thevelein, J.M. & Hohmann, S. (1999). Fps1p controls the accumulation and release of the compatible solute glycerol in yeast osmoregulation. *Mol Microbiol*, Vol.31, No.4, pp. 1087-1104, ISSN 0950-382X

Tanghe, A., van Dijck, P., Colavizza, D. & Thevelein, J.M. (2004). Aquaporin-mediated improvement of freeze tolerance of *Saccharomyces cerevisiae* is restricted to rapid freezing conditions. *Appl Environ Microbiol*, Vol.70, No.6, pp. 3377-3382

Tanghe, A., van Dijck, P., Dumortier, F., Teunissen, A., Hohmann, S. & Thevelein, J.M. (2002). Aquaporin expression correlates with freeze tolerance in baker's yeast, and overexpression improves freeze tolerance in industrial strains. *Appl Environ Microbiol*, Vol.68, No.12, pp. 5981-5989

Terao, Y., Nakamori, S. & Takagi, H. (2003). Gene dosage effect of L-proline biosynthetic enzymes on L-proline accumulation and freeze tolerance in *Saccharomyces cerevisiae*. *Appl Environ Microbiol*, Vol.69, No.11, pp. 6527-6532

Trûper, H.G. & De' Clari, L. (1997). Taxonomic note: Necessary correction of specific epithets formed as substantives (Nouns) "in Apposition". *International Journal of Systematic Bacteriology*, Vol.47, No.3, pp. 908-909

Tulha, J., Lima, A., Lucas, C. & Ferreira, C. (2010). *Saccharomyces cerevisiae* glycerol/H⁺ symporter Stl1p is essential for cold/near-freeze and freeze stress adaptation. A simple recipe with high biotechnological potential is given. *Microb Cell Fact*, Vol.9, No.82, ISSN 1475-2859

Valmorri, S., Tofalo, R., Settanni, L., Corsetti, A. & Suzzi, G. (2010). Yeast microbiota associated with spontaneous sourdough fermentations in the production of traditional wheat sourdough breads of the Abruzzo region (Italy). *Antonie Van Leeuwenhoek*, Vol.97, No.2, pp. 119-129, ISSN 1572-9699

van Dijken, J.P. & Scheffers, W.A. (1986). Redox balances in the metabolism of sugars by yeasts. *FEMS Microbiology Letters*, Vol.32, No.3-4, pp. 199-224, ISSN 0378-1097

van Urk H., Voll W.S.L., Scheffers W.A. & van Dijken J. (1990). Transient-state analysis of metabolic fluxes in Crabtree-positive and Crabtree-negative yeasts. *Appl. Environ. Microbiol*, Vol.56, No.1, pp. 281-287

Veiga, A., Arrabaca, J.D. & Loureiro-Dias, M.C. (2003). Stress situations induce cyanide-resistant respiration in spoilage yeasts. *J Appl Microbiol*, Vol.95, No.2, pp. 364-371, ISSN 1364-5072

Vernocchi, P., Valmorri, S., Dalai, I., Torriani, S., Gianotti, A., Suzzi, G., Guerzoni, M.E., Mastrocola, D. & Gardini, F. (2004a). Characterization of the yeast population involved in the production of a typical Italian bread. *Journal of Food Science*, Vol.69, No.7, pp. 182-186, ISSN 1750-3841

Vernocchi, P., Valmorri, S., Gatto, V., Torriani, S., Gianotti, A., Suzzi, G., Guerzoni, M.E. & Gardini, F. (2004b). A survey on yeast microbiota associated with an Italian traditional sweet-leavened baked good fermentation. *Food Research International*, Vol.37, No.5, pp. 469-476, ISSN 0963-9969

Vogel, R. (1997). Microbial Ecology of Cereal Fermentations. *Food Tecnhol. Biotechnol.*, Vol.35, pp. 51-54, ISSN 1330-9862

Vrancken, G., de Vuyst, L., van der Meulen, R., Huys, G., Vandamme, P. & Daniel, H.M. (2010). Yeast species composition differs between artisan bakery and spontaneous laboratory sourdoughs. *FEMS Yeast Res*, Vol.10, No.4, pp. 471-481, ISSN 1567-1364

Xu, Z.-Y., Zhang, X., Schläppi, M. & Xu, Z.-Q. (2011). Cold-inducible expression of *AZI1* and its function in improvement of freezing tolerance of *Arabidopsis thaliana* and *Saccharomyces cerevisiae*. *Journal of Plant Physiology*, Vol.168, No.13, pp. 1576-1587, ISSN 0176-1617

Yazdani, S.S. & Gonzalez, R. (2007). Anaerobic fermentation of glycerol: a path to economic viability for the biofuels industry. *Curr Opin Biotechnol*, Vol.18, No.3, pp. 213-219, ISSN 0958-1669

Yokoigawa, K., Sato, M. & Soda, K. (2006). Simple improvement in freeze-tolerance of bakers' yeast with poly-γ-glutamate. *J Biosci Bioeng*, Vol.102, No.3, pp. 215-219, ISSN 1389-1723

Zhang, L., Ohta, A., Horiuchi, H., Takagi, M. & Imai, R. (2001). Multiple mechanisms regulate expression of low temperature responsive (*LOT*) genes in *Saccharomyces cerevisiae*. *Biochem Biophys Res Commun*, Vol.283, No.2, pp. 531-535, ISSN 0006-291X

Zhang, X., Shanmugam, K.T. & Ingram, L.O. (2010). Fermentation of glycerol to succinate by metabolically engineered strains of *Escherichia coli*. *Appl. Environ. Microbiol.*, Vol.76, No.8, pp. 2397-2401

2

Epidemiology of Foodborne Illness

Saulat Jahan
Research and Information Unit
Primary Health Care Administration, Qassim
Ministry of Health
Kingdom of Saudi Arabia

1. Introduction

Foodborne illnesses comprise a broad spectrum of diseases and are responsible for substantial morbidity and mortality worldwide. It is a growing public health problem in developing as well as developed countries. It is difficult to determine the exact mortality associated with foodborne illnesses (Helms et al., 2003). However, worldwide an estimated 2 million deaths occurred due to gastrointestinal illness, during the year 2005 (Fleury et al., 2008). More than 250 different foodborne illnesses are caused by various pathogens or by toxins (Linscott, 2011). Foodborne illnesses result from consumption of food containing pathogens such as bacteria, viruses, parasites or the food contaminated by poisonous chemicals or bio-toxins (World Health Organization [WHO], 2011c). Although majority of the foodborne illness cases are mild and self-limiting, severe cases can occur in high risk groups resulting in high mortality and morbidity in this group. The high risk groups for foodborne diseases include infants, young children, the elderly and the immunocompromised persons (Fleury et al., 2008).

There are changes in the spectrum of foodborne illnesses along with demographic and epidemiologic changes in the population. A century ago, cholera and typhoid fever were prevalent foodborne illnesses, globally. During last few decades, other foodborne infections have emerged, such as diarrheal illness caused by the parasite *Cyclospora*, and the bacterium *Vibrio parahemolyticus*. The newly identified microbes pose a threat to public health as they can easily spread globally and can mutate to form new pathogens. In the United States, 31 different pathogens are known to cause foodborne illness, however, numerous episodes of foodborne illnesses and hospitalizations are caused by unspecified agents (Centers for Disease Control and Prevention [CDC], 2011a).

Although foodborne illnesses cause substantial morbidity in the developed countries, the main burden is borne by developing countries. These illnesses are an obstacle to global development efforts and in the achievement of the Millennium Development Goals (MDGs) (WHO, 2011c). There is an impact of foodborne illnesses on four out of the eight MDGs. These include MDG 1(Eradication of extreme poverty); MDG 3 (Reduction in child mortality); MDG 5 (Improvement of maternal health); MDG 6 (Combating HIV/AIDS and other illnesses). The population in developing countries is more prone to suffer from foodborne illnesses because of multiple reasons, including lack of access to clean water for food preparation; inappropriate transportation and storage of foods; and lack of awareness regarding safe and hygienic food practices (WHO, 2011c). Moreover, majority of the developing countries have limited capacity to implement rules and regulations regarding food safety. Also, there is lack of effective surveillance and monitoring systems for foodborne illness, inspection systems for food safety, and educational programs regarding awareness of food hygiene (WHO, 2011a).

Foodborne illnesses have an impact on the public health as well as economy of a country (Helms et al., 2003). They have a negative impact on the trade and industries of the affected countries. Identification of a contaminated food product can result in recalling of that specific food product leading to economic loss to the industry. Foodborne outbreaks may lead to closure of the food outlets or food industry resulting in job losses for workers, affecting the individuals as well as the communities. Moreover, local foodborne illness outbreaks may become a global threat. The health of people in many countries can be affected by consuming contaminated food products, and may negatively impact a country's tourist industry. The foodborne illness outbreaks are reported frequently at national as well as international level underscoring the importance of food safety(WHO, 2011c).

Increasing commercialization of food production has resulted in the emergence and dissemination of previously unknown pathogens, and resulted in diseases such as bovine spongiform encephalopathy (BSE). BSE is a variant of Creutzfeldt-Jakob disease (vCJD) which affected human population in UK during the 1990s (WHO, 2011c).

During last few decades there are advances in technology, regulation, and awareness regarding food safety but new challenges have emerged because of mass production, distribution, and importation of food and emerging foodborne pathogens (Scallan, 2007; Taormina & Beuchat, 1999). One of the major issues of public health importance is the increasing resistance of foodborne pathogens to antibiotics (WHO, 2011c).

For the sake of public health, it is important to understand the epidemiology of foodborne illnesses because it will help in prevention and control efforts, appropriately allocating resources to control foodborne illness, monitoring and evaluation of food safety measures, development of new food safety standards, and assessment of the cost-effectiveness of interventions. The purpose of this chapter is to describe the epidemiology of foodborne illness in order to determine the magnitude of the problem, risk factors, monitoring and surveillance and measures of control.

2. Impact of foodborne illness

2.1 Public health impact

Foodborne illnesses are prevalent (Hoffman et al., 2005) but the magnitude of illness and associated deaths are not accurately reflected by the data available in both developed and developing countries. To fill the current data gap, the World Health Organization (WHO) has taken initiative for estimation of the global burden of foodborne illnesses (Kuchenmüller et al., 2009). World Health Organization and the US Centers for Disease Control and Prevention (CDC) report every year a large number of people affected by foodborne illnesses (Busani, Scavia, Luzzi, & Caprioli, 2006). Globally, an estimated 2 million people died from diarrheal diseases in 2005; approximately 70% of diarrheal diseases are foodborne. It is estimated that up to 30% of the population suffer from foodborne illnesses each year in some industrialized countries (WHO, 2011c). According to the estimation by CDC in 1999, around 76 million foodborne illnesses occur annually, resulting in 325,000 hospitalizations and 5200 deaths in the United States (Buzby & Roberts, 2009). However, a decrease in the incidence rates of notified foodborne illness was noticed from 1996 to 2005, but these rates have remained static since 2005 (Anderson, Verrill, & Sahyoun, 2011). There is a 20% reduction in illnesses caused by the specific pathogens tracked by FoodNet system, over the past 10 years. There are many explanations for this decrease in foodborne illness. It may be due to improved food safety because of regulatory and industry efforts or because of better detection, prevention, education, and control efforts (CDC, 2011a). According to 2011 estimates of CDC, annually 48 million Americans get sick, there are 128,000 hospitalizations, and 3,000 deaths due to foodborne illnesses in the US (CDC, 2011b). In Canada, an estimated 1.3 episodes per person-year of enteric disease occur (Fleury et al., 2008). In New Zealand, there are an estimated 119,320 episodes of foodborne illnesses each year, accounting for a rate of 3,241 per 100,000 population (Scott et al., 2000).

Foodborne-illness outbreaks are under-reported and it is estimated that 68% of foodborne-illness outbreaks are notified to the Centers for Disease Control and Prevention (Jones et al., 2004). Even during foodborne illness outbreaks, only a small proportion of the total number of cases is reported. In the United States, during 1993–1997, an average of 550 foodborne illness outbreaks was reported annually. Each outbreak had an average of 31 cases (Jones et al., 2004). Foodborne illnesses also play an important role in new and emerging infections. It is estimated that during past 60 years, an estimated 30% of all emerging infections comprised of pathogens transmitted through food leading to foodborne illness (Kuchenmüller et al., 2009).

2.2 Economic impact of foodborne illness

Every illness has an economic cost and same is the case with foodborne illness. However, the economic cost of health losses related to foodborne illnesses has not been extensively studied. There are few studies available which provide either incomplete cost estimates or their estimates are based on limiting assumptions (Buzby & Roberts, 2009). In the United States, data from Foodborne Diseases Active Surveillance Network (FoodNet) and other related studies contributed to estimates of the economic cost of foodborne illness (Angulo & Scallan, 2007).

The annual economic cost of foodborne illness is calculated by multiplying the cost per case with the expected annual number of foodborne illnesses experienced. It was estimated that in 1999, the US government spent $1 billion on food safety efforts at federal level, an additional $300 million were spent by state governments. Moreover, it is estimated that a total of $152 billion a year is spent on foodborne illness in the U.S. (Scharff, 2010). The foodborne illness also bears substantial economic burden at regional level. The annual estimated economic cost of foodborne illness for Ohio is between $1.0 and $7.1 billion i-e., cost of $91 to $624 per Ohio resident (Scharff, McDowell, & Medeiros, 2009).

A retrospective study performed in Uppsala, Sweden during 1998–99, estimated average costs per foodborne illness as $246 to society and $57 to the patient. An estimated $123 million was the annual cost of foodborne illnesses in Sweden (Lindqvist et al., 2001).

In New Zealand, the total cost of foodborne illness cases was estimated to be $55.1 million, accounting for $462 per case. The direct medical costs were calculated as $2.1 million while direct non-medical costs were $0.2 million (Scott et al., 2000). The estimated total costs were $161.9 million including government outlays of $16.4 million, industry costs of $12.3 million and $133.2 million for incident case costs of disease associated with treatment, loss of output and residual lifestyle loss (New Zealand Food Safety Authority, 2010).

2.3 The unknown burden

Data from surveillance systems and sentinel sites show only the tip of the iceberg and do not reflect true disease burden of foodborne illness (Figure 1). The reasons are that the sick persons may not seek medical care, may not be tested at laboratory or may not be notified to the relevant health authorities (WHO, 2011b).

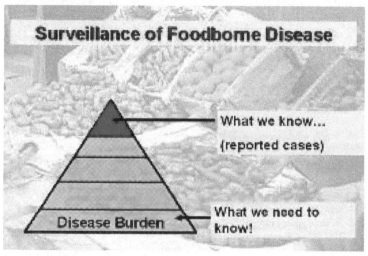

Fig. 1. The Unknown burden of foodborne illness. Adapted from WHO. Available at: http://www.who.int/foodsafety/foodborne_disease/ferg/en/index1.html Accessed August 04, 2011.

3. Risk factors

Factors such as socioeconomic status, malnutrition, micronutrient deficiencies, genetic factors and immunity play an important role in the causation of foodborne illnesses. The risk factors may be related to the host as well as to the environment.

3.1 Host risk factors

The risk factors for the foodborne illness caused by common pathogens have been determined by various studies. A case-control study conducted in the United States determined risk factors for Campylobacter infection. Traveling abroad and consumption of chicken or non-poultry meat prepared at a restaurant were stated as the main risk factors (Friedman et al., 2004). In Australia, annually an estimated 50,500 cases of *Campylobacter* infection in persons more than 5 years of age are associated with consumption of chicken (Stafford et al., 2008).

The risk factors for *Salmonella* Enteritidis have been explored in various studies conducted in different regions of the world. Foodborne Diseases Active Surveillance Network (FoodNet) in US, conducted a population-based case-control study for 12-month period during 2002–2003, to determine the risk factors associated with sporadic *Salmonella* Enteritidis infection. The study enrolled 218 cases and 742 controls. Travelling outside US five days before the onset of illness was found to be strongly associated with *Salmonella* Enteritidis infection [odds ratio (OR) 53, 95% Confidence Interval (CI) 23–125]. In addition, eating chicken cooked outside the home, consuming undercooked eggs and contact with birds and reptiles were associated with *Salmonella* Enteritidis infections (Marcus et al., 2007). In Denmark, a prospective case-control study of sporadic *Salmonella* Enteritidis infection was conducted in 1997–1999. It was revealed that foreign travel had a positive association with the disease. Foreign travelers were approximately three times more likely to suffer from *Salmonella* Enteritidis infection as compared to the non-travelers. Among non-travelers, the risk factors were eating eggs or dishes containing raw or undercooked eggs. The study concluded that eggs are the main source of *Salmonella* Enteritidis in Denmark (Mølbak & Neimann, 2002). A case-control study to determine risk factors for salmonellosis was conducted in the Netherlands. Consumption of raw eggs (OR 3.1, 95% CI 1.3–7.4) and products containing raw eggs (OR 1.8, 95% CI 1.1–3.0) were found to be the risk factors associated with *Salmonella* Enteritidis infection. The risk factors for *Salmonella* Typhimurium infection included contact with raw meat (OR 3.0, 95% CI 1.1–7.9), consumption of undercooked meat (OR 2.2, 95% CI 1.1–4.1) and recent antibiotic use (OR 1.9, 95% CI 1.0–3.4) (Doorduyn, Van Den Brandhof, Van Duynhoven, Wannet, & Van Pelt, 2006). In Canada, a matched case-control study was conducted to determine risk factors for sporadic *Salmonella* Typhimurium infection. The study was conducted during 1999 and 2000. *Salmonella* Typhimurium phage-type 104 (DT104) and non-DT104 diarrheal illness in Canada, were studied. The risk factors associated with DT104 illness were recent antibiotic use (OR 5.2, 95% CI 1.8–15.3), residence on a livestock farm (OR 4.9, 95% CI 1.9–18.9), and travel outside Canada in the recent past (OR 4.1, 95% CI 1.2–13.8) (Doré et al., 2004). To determine the risk factors for typhoid fever, a community-based case-control study was conducted during 2001- 2003, in Jakarta, Indonesia. The risk factors for typhoid fever were recent typhoid fever in the household (OR 2.38, 95% CI 1.03-5.48); not using soap for hand washing (OR 1.91, 95% CI 1.06-3.46); eating and sharing food in the same dish (OR 1.93, 95% CI 1.10-3.37); and non availability of appropriate sanitation services (OR 2.20, 95% CI 1.06-4.55). Typhoid fever was also found to be associated with young age and female gender (Vollaard et al., 2004).

There is a well-established association between *E. coli* O157:H7 infections and consumption of contaminated ground beef. Sporadic *E. coli* O157:H7 infection has been found to be linked to eating undercooked hamburgers and exposure to cattle on farms (Allos, Moore, Griffin, & Tauxe, 2004). A case-control study of sporadic Shiga toxin-producing *Escherichia coli* O157 (STEC O157) infections was conducted in 1999–2000, in US. In this 12-month study, 283 patients and 534 controls participated. The risk factors for STEC O157 infection included consumption of pink hamburgers, intake of unsafe surface water, and exposure to cattle (Voetsch et al., 2007).

3.2 Environmental risk factors
The environmental risk factors, food industry, food manufacturing, retail and catering sectors play important role in foodborne illness. Various studies of both sporadic and outbreak-associated illness involving different geographic areas, varied study designs, and a variety of pathogens have revealed that restaurants play an important role as source of foodborne illness in the US. Many factors may contribute to an increased risk of foodborne illness due to foods consumed in restaurants such as cross-contamination events within restaurants (Jones & Angulo, 2006). Research has shown that populations with low socioeconomic status have less access to high-quality food products, resulting in reliance on small markets that may sell foods of poorer quality. These populations have less access to supermarkets selling a variety of fresh fruits and vegetables and low-fat foods. Consumption of a poorer- quality diet along with other risk factors may lead to foodborne illness.

4. Causative agents

A variety of agents can cause foodborne illnesses. Although most of the foodborne illnesses are caused by bacterial or viral pathogens, there are certain non-infectious causes such as, chemicals and toxins. Approximately 67% of all foodborne illnesses caused by pathogens have viral etiology. Most commonly implicated viruses in foodborne illnesses are norovirus, hepatitis A virus, hepatitis E virus, rotavirus, and astrovirus (Atreya, 2004).

4.1 Foodborne pathogens

Various bacterial, viral and parasitic agents causing foodborne illness of public health importance are listed below (Nelson & Williams, 2007; The Food Safety and Inspection Service [FSIS], 2008):

Bacterial Agents
- *Listeria monocytogenes*
- *Staphylococcus aureus*
- *Bacillus cereus*
- *Bacillus anthracis*
- *Clostridium botulinum*
- *Clostridium perfringens*
- *Clostridium difficile*
- *Salmonella* spp
- *Shigella* spp
- *Campylobacter* spp
- *Escherichia coli* 0157:H7
- *Yersinia enterocolitica*
- *Brucella* spp
- *Vibrio Cholerae*

Viral Agents
- *Norovirus*
- *Hepatitis A*
- *Hepatitis E*
- *Adenovirus (Enteric)*
- *Rotaviruses*

Parasitic Agents
- *Cryptosporidium sp.*
- *Cyclospora cayetanensis*
- *Giardia lamblia*
- *Entamoeba histolytica*
- *Balantadium coli*

5. Clinical manifestations

The clinical features of foodborne illness include nausea, vomiting, diarrhea, abdominal pain and fever (McCulloch, 2000). Symptoms of foodborne illnesses may persist for two to three days. In majority of patients, the illness is mild in nature; however, severe complications may occur in some cases. The serious complications of foodborne illness include hypovolemic shock, septicemia, hemolytic uremic syndrome, reactive arthritis, and

Guillan- Barré syndrome. Severe complications may occur in immunocompromised cases, patients with co-morbidities, and in very young or elderly patients. Several cases of foodborne illnesses remain undiagnosed as the patient may not seek medical attention. If a person reports to a health care facility, generally, stool specimens are taken and tested for bacteria and parasites (Linscott, 2011).

Table 1 displays the common causative agents of foodborne illnesses along with their incubation period, clinical features, possible contaminants, diagnostic procedures and steps for prevention (Iowa State University, 2010; Linscott, 2011; Nelson & Williams, 2007).

Organism	Incubation Period	Clinical Features	Possible contaminants	Laboratory diagnosis	Preventive measures
Bacillus cereus	30 minutes to 15 hours	Sudden onset of nausea and vomiting, abdominal cramps with or without diarrhea	Cooked but not properly refrigerated foods, such as vegetables, fish, rice, potatoes and pasta.	Stool test Food sources may be tested.	Careful attention to food preparation, cooking and storage standards
Brucella sp.	7 to 21 days	Fever, night sweats, backache, muscle aches, diarrhea	Unpasteurized dairy foods and meat	Serology, blood culture.	Consumption of pasteurized dairy products. Cooking meat thoroughly
Campylobacter jejuni	1 to 7 days	Fever, headache, nausea, abdominal cramps, diarrhea. Guillian-Barre syndrome may occur in some patients.	Raw and undercooked poultry, eggs, raw beef, unpasteurized milk, contaminated water	Stool culture, rapid immune-chromogenic tests, molecular assays.	Pasteurization of milk, cooking foods properly, prevention of cross-contamination
Clostridium botulinum – preformed toxin	12 to 72 hours	Abdominal cramps, nausea, vomiting, diarrhea, diplopia, blurred vision, headache, dryness of mouth, muscle paralysis, respiratory failure, nerve damage.	Inappropriately canned foods, meats, sausage, fish.	Detection of botulinum toxin in serum, stool, or patient's food	Proper canning of foods, proper cooking of foods
Clostridium perfringen-toxin	8 to 22 hours	Nausea, abdominal cramps and diarrhea, dehydration in some cases	Meat, poultry, gravy, inadequately reheated food	Stool test for enterotoxin	Maintenance of proper cooking temperatures

Organism	Incubation Period	Clinical Features	Possible contaminants	Laboratory diagnosis	Preventive measures
Cryptosporidium sp.	2 to 10 days	Nausea, loss of appetite, watery diarrhea accompanied by mild abdominal cramp; severity depends on host immune status.	Undercooked food or food contaminated by ill food handler, contaminated drinking water or milk	Stool test for detecting oocysts by modified acid-fast stain, direct fluorescent antibodies, or with immunoassays	Avoidance of contaminated water or food, washing hands after using the toilet and before handling food.
Cyclospora cayetanensis	1 to 14 days	Watery diarrhea, abdominal cramps, nausea, anorexia, and weight loss	Fresh fruits and vegetables	Stool test for detecting oocysts by modified acid-fast stain	Hygienic practices in the agricultural setting and in the individual's environment.
Entero-hemorrhagic E.coli – 0157;H7 and other Shiga toxins	1 to 8 days	Bloody diarrhea, vomiting, abdominal cramps, fever, hemorrhagic colitis, hemolytic uremic syndrome	Undercooked ground beef, unpasteurized milk, fruit juices, raw fruits, and vegetables	Stool culture for isolating the organism (Sorbitol MacConkey or CHROMagar media), Antisera or latex agglutination, Immunoassays.	Cooking meat thoroughly, prevention of cross-contamination
Hepatitis A virus (HAV)	15 to 50 days	Anorexia, nausea, abdominal discomfort, malaise, diarrhea, fever, dark colored urine, and jaundice. Arthritis, urticarial rash and aplastic anemia in rare cases	Consumption of contaminated water or food. Shellfish, clams, oysters, fruits, vegetables, iced drinks, lettuce, and salads.	Serum test for IgM antibodies to HAV, Serum test for Alanine transaminase (ALT), Aspartate aminotransferase (AST) and bilirubin	Washing hands after using toilet and before preparing food
Listeria monocytogenes	2 days to 3 weeks	Diarrhea, nausea, fever, muscular pain, flu-like symptoms in pregnant females (may result in preterm birth or still birth). Meningitis and septicemia may occur in older age or immunocompromised patients	Unpasteurized milk, cheese prepared with unpasteurized milk, vegetables, meats, hotdogs, and seafood	Blood culture, cerebrospinal fluid cultures, detection of antibodies to listerolysin O.	Pasteurization of dairy products, cooking foods properly, prevention of cross-contamination.

Organism	Incubation Period	Clinical Features	Possible contaminants	Laboratory diagnosis	Preventive measures
Noroviruses	Between 12 and 48 hours (average, 36 hours); duration, 12-60 hours	Nausea, vomiting, diarrhea, abdominal cramps, headache, and fever	Raw food products, contaminated food or drinking water. Shellfish, oysters and salads.	Stool culture, nucleic acid hybridization, Reverse transcription polymerase chain reaction (PCR), Electron microscopy, Enzyme , and Immunoassays (ELISA).	Proper sewage disposal, water chlorination, restricting infected food handlers from handling food
Salmonella sp.	Non-Typhi: 1 to 3 days Typhi: 3 to 60 days	Non- Tyhi: Nausea, diarrhea, abdominal pain, and fever Typhi: fever, headache, shivering, loss of appetite, malaise, constipation, and muscular pain	Contaminated egg; poultry; meat; unpasteurized milk, dairy foods and fruit juice; raw fruits and vegetables.	Non –Typhi: Stool culture Typhi: Stool culture and blood culture	Cooking food thoroughly, prevention of cross-contamination.
Shigella sp.	12 to 50 hours	Vomiting, abdominal pain, diarrhea with blood and mucus, and fever	Contaminated food or drinking water. Raw vegetables, salads, dairy foods, and poultry.	Stool culture	Practicing proper washing and hygienic techniques in food preparation
Staphylococcus aureus (preformed enterotoxin)	1 to 6 hours	Nausea, vomiting, abdominal pain, fever and diarrhea.	Inappropriately refrigerated foods such as meat, salads, salad dressing, cream pastries, cream-filled baked products, poultry, gravy, and sandwich fillings	Stool or vomitus test for detection of toxin. Suspected food may be tested to detect toxin	Refrigerating foods properly, using hygienic practices
Vibrio cholerae (O1, O139) (non- O1 or non O139)	4 hours to 4 days	Severe watery diarrhea, abdominal cramps, nausea vomiting, headache, fever, and chills. Severe dehydration and death may occur.	Contaminated water, shellfish and crustaceans	Stool culture	Cooking food thoroughly

Organism	Incubation Period	Clinical Features	Possible contaminants	Laboratory diagnosis	Preventive measures
Vibrio parahaemolyticus	2 to 48 h	Watery diarrhea, abdominal pain, nausea, and vomiting	Raw or undercooked seafood from contaminated seawater	Stool culture	Cooking food thoroughly, practicing hygienic techniques in food preparation and storing food at the appropriate temperatures
Yersinia enterocolitica	1 to 3 days	Diarrhea, vomiting, fever, and abdominal cramps	Contaminated meat and milk, undercooked pork, unpasteurized milk, and contaminated water.	Stool culture, blood culture.	Pasteurization of milk; Cooking food thoroughly, prevention of cross-contamination, practicing hygienic techniques in food preparation

Table 1. Causative agents, clinical features and preventive measures for common foodborne illnesses

6. Foodborne illness outbreaks

Manifestation of the same symptoms or illness by two or more of the individuals after consumption of the same contaminated food, is labeled as an outbreak of foodborne illness (The Food Safety and Inspection Service (FSIS), 2008). The outbreak can be recognized when several people who ate together at an occasion become sick and have similar clinical manifestations. Most of the outbreaks are localized, such as an outbreak after eating a catered meal or at a restaurant. However, more widespread outbreaks also occur that affect people in various places, and may last for several weeks (FSIS, 2008).

An outbreak is investigated by describing it systematically and trying to find out the causative agent. The description of outbreak includes time, place, and person distribution. Data is collected by interviews, by testing suspected food source, and by gathering other related information (FSIS, 2008). It is important that foodborne illness outbreaks are investigated timely and proper environmental assessments are done so that appropriate prevention strategies are identified (Lynch, Tauxe, & Hedberg, 2009). According to CDC, the etiology of majority (68%) of reported foodborne-illness outbreaks is unknown. The causative agent of many of the outbreaks cannot be determined because of certain issues related to outbreak investigations e.g. lack of timely reporting, lack of resources for investigations and other priorities in health departments. In addition, ill persons who do not seek health care and limited testing of specimens are also the contributory factors in failure to determine the cause of foodborne illness outbreak (Lynch et al., 2009).

A number of foodborne illness outbreaks are reported from various parts of the world. Analysis of foodborne outbreak data helps in the estimation of the proportion of human cases of specific enteric diseases attributable to a specific food item (Greig & Ravel, 2009). Worldwide, a total of 4093 foodborne outbreaks occurred between 1988 and 2007. It was found that *Salmonella* Enteritidis outbreaks were more common in the EU states and eggs were the most frequent vehicle of infection. Poultry products in the EU and dairy products in the United States, were related to *Campylobacter* associated outbreaks. In Canada, *Escherichia coli* outbreaks were associated with beef. In Australia and New Zealand, *Salmonella* Typhiumurium outbreaks were more common (Greig & Ravel, 2009). Daniels and colleague (2002) conducted a study in the United States, to describe the epidemiology of foodborne illness outbreaks in schools, colleges and universities. The data from January 1, 1973, to December 31, 1997 was reviewed. Characteristics of ill persons, the magnitude of foodborne illness outbreaks, causative agents, food vehicles for transmitting infection, place of preparing food and contributory factors for occurrence of outbreaks, were examined. A total of 604 outbreaks of foodborne illness were reported during the study period. In majority (60%) of the outbreaks the etiology was unknown. Among the outbreaks with a known etiology, in 36% of outbreak reports Salmonella was the most commonly identified pathogen. Foods containing poultry, salads, Mexican-style food, beef and dairy foods were the most commonly implicated vehicles for transmission (Daniels et al., 2002). A total of 6,647 outbreaks of foodborne illness were reported during 1998-2002 in U.S (Lynch, Painter, Woodruff, & Braden, 2006). As a result of these outbreaks, 128,370 persons were reported to become ill. The etiology was identified in 33% outbreaks. The largest percentage of outbreaks was caused by bacterial pathogens and *Salmonella* Enteritidis was the most common causative agent. However, the highest mortality was caused by *Listeria monocytogenes*. Viral pathogens were responsible for 33% of the outbreaks. Among the viral pathogens, norovirus was the most common causative agent (Lynch et al., 2006). In 2002, a salmonellosis outbreak occurred in five states of U.S. It occurred after consuming ground beef. During this outbreak, forty seven cases were reported; out of which 17 people were hospitalized and one death was reported (FSIS, 2008).

Systematic surveillance systems are very important to get information about causative organisms, sources of infection and modes of transmission in foodborne illness (O'Brien et al., 2006). An electronic surveillance system (SurvNet) was established for monitoring and investigation of infectious disease outbreaks, in 2001, in Germany (Krause et al., 2007). During 2001–2005, a total of 30,578 outbreaks were reported. These outbreaks ranged in size from 2 to 527 cases. The most common settings for outbreaks in 2004 and 2005 were households, nursing homes, and hospitals (Krause et al., 2007). In England and Wales, systematic national surveillance of outbreaks of infectious intestinal disease (IID) was introduced in 1992. This system provides information on etiologic agent, sources of infection and modes of transmission. Between January 1, 1992 and January 31, 2003, a total of 1763 outbreaks of IID were reported (O'Brien et al., 2006). From 1992 to 2008, 2429 foodborne outbreaks were reported in England and Wales. Approximately half of the outbreaks were caused by *Salmonella* spp. Poultry and red meat was the most commonly implicated foods in the causation of outbreaks. The associated factors in most outbreaks were cross-contamination, lack of adequate heat treatment and improper food storage (Gormley et al., 2011). In central Taiwan, 274 outbreaks of foodborne illness including 12,845 cases and 3 deaths were reported during 1991 to 2000. Majority (62.4%) of the outbreaks were caused by bacterial pathogens. The main etiologic agents were *Bacillus cereus*, *Staphylococcus aureus*, and *Vibrio parahaemolyticus*. The important contributing factor was improper handling of food. The implicated foods included seafood, meat products and cereal products (Chang & Chen, 2003).

A review of *E. coli* O157 outbreaks reported to CDC from 1982 to 2002, was conducted (Sparling, Crowe, Griffin, Swerdlow, & Rangel, 2005). During the study period, 49 states reported 350 outbreaks. During these outbreaks 8,598 cases were reported; out of which 1,493 (17%) were hospitalized, and 40 (0.5%) died. Transmission route for 183 (52%) was foodborne and the food vehicle for 75 (41%) foodborne outbreaks was ground beef (Sparling et al., 2005).

In a review of all confirmed foodborne outbreaks reported to the Centers for Disease Control and Prevention (CDC) from 1982 to 1997, a total of 2,246 foodborne outbreaks were reported; 697 (31%) were of known etiology. *Salmonella* was responsible for 65% of outbreaks with a known etiology (Hedberg et al., 2008). An analysis of national foodborne outbreak data from 1973 through 2001for determining the proportion of *Salmonella* Heidelberg outbreaks and its causes was conducted (Chittick, Sulka, Tauxe, & Fry, 2006). Out of 6,633 outbreaks with known etiology, *Salmonella* Heidelberg was responsible for 184 (3%) outbreaks. 53 outbreaks were poultry or egg-related (Chittick et al., 2006). Because of increasing consumption of fresh produce, changes in production and distribution, there is an increase in foodborne outbreaks caused by contaminated fresh produce (Lynch et al., 2009; Sivapalasingam et al., 2004). An analysis of data for 1973 through 1997 from the Foodborne Outbreak Surveillance System was conducted. A total of 190 outbreaks were reported including 16,058 illnesses, 598 hospitalizations, and eight deaths. There was an increase in the proportion of produce-associated outbreaks, rising from 0.7% in the 1970s to 6% in the 1990s. The food items most frequently implicated in the outbreaks associated with fresh produce include salad, lettuce, juice, melon, sprouts, and berries. In this analysis, the bacterial pathogens were the most common causative agents. Among the bacterial pathogens *Salmonella* was the commonest causative agent (Sivapalasingam et al., 2004). In a study carried out from October 2004 to October 2005 in Catalonia, Spain, 181 outbreaks were reported; 72 were caused by *Salmonella* and 30 by norovirus (NoV); 66.7% of NoV outbreaks occurred in restaurants. Hospitalizations were reported more commonly in outbreaks caused by bacterial pathogens as compared to those caused by NoV (Martinez et al., 2008).

In Spain, 971 and 1227 outbreaks were reported in 2002 and 2003, respectively. A substantial proportion of outbreaks were associated with consumption of eggs and egg products (Crespo et al., 2005). A study was conducted in northeastern Spain to investigate the trend of foodborne *Salmonella*-caused outbreaks and number of cases, hospitalizations, and deaths. A review of the information on reported outbreaks of foodborne disease from 1990 to 2003 found a total of 1,652 outbreaks. 1,078 outbreaks had a known etiologic agent; out of which 871 (80.8%) were caused by *Salmonella*, with 14,695 cases, 1,534 hospitalizations, and 4 deaths. Forty-eight percent of *Salmonella*-caused outbreaks were associated with consumption of eggs. The study concluded that to prevent and control these outbreaks, health education programs are needed to create awareness regarding risk of consuming raw or undercooked eggs (Domínguez et al., 2007).

In 2002, in the Netherlands a national study of foodborne illness outbreaks was performed (van Duynhoven et al., 2005). A total of 281 foodborne illness outbreaks were included. Most of these outbreaks were reported from nursing homes, restaurants, hospitals and day-care centers. In restaurant outbreaks, food was the mode of transmission in almost 90% outbreaks. The causative agents included norovirus (54%), Salmonella spp. (4%), rotavirus (2%), and Campylobacter spp. (1%) (van Duynhoven et al., 2005). The Danish and Dutch investigation reported foodborne outbreaks caused by the same strain of *Salmonella* serotype Typhimurium DT104 during the year 2005. These were two distinct outbreaks caused by the same organism. The outbreaks were traced to two kinds of raw beef provided by the same supplier (Ammon & Tauxe, 2007).

A study conducted in Qassim province, Saudi Arabia, analyzed the foodborne illness surveillance data for the year 2006. During the study period, 31 foodborne illness outbreaks comprising of 251 cases, were reported. The most common etiologic agent was *Salmonella* spp, followed by *Staphylococcus* aureus. Commercially prepared foods were consumed by the majority (68.9%) of the cases. Meat and Middle Eastern meat sandwich were the commonly implicated food vehicles (Al-Goblan & Jahan, 2010).

The globalization of the food industry has resulted in occurrence of international viral foodborne outbreaks (Verhoef et al., 2009). Although mostly the bacterial pathogens are targeted for prevention of foodborne illness, yet it is suspected that noroviruses (NoV) are the most common cause of gastroenteritis. The role of NoV in foodborne illness has become evident because of new molecular assays. An analysis of 8, 271 foodborne outbreaks reported to CDC from 1991 to 2000 showed an increase in the proportion of NoV-confirmed outbreaks. It rose from 1% in 1991 to 12% in 2000. In addition, NoV outbreaks were larger in magnitude as compared to the bacterial outbreaks (Widdowson et al., 2005).

The Belgian data for foodborne norovirus (NoV) outbreaks showed that in 2007, 10 NoV foodborne outbreaks were reported accounting for 392 cases in Belgium. NoV was the most commonly identified agent in foodborne outbreaks. Sandwiches were the most commonly implicated food. A review of forty outbreaks due to NoV, from 2000 to 2007 revealed that the food handler was responsible for the outbreak in 42.5% of the cases. The source of transmission was water (27.5%), shellfish (17.5%) and raspberries (10.0%) in remaining cases (Baert et al., 2009).

The contributory factors observed in most of the foodborne illness outbreaks included poor personal hygiene, inadequate holding times and temperatures, cross-contamination, lack of adequate heat treatment and improper food storage (Chang & Chen, 2003; Gormley et al., 2011; Hedberg et al., 2008).

7. Emerging foodborne infections

The changes in the methods of food production have affected the epidemiology of foodborne infection (Tassios & Kerr, 2010). Various new pathogens have emerged due to changing production processes in food industry. Some of these are new pathogens and were unknown previously, others are emerging pathogens for foodborne infections, and some others are evolving pathogens that have become more potent (Mor-Mur & Yuste, 2009). These pathogens include *Campylobacter jejuni, Salmonella* Typhimurium DT104, enterohemorrhagic *Escherichia coli, Listeria monocytogenes, Arcobacter butzleri, Mycobacterium avium* subsp. *Paratuberculosis,* etc. *Campylobacter jejuni* O:19 and other serotypes may cause neuropathy called Guillain–Barré syndrome. Multi-drug resistant *Salmonella* Typhimurium DT104 and other serotypes may cause salmonellosis leading to chronic reactive arthritis. Enterohemorrhagic *Escherichia coli* infection can cause complications such as hemolytic uremic syndrome and thrombotic thrombocytopenic purpura. *Listeria monocytogenes* causes listeriosis, a public health problem of major concern due to severe non-enteric nature of the disease i-e, meningitis or meningoencephalitis, and septicemia. *Arcobacter butzleri* is isolated from raw poultry, meat, and meat products and is considered a potential foodborne pathogen. *Mycobacterium avium* subsp. *paratuberculosis* may contribute to Crohn's disease (Mor-Mur & Yuste, 2009).

The new emerging foodborne infections are also related to food handling practices. Foodborne infections caused by *Escherichia coli* O157, *Yersinia enterocolitica, Campylobacter* and *Vibrio*, are associated with processing and packaging of food, or the importation of certain food from a new geographical area (Robinson, 2007). The well known foodborne pathogens may develop antimicrobial resistance and generate new public health challenges (Newell et al., 2010). Antimicrobial resistance has developed in many bacterial foodborne pathogens, such as *Salmonella, Shigella, Vibrio* spp., *Campylobacter*, methicillin resistant *Staphylcoccus aureas, E. coli* and *Enterococci*.

Previously un-recognized foodborne pathogens are constantly emerging. The emerging viral foodborne infections include norovirus, hepatitis A, rotaviruses and SARS (Newell et al., 2010). Currently, the norovirus (NoV) and hepatitis A virus (HAV) are considered the most important human foodborne pathogens. NoV and HAV are highly infectious and may cause large outbreaks (Koopmans & Duizer, 2004). Hepatitis E causes substantial morbidity in many developing countries while in developed countries, it was previously thought to be confined to travelers returning from endemic areas (Dalton, Bendall, Ijaz, & Banks, 2008). However, this concept is changing based on the evidence available. In developed countries, Autochthonous hepatitis E is found to be quite common and affects older age groups leading to morbidity and mortality (Dalton et al., 2008).

8. Monitoring and surveillance of foodborne illness

Worldwide, there are various surveillance systems to monitor, investigate, control and prevent illness. To assess and monitor morbidity and mortality in the United States, surveillance activities are conducted by several systems in collaboration with federal agencies and health departments (McCabe-Sellers & Beattie, 2004). Some surveillance systems are specific for foodborne illnesses. In addition to monitoring the foodborne illness, these surveillance activities also help in evaluating the safety of the food supply (Allos et al., 2004). Some of these surveillance systems are discussed below:

8.1 FoodNet (Foodborne disease active surveillance network)

FoodNet is the surveillance system in the United States. For Foodnet, CDC has collaborated with ten Emerging Infections Program (EIP) sites (California, Colorado, Connecticut, Georgia, New York, Maryland, Minnesota, Oregon, Tennessee and New Mexico), the US Department of Agriculture, and the Food and Drug Administration. It performs active surveillance for foodborne illnesses and also conducts epidemiologic studies to determine the changing epidemiology of foodborne illnesses. It responds to new and emerging foodborne illnesses, monitors the burden of foodborne illnesses, and identifies their sources (FSIS, 2008).

Figure 2 shows the burden of illness pyramid. It helps in understanding foodborne disease reporting in FoodNet surveillance system. It shows steps involved in the registration of an episode of foodborne illness in the population. Moving from bottom to top of pyramid, the steps are: exposure of some individuals in the general population to an organism; out of the exposed some persons become ill; out of all ill some seek medical care; a specimen is obtained from some of these persons and sent to a clinical laboratory; some of these specimens are tested for a specific pathogen; the causative organism is identified in some of these tested specimens; a local or state health department receives the report of the laboratory-confirmed case (CDC, 2011c).

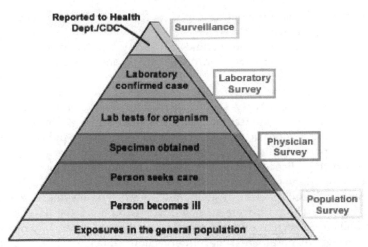

Fig. 2. FoodNet surveillance, burden of illness pyramid. Adapted from CDC. Available at: http://www.cdc.gov/foodnet/surveillance_pages/burden_pyramid.htm. Accessed August 04, 2011.

8.2 PulseNet (The molecular subtyping network for foodborne bacterial disease surveillance)

In the United States, PulseNet has created a national framework for pathogen-specific surveillance (Li et al., 2010). PulseNet is responsible for molecular subtyping of foodborne illness surveillance. It helps in detecting widespread foodborne outbreaks by comparing strains of bacterial pathogens from all over the United States. It performs DNA "fingerprinting" on foodborne bacteria by pulsed-field gel electrophoresis. By identifying and labeling each "fingerprint" pattern, it is possible to rapidly compare these patterns through an electronic database at the CDC, thus identifying related strains (FSIS, 2008).

8.3 Electronic Foodborne Outbreak Reporting System (eFORS)

The Electronic Foodborne Outbreak Reporting System's (eFORS) database is a surveillance system that collects reports on foodborne outbreaks. It requires specialized knowledge and expertise to appropriately analyze and interpret the data (Middaugh, Hammond, Eisenstein, & Lazensky, 2010). Various studies are conducted by analyzing the data collected within the Electronic Foodborne Outbreak Reporting System (eFORS) in various settings, such as schools in order to examine the magnitude of foodborne illness, their etiologies and to provide recommendations for prevention of foodborne illness (Venuto, Halbrook, Hinners, & Mickelson, 2010).

9. Strategies for control of foodborne illness

The contamination of food is influenced by multiple factors and may occur anywhere in the food production process (Newell et al., 2010). However, most of the foodborne illnesses can be traced back to infected food handlers. Therefore, it is important that strict personal hygiene measures should be adopted during food preparation. To prevent foodborne infections in children, educational measures are needed for parents and care-takers. The interventions should focus on avoiding exposure to infectious agents and on preventing cross-contamination (Marcus, 2008).

Good agriculture practice and good manufacturing practice should be adopted to prevent introduction of pathogens into food products (Koopmans & Duizer, 2004). In order to control foodborne viral infections, it is important to increase awareness of food handlers regarding the presence and spread of these viruses. In addition, standardized methods for the detection of foodborne viruses should be utilized and laboratory-based surveillance should be established for early detection of outbreaks (Koopmans & Duizer, 2004).

To prevent food-related zoonotic diseases, collaboration between public health, veterinary and food safety experts should be established. This collaboration will help in monitoring trends in the existing diseases and in detecting emerging pathogens. It will help in developing effective prevention and control strategies (Newell et al., 2010). The control strategies should be based on creating awareness among the consumers, farmers and those raising farm animals. The improvement of farming conditions, the development of more sensitive methods for detection of pathogens in slaughtered animals and in food products, and proper sewage disposal are other intervention strategies (Pozio, 2008). Hygienic measures are required throughout the continuum from "farm to fork". Further research is also required to explore pathways of the foodborne illness and to determine the vehicles of the greatest importance (Unicomb, 2009).

In a study conducted in Turkey, knowledge, attitudes, and practices about food safety among food handlers, were explored. The study revealed that food handlers in Turkish food industry often lacked knowledge regarding basic food hygiene. The authors concluded that the food handlers must be educated regarding safe food handling practices (Bas, Safak Ersun, & KIvanç, 2006). For the prevention of foodborne outbreaks, training of food handlers, regarding appropriate preparation and storage of food is required. In addition, effective environmental cleaning and disinfection, excluding infected staff, implementing hand hygiene principles, and preventing cross-contamination are recommended (Greig & Lee, 2009).

Proper processing of food is necessary to ensure the reduction or elimination of the growth of harmful microorganisms. Pasteurization of milk and dairy products and hygienic manufacturing processes for canned foods will help reduce the cases of food-borne illnesses. Food irradiation is a recent technology for prevention of food-borne illnesses. The food irradiation methods include Gamma irradiation, Electron beam irradiation, and X-irradiation. Irradiation destroys the organism's DNA and prevents DNA replication. Food irradiation could eliminate *E. coli* in ground beef, *Campylobacter* in poultry, *Listeria* in food and dairy products, and *Toxoplasma gondii* in meat. However, all food products cannot be irradiated (Linscott, 2011).

The consumers should also take precautions to prevent foodborne illnesses. These include cooking meat, poultry, and eggs at appropriate temperatures; proper refrigeration and storage of foods at recommended temperatures; prevention of cross-contamination of food; use of clean slicing boards and utensils while cooking; and washing hands often while preparing food (Linscott, 2011).

10. Conclusion

Globally, foodborne illnesses are responsible for substantial morbidity and mortality. Although it is difficult to determine the exact mortality associated with foodborne illnesses, worldwide an estimated 2 million deaths occurred due to gastrointestinal illness during the year 2005. Foodborne illnesses result from consumption of food containing pathogens or the food contaminated by poisonous chemicals or toxins. Foodborne illnesses also play an important role in new and emerging infections. Various new pathogens have emerged due to changing dynamics of food industry. It is estimated that during past 60 years, approximately 30% of all emerging infections comprised of pathogens transmitted through food.

It is important to monitor and investigate the foodborne illnesses in order to control and prevent them. For this purpose various surveillance systems are established in different parts of the world. FoodNet, PulseNet and EFORS are the surveillance systems for monitoring foodborne illness in the United States. These surveillance systems play vital role in prevention, early detection and control of foodborne illness outbreaks.

The contamination of food is influenced by multiple factors and may occur anywhere along the food chain. Good agriculture practice and good manufacturing practice should be adopted to prevent introduction of pathogens into food products. Most foodborne illnesses can be tracked to infected food handlers. Therefore, it is important that strict personal hygiene measures should be adopted during food preparation. The consumers should also take precautions for prevention of foodborne illness. These include cooking food at appropriate temperatures and following standard hygiene practices, proper storage and prevention of cross-contamination of food. Thus integrated intervention strategies are required to prevent foodborne illness at community level. Successful implementation of these interventions requires inter-sectoral collaboration including agriculture industry, food industry and health care sector.

11. References

Al-Goblan, A. S., & Jahan, S. (2010). Surveillance for foodborne illness outbreaks in Qassim, Saudi Arabia, 2006. *Foodborne Pathogens and Disease, 7*(12), 1559-1562. doi:10.1089/fpd.2010.0638

Allos, B. M., Moore, M. R., Griffin, P. M., & Tauxe, R. V. (2004). Surveillance for Sporadic Foodborne Disease in the 21st Century: The FoodNet Perspective. *Clinical Infectious Diseases, 38*(s3), S115-S120. doi:10.1086/381577

Anderson, A. L., Verrill, L. A., & Sahyoun, N. R. (2011). Food Safety Perceptions and Practices of Older Adults. *Public Health Reports, 126*(2), 220-227.

Ammon. A., & Tauxe. R. V. (2007). Investigation of multi-national foodborne outbreaks in Europe: some challenges remain, *135*(6), 887-889. doi:10.1017/S0950268807008898

Angulo, F. J., & Scallan, E. (2007). Activities, Achievements, and Lessons Learned during the First 10 Years of the Foodborne Diseases Active Surveillance Network: 1996–2005. *Clinical Infectious Diseases, 44*(5), 718 -725. doi:10.1086/511648

Atreya, C. D. (2004). Major foodborne illness causing viruses and current status of vaccines against the diseases. *Foodborne Pathogens and Disease, 1*(2), 89-96. doi:10.1089/153531404323143602

Baert, L., Uyttendaele, M., Stals, A., VAN Coillie, E., Dierick, K., Debevere, J., & Botteldoorn, N. (2009). Reported foodborne outbreaks due to noroviruses in Belgium: the link between food and patient investigations in an international context. *Epidemiology and Infection, 137*(3), 316-325. doi:10.1017/S0950268808001830

Bas, M., Safak Ersun, A., & KIvanç, G. (2006). The evaluation of food hygiene knowledge, attitudes, and practices of food handlers' in food businesses in Turkey. *Food Control*, *17*(4), 317-322. doi:16/j.foodcont.2004.11.006

Busani, L., Scavia, G., Luzzi, I., & Caprioli, A. (2006). Laboratory surveillance for prevention and control of foodborne zoonoses. *Annali dell'Istituto Superiore Di Sanità*, *42*(4), 401-404.

Buzby, J. C., & Roberts, T. (2009). The Economics of Enteric Infections: Human Foodborne Disease Costs. *Gastroenterology*, *136*(6), 1851-1862. doi:10.1053/j.gastro.2009.01.074

Centers for Disease Control and Prevention [CDC]. (2011a). CDC 2011 Estimates: Findings. Retrieved June 26, 2011, from http://www.cdc.gov/foodborneburden/2011-foodborne-estimates.html

Centers for Disease Control and Prevention [CDC]. (2011b). Estimates of Foodborne Illness in the United States. Retrieved June 26, 2011, from http://www.cdc.gov/foodborneburden/index.html

Centers for Disease Control and Prevention [CDC]. (2011c). Foodnet surveillance: Burden of illness. Retrieved August 18, 2011, from http://www.cdc.gov/foodnet/surveillance_pages/burden_pyramid.htm

Chang, J. M., & Chen, T. H. (2003). Bacterial Foodborne Outbreaks in Central Taiwan, 1991-2000. *Journal of Food and Drug Analysis*, *11*(1), 53-59.

Chittick, P., Sulka, A., Tauxe, R. V., & Fry, A. M. (2006). A Summary of National Reports of Foodborne Outbreaks of Salmonella Heidelberg Infections in the United States: Clues for Disease Prevention. *Journal of Food Protection*, *69*, 1150-1153.

Crespo, P. S., Hernández, G., Echeíta, A., Torres, A., Ordóñez, P., & Aladueña, A. (2005). Surveillance of foodborne disease outbreaks associated with consumption of eggs and egg products: Spain, 2002-2003. *Euro Surveillance: Bulletin Européen Sur Les Maladies Transmissibles = European Communicable Disease Bulletin*, *10*(6), E050616.2.

Dalton, H. R., Bendall, R., Ijaz, S., & Banks, M. (2008). Hepatitis E: an emerging infection in developed countries. *The Lancet Infectious Diseases*, *8*(11), 698-709. doi:16/S1473-3099(08)70255-X

Daniels, N. A., MacKinnon, L., Rowe, S. M., Bean, N. H., Griffin, P. M., & Mead, P. S. (2002). Foodborne disease outbreaks in United States schools. *The Pediatric Infectious Disease Journal*, *21*(7), 623-628. doi:10.1097/01.inf.0000019885.31694.9e

Domínguez, A., Torner, N., Ruiz, L., Martínez, A., Bartolomé, R., Sulleiro, E., Teixidó, A., et al. (2007). Foodborne Salmonella-Caused Outbreaks in Catalonia (Spain), 1990 to 2003. *Journal of Food Protection*, *70*, 209-213.

Doorduyn, Y., Van Den Brandhof, W. E., Van Duynhoven, Y. T. H. P., Wannet, W. J. B., & Van Pelt, W. (2006). Risk Factors for Salmonella Enteritidis and Typhimurium (DT104 and Non-DT104) Infections in The Netherlands: Predominant Roles for Raw Eggs in Enteritidis and Sandboxes in Typhimurium Infections. *Epidemiology and Infection*, *134*(03), 617-626. doi:10.1017/S0950268805005406

Doré, K., Buxton, J., Henry, B., Pollari, F., Middleton, D., Fyfe, M., Ahmed, R., et al. (2004). Risk Factors for Salmonella Typhimurium DT104 and Non-DT104 Infection: A Canadian Multi-Provincial Case-Control Study. *Epidemiology and Infection*, *132*(03), 485-493. doi:10.1017/S0950268803001924

Fleury, M.D., Stratton, J., Tinga, C., Charron, D.F., & Aramini J. (2008). A descriptive analysis of hospitalization due to acute gastrointestinal illness in Canada, 1995-2004. *Canadian Journal of Public Health*, 99 (6), 489-93.

Friedman, C. R., Hoekstra, R. M., Samuel, M., Marcus, R., Bender, J., Shiferaw, B., Reddy, S., et al. (2004). Risk Factors for Sporadic *Campylobacter* Infection in the United States: A Case-Control Study in FoodNet Sites. *Clinical Infectious Diseases*, 38(s3), S285-S296. doi:10.1086/381598

Gormley, F. J., Little, C. L., Rawal, N., Gillespie, I. A., Lebaigue, S., & Adak, G. K. (2011). A 17-year review of foodborne outbreaks: describing the continuing decline in England and Wales (1992-2008). *Epidemiology and Infection*, 139(5), 688-699. doi:10.1017/S0950268810001858

Greig, J. D., & Lee, M. B. (2009). Enteric outbreaks in long-term care facilities and recommendations for prevention: a review. *Epidemiology and Infection*, 137(2), 145-155. doi:10.1017/S0950268808000757

Greig, J. D., & Ravel, A. (2009). Analysis of foodborne outbreak data reported internationally for source attribution. *International Journal of Food Microbiology*, 130(2), 77-87. doi:16/j.ijfoodmicro.2008.12.031

Hedberg, C. W., Palazzi-Churas, K. L., Radke, V. J., Selman, C. A., & Tauxe, R. V. (2008). The use of clinical profiles in the investigation of foodborne outbreaks in restaurants: United States, 1982–1997. *Epidemiology and Infection*, 136(01), 65-72. doi:10.1017/S0950268807008199

Helms, M., Vastrup, P., Gerner-Smidt, P., & Mølbak, K. (2003). Short and long term mortality associated with foodborne bacterial gastrointestinal infections: registry based study. *BMJ (Clinical Research Ed.)*, 326(7385), 357.

Hoffman, R. E., Greenblatt, J., Matyas, B. T., Sharp, D. J., Esteban, E., Hodge, K., & Liang, A. (2005). Capacity of State and Territorial Health Agencies To Prevent Foodborne Illness. *Emerging Infectious Diseases*, 11(1), 11-16.

Jones, T. F., & Angulo, F. J. (2006). Eating in restaurants: a risk factor for foodborne disease? *Clinical Infectious Diseases: An Official Publication of the Infectious Diseases Society of America*, 43(10), 1324-1328. doi:10.1086/508540

Iowa State University. (2010). Common foodborne pathogens. Retrieved July 23, 2011, from http://www.extension.iastate.edu/foodsafety/pathogens/index.cfm?parent=37

Jones, T. F., Imhoff, B., Samuel, M., Mshar, P., McCombs, K. G., Hawkins, M., Deneen, V., et al. (2004). Limitations to Successful Investigation and Reporting of Foodborne Outbreaks: An Analysis of Foodborne Disease Outbreaks in FoodNet Catchment Areas, 1998–1999. *Clinical Infectious Diseases*, 38(s3), S297-S302. doi:10.1086/381599

Krause, G., Altmann, D., Faensen, D., Porten, K., Benzler, J., Pfoch, T., Ammon, A., et al. (2007). SurvNet Electronic Surveillance System for Infectious Disease Outbreaks, Germany, 13(10), 1548-1555. doi:10.3201/eid1310.070253

Koopmans, M., & Duizer, E. (2004). Foodborne viruses: an emerging problem. *International Journal of Food Microbiology*, 90(1), 23-41. doi:16/S0168-1605(03)00169-7

Kuchenmüller, T., Hird, S., Stein, C., Kramarz, P., Nanda, A., & Havelaar, A. H. (2009). Estimating global burden of foodborne diseases – a collaborative effort. *Eurosurveillance*, 14 (18), 191-95.

Li, J., Smith, K., Kaehler, D., Everstine, K., Rounds, J., & Hedberg, C. (2010). Evaluation of a Statewide Foodborne Illness Complaint Surveillance System in Minnesota, 2000 through 2006. *Journal of Food Protection*, *73*(11), 2059-2064.

Lindqvist, R., Andersson, Y., Lindbäck, J., Wegscheider, M., Eriksson, Y., Tideström, L., Lagerqvist-Widh, A., et al. (2001). A one-year study of foodborne illnesses in the municipality of Uppsala, Sweden. *Emerging Infectious Diseases*, *7*(3 Suppl), 588-592.

Linscott, A. J. (2011). Food-Borne Illnesses. *Clinical Microbiology Newsletter*, *33*(6), 41-45. doi:10.1016/j.clinmicnews.2011.02.004

Lynch, M. F., Tauxe, R. V., & Hedberg, C. W. (2009). The Growing Burden of Foodborne Outbreaks Due to Contaminated Fresh Produce: Risks and Opportunities. *Epidemiology and Infection*, *137*(Special Issue 03), 307-315. doi:10.1017/S0950268808001969

Lynch, M., Painter, J., Woodruff, R., & Braden, C. (2006). Surveillance for foodborne-disease outbreaks--United States, 1998-2002. *MMWR. Surveillance Summaries: Morbidity and Mortality Weekly Report. Surveillance Summaries / CDC*, *55*(10), 1-42.

Marcus, R. (2008). New information about pediatric foodborne infections: the view from FoodNet. *Current Opinion in Pediatrics*, *20*(1), 79-84. doi:10.1097/MOP.0b013e3282f43067

Marcus, R., Varma, J. K., Medus, C., Boothe, E. J., Anderson, B. J., Crume, T., Fullerton, K. E., et al. (2007). Re-Assessment of Risk Factors for Sporadic Salmonella Serotype Enteritidis Infections: A Case-Control Study in Five FoodNet Sites, 2002–2003. *Epidemiology and Infection*, *135*(01), 84-92. doi:10.1017/S0950268806006558

Martinez, A., Dominguez, A., Torner, N., Ruiz, L., Camps, N., Barrabeig, I., Arias, C., et al. (2008). Epidemiology of foodborne Norovirus outbreaks in Catalonia, Spain. *BMC Infectious Diseases*, *8*(1), 47. doi:10.1186/1471-2334-8-47

McCulloch, J.E. (ed) (2000). Infection Control: Science, Management and Practice. London: Whurr Publishers.

McCabe-Sellers, B. J., & Beattie, S. E. (2004). Food safety: Emerging trends in foodborne illness surveillance and prevention. *Journal of the American Dietetic Association*, *104*(11), 1708-1717. doi:10.1016/j.jada.2004.08.028

Middaugh, J. P., hammond, R. M., Eisenstein, L., & Lazensky, R. (2010). Advancement of the science. Using the Electronic Foodborne Outbreak Reporting System (eFORS) to Improve Foodborne Outbreak Surveillance, Investigations, and Program Evaluation. *Journal of Environmental Health*, *73*(2), 8-11.

Mølbak, K., & Neimann, J. (2002). Risk Factors for Sporadic Infection with Salmonella Enteritidis, Denmark, 1997–1999. *American Journal of Epidemiology*, *156*(7), 654-661. doi:10.1093/aje/kwf096

Mor-Mur, M., & Yuste, J. (2009). Emerging Bacterial Pathogens in Meat and Poultry: An Overview. *Food and Bioprocess Technology*, *3*(1), 24-35. doi:10.1007/s11947-009-0189-8

Nelson, K., & Williams, C. (Ed.) Second Edition, 2007. *Infectious Disease Epidemiology, Theory and Practice*. New York: Aspen Publishers.

Newell, D. G., Koopmans, M., Verhoef, L., Duizer, E., Aidara-Kane, A., Sprong, H., Opsteegh, M., et al. (2010). Food-borne diseases -- The challenges of 20 years ago still persist while new ones continue to emerge. *International Journal of Food Microbiology*, *139*(Supplement 1), S3-S15. doi:16/j.ijfoodmicro.2010.01.021

New Zealand Food Safety Authority. (2010). The economic cost of foodborne disease in New Zealand. Retrieved July 19, 2011, from http://www.foodsafety.govt.nz/elibrary/industry/economic-cost-foodborne-disease/foodborne-disease.pdf

O'Brien, S. J., Gillespie, I. A., Sivanesan, M. A., Elson, R., Hughes, C., & Adak, G. K. (2006). Publication Bias in Foodborne Outbreaks of Infectious Intestinal Disease and Its Implications for Evidence-Based Food Policy. England and Wales 1992–2003. *Epidemiology and Infection, 134*(04), 667-674. doi:10.1017/S0950268805005765

Pozio, E. (2008). Epidemiology and control prospects of foodborne parasitic zoonoses in the European Union. *Parassitologia, 50*(1-2), 17-24.

Robinson, R. K. (2007). Emerging Foodborne Pathogens. *International Journal of Dairy Technology, 60*(4), 305-306. doi:10.1111/j.1471-0307.2007.00331.x

Scallan, E. (2007). Activities, Achievements, and Lessons Learned during the First 10 Years of the Foodborne Diseases Active Surveillance Network: 1996-2005. *Clinical Infectious Diseases, 44*(5), 718-725. doi:10.1086/511648

Scharff, R, L. (2010). Health-related costs from foodborne illness in the United States. Retrieved July 19, 2011, from http://www.marlerblog.com/uploads/image/PSP-Scharff%20v9.pdf

Scharff, R. L., McDowell, J., & Medeiros, L. (2009). Economic Cost of Foodborne Illness in Ohio. *Journal of Food Protection, 72*, 128-136.

Scott, W. G., Scott, H. M., Lake, R. J., & Baker, M. G. (2000). Economic cost to New Zealand of foodborne infectious disease. *The New Zealand Medical Journal, 113*(1113), 281-284.

Sivapalasingam, S., Friedman, C. R., Cohen, L., & Tauxe, R. V. (2004). Fresh produce: a growing cause of outbreaks of foodborne illness in the United States, 1973 through 1997. *Journal of Food Protection, 67*(10), 2342-2353.

Sparling, P. H., Crowe, C., Griffin, P. M., Swerdlow, D. L., & Rangel, J. M. (2005). Epidemiology of *Escherichia coli* O157:H7 Outbreaks, United States, 1982–2002. *Public Health Resources.* Retrieved from http://digitalcommons.unl.edu/publichealthresources/73

Stafford, R. J., Schluter, P. J., Wilson, A. J., Kirk, M. D., Hall, G., & Unicomb, L. (2008). Population-Attributable Risk Estimates for Risk Factors Associated with Campylobacter Infection, Australia. *Emerging Infectious Diseases, 14*(6), 895-901. doi:10.3201/eid1406.071008

Taormina, P. J., & Beuchat, L. R. (1999). Infections Associated with Eating Seed Sprouts: An International Concern. *Emerging Infectious Diseases, 5*(5), 626-634.

Tassios, P. T., & Kerr, K. G. (2010, January). Hard to swallow—emerging and re-emerging issues in foodborne infection. *Clinical Microbiology & Infection,* 1-2.

The Food Safety and Inspection Service [FSIS].(2008). Disposition/Food Safety: Overview of Food Microbiology. Retrieved July 23, 2011, from www.fsis.usda.gov/PDF/PHVt-Food_Microbiology.pdf

Unicomb, L. E. (2009). Food Safety: Pathogen Transmission Routes, Hygiene Practices and Prevention, *27*(5), 599-601.

van Duynhoven, Y. T. H. P., de Jager, C. M., Kortbeek, L. M., Vennema, H., Koopmans, M. P. G., van Leusden, F., van der Poel, W. H. M., et al. (2005). A one-year intensified study of outbreaks of gastroenteritis in The Netherlands. *Epidemiology and Infection, 133*(1), 9-21.

Venuto, M., Halbrook, B., Hinners, M., & Mickelson, A. L. S. (2010). Analyses of the eFORS (Electronic Foodborne Outbreak Reporting System) Surveillance Data (2000--2004) in School Settings. *Journal of Environmental Health, 72*(7), 8-13.

Verhoef, L. P. B., Kroneman, A., van Duynhoven, Y., Boshuizen, H., van Pelt, W., & Koopmans, M. (2009). Selection Tool for Foodborne Norovirus Outbreaks. *Emerging Infectious Diseases, 15*(1), 31-38. doi:10.3201/eid1501.080673

Widdowson, M.-A., Sulka, A., Bulens, S. N., Beard, R. S., Chaves, S. S., Hammond, R., Salehi, E. D. P., et al. (2005). Norovirus and foodborne disease, United States, 1991-2000. *Emerging Infectious Diseases, 11*(1), 95-102.

Vollaard, A. M., Ali, S., van Asten, H. A. G. H., Widjaja, S., Visser, L. G., Surjadi, C., & van Dissel, J. T. (2004). Risk Factors for Typhoid and Paratyphoid Fever in Jakarta, Indonesia. *JAMA: The Journal of the American Medical Association, 291*(21), 2607 -2615. doi:10.1001/jama.291.21.2607

Voetsch, A. C., Kennedy, M. H., Keene, W. E., Smith, K. E., Rabatsky-Ehr, T., Zansky, S., Thomas, S. M., et al. (2007). Risk Factors for Sporadic Shiga Toxin-Producing Escherichia Coli O157 Infections in FoodNet Sites, 1999–2000. *Epidemiology and Infection, 135*(06), 993-1000. doi:10.1017/S0950268806007564

World Health Organization [WHO]. (2011a). Food Safety. Retrieved June 26, 2011, from http://www.who.int/foodsafety/foodborne_disease/ferg1/en/index.html

World Health Organization [WHO]. (2011b). Initiative to estimate the Global Burden of Foodborne Diseases. Retrieved June 26, 2011, from http://www.who.int/foodsafety/foodborne_disease/ferg/en/index1.html

World Health Organization [WHO]. (2011c). Initiative to estimate the Global Burden of Foodborne Diseases: Information and publications. Retrieved June 26, 2011, from http://www.who.int/foodsafety/foodborne_disease/ferg/en/index7.html

3

Starch: From Food to Medicine

Emeje Martins Ochubiojo[1] and Asha Rodrigues[2]
[1]National Institute for Pharmaceutical Research and Development,
[2]Physical and Materials Chemistry Division,
National Chemical Laboratory,
[1]Nigeria
[2]India

1. Introduction

Starch is a natural, cheap, available, renewable, and biodegradable polymer produced by many plants as a source of stored energy. It is the second most abundant biomass material in nature. It is found in plant leaves, stems, roots, bulbs, nuts, stalks, crop seeds, and staple crops such as rice, corn, wheat, cassava, and potato. It has found wide use in the food, textiles, cosmetics, plastics, adhesives, paper, and pharmaceutical industries. In the food industry, starch has a wide range of applications ranging from being a thickener, gelling agent, to being a stabilizer for making snacks, meat products, fruit juices (Manek, et al., 2005). It is either used as extracted from the plant and is called "native starch", or it undergoes one or more modifications to reach specific properties and is called "modified starch". Worldwide, the main sources of starch are maize (82%), wheat (8%), potatoes (5%), and cassava (5%). In 2000, the world starch market was estimated to be 48.5 million tons, including native and modified starches. The value of the output is worth €15 billion per year (Le Corre, et al., 2010). As noted by Mason (2009), as far back as the first century, Celsus, a Greek physician, had described starch as a wholesome food. Starch was added to rye and wheat breads during the 1890s in Germany and to beer in 1918 in England. Also, Moffett, writing in 1928, had described the use of corn starch in baking powders, pie fillings, sauces, jellies and puddings. The 1930s saw the use of starch as components of salad dressings in mayonnaise. Subsequently, combinations of corn and tapioca starches were used by salad dressing manufacturers. (Mason, 2009). Starch has also find use as sweetners; sweeteners produced by acid-catalyzed hydrolysis of starch were used in the improvement of wines in Germany in the 1830s. Between 1940 and 1995, the use of starch by the US food industry was reported to have increased from roughly 30 000 to 950 000 metric tons. The leading users of starch were believed to be the brewing, baking powder and confectionery industries. Similar survey in Europe in 1992, showed that, 2.8 million metric tons of starch was used in food. Several uses of starch abound in literature and the reader is advised to refer to more comprehensive reviews on the application of starch in the food industry. In fact, the versatility of starch applications is unparalleled as compared to other biomaterials.

It is obvious that, the need for starch will continue to increase especially as this biopolymer finds application in other industries including medicine and Pharmacy. From serving as food for man, starch has been found to be effective in drying up skin lesions (dermatitis), especially where there are watery exudates. Consequently, starch is a major component of dusting powders, pastes and ointments meant to provide protective and healing effect on skins. Starch mucilage has also performed well as emollient and major base in enemas. Because of its ability to form complex with iodine, starch has been used in treating iodine poisoning. Acute diarrhea has also been effectively prevented or treated with starch based solutions due to the excellent ability of starch to take up water. In Pharmacy, starch appears indispensable; It is used as excipients in several medicines. Its traditional role as a disintegrant or diluent is giving way to the more modern role as drug carrier; the therapeutic effect of the starch-adsorbed or starch-encapsulated or starch-conjugated drug largely depends on the type of starch.

2. The role of excipients in drug delivery

The International Pharmaceutical Excipient Council (IPEC) defines excipients as substances, other than the active pharmaceutical ingredient (API) in finished dosage form, which have been appropriately evaluated for safety and are included in a drug delivery system to either aid the processing or to aid manufacture, protect, support, enhance stability, bioavailability or patient acceptability, assist in product identification, or enhance other attributes of the overall safety and effectiveness of the drug delivery system during storage or use (Robertson, 1999). They can also be defined as additives used to convert active pharmaceutical ingredients into pharmaceutical dosage forms suitable for administration to patients. Excipients no longer maintain the initial concept of —Inactive support; because of the influence they have over both biopharmaceutical aspects and technological factors (Jansook and Loftsson, 2009; Killen and Corrigan, 2006; Langoth, et al., 2003; Lemieux, et al., 2009; Li, et al., 2003; Massicotte, et al., 2008; Munday and Cox, 2000; Nykänen, et al., 2001; Williams, et al.) The desired activity, the excipient's equivalent of the active ingredients efficacy, is called its functionality. The inherent property of an excipient is its functionality in the dosage form. In order to deliver a stable, uniform and effective drug product, it is essential to know the properties of the active pharmaceutical ingredient alone and in combination with all other ingredients based on the requirements of the dosage form and process applied. This underscores the importance of excipients in dosage form development.

The ultimate application goal of any drug delivery system including nano drug delivery, is to develop clinically useful formulations for treating diseases in patients (Park, 2007). Clinical applications require approval from FDA. The pharmaceutical industry has been slow to utilize the new drug delivery systems if they include excipients that are not generally regarded as safe. This is because, going through clinical studies for FDA approval of a new chemical entity is a long and costly process; there is therefore, a very strong resistance in the industry to adding any untested materials that may require seeking approval. To overcome this reluctant attitude by the industry, scientists need to develop not only new delivery systems that are substantially better than the existing delivery systems (Park, 2007), but also seek for new ways of using old biomaterials. The use of starch (native or modified) is an important strategy towards the attainment of this objective. This is because starch unlike synthetic products is biocompatible, non toxic, biodegradable, eco-friendly and of low prices. It is generally a non-polluting renewable source for sustainable supply of cheaper pharmaceutical products.

3. What is starch?

Starch, which is the major dietary source of carbohydrates, is the most abundant storage polysaccharide in plants, and occurs as granules in the chloroplast of green leaves and the amyloplast of seeds, pulses, and tubers (Sajilata, et al., 2006). Chemically, starches are polysaccharides, composed of a number of monosaccharides or sugar (glucose) molecules linked together with α-D-(1-4) and/or α-D-(1-6) linkages. The starch consists of 2 main structural components, the amylose, which is essentially a linear polymer in which glucose residues are α-D-(1-4) linked typically constituting 15% to 20% of starch, and amylopectin, which is a larger branched molecule with α-D-(1-4) and α-D-(1-6) linkages and is a major component of starch. Amylose is linear or slightly branched, has a degree of polymerization up to 6000, and has a molecular mass of 105 to 106 g/mol. The chains can easily form single or double helices. Amylopectin on the other hand has a molecular mass of 107 to 109 g/mol. It is highly branched and has an average degree of polymerization of 2 million, making it one of the largest molecules in nature. Chain lengths of 20 to 25 glucose units between branch points are typical. About 70% of the mass of starch granule is regarded as amorphous and about 30% as crystalline. The amorphous regions contain the main amount of amylose but also a considerable part of the amylopectin. The crystalline region consists primarily of the amylopectin (Sajilata, et al., 2006).

Starch in the pharmaceutical industry

During recent years, starch has been taken as a new potential biomaterial for pharmaceutical applications because of the unique physicochemical and functional characteristics (Cristina Freire, et al., 2009; Freire, et al., 2009; Serrero, et al.).

3.1 Starch as pharmaceutical excipient

Native starches were well explored as binder and disintegrant in solid dosage form, but due to poor flowability their utilization is restricted. Most common form of modified starch i.e. Pre-gelatinized starch marketed under the name of starch 1500 is now a day's most preferred directly compressible excipients in pharmaceutical industry. Recently modified rice starch, starch acetate and acid hydrolyzed dioscorea starch were established as multifunctional excipient in the pharmaceutical industry. The International Joint Conference on Excipients rated starch among the top ten pharmaceutical ingredients (Shangraw, 1992).

3.2 Starch as tablet disintegrant

They are generally employed for immediate release tablet formulations, where drug should be available within short span of time to the absorptive area. Sodium carboxymethyl starch, which is well established and marketed as sodium starch glycolate is generally used for immediate release formulation. Some newer sources of starch have been modified and evaluated for the same.

3.3 Starch as controlled/sustained release polymer for drugs and hormones

Modified starches in different forms such as Grafted, acetylated and phosphate ester derivative have been extensively evaluated for sustaining the release of drug for better patient compliances. Starch-based biodegradable polymers, in the form of microsphere or hydrogel, are suitable for drug delivery (Balmayor, et al., 2008), (Reis, et al., 2008). For example, high amylose corn starch has been reported to have good sustained release properties and this has been attributed to its excellent gel-forming capacity (Rahmouni, et al., 2003; Te Wierik, et al., 1997). Some authors (Efentakis, et al., 2007; Herman and Remon, 1989; Michailova, et al., 2001) have explained the mechanism of drug release from such gel-forming matrices to be a result of the controlled passage of drug molecules through the obstructive gel layer, gel structure and matrix.

3.4 Starch as plasma volume expander

Acetylated and hydroxyethyl starch are now mainly used as plasma volume expanders. They are mainly used for the treatment of patients suffering from trauma, heavy blood loss and cancer.

3.5 Starch in bone tissue engineering

Starch-based biodegradable bone cements can provide immediate structural support and degrade from the site of application. Moreover, they can be combined with bioactive particles, which allow new bone growth to be induced in both the interface of cement-bone and the volume left by polymer degradation (Boesel, et al., 2004). In addition, starch-based biodegradeable polymer can also be used as bone tissue engineering scaffold (Gomes, et al., 2003).

3.6 Starch in artificial red cells

Starch has also been used to produce a novel and satisfactory artificial RBCs with good oxygen carrying capacity. It was prepared by encapsulating hemoglobin (Hb) with long-chain fatty-acids-grafted potato starch in a self-assembly way (Xu, et al., 2011).

3.7 Starch in nanotechnology

Starch nanoparticles, nanospheres, and nanogels have also been applied in the construction of nanoscale sensors, tissues, mechanical devices, and drug delivery system. (Le Corre, et al., 2010).

3.7.1 Starch microparticles

The use of biodegradable microparticles as a dosage form for the administration of active substances is attracting increasing interest, especially as a means of delivering proteins. Starch is one of the polymers that is suitable for the production of microparticles. It is biodegradable and has a long tradition as an excipient in drug formulations. Starch microparticles have been used for the nasal delivery of drugs and for the delivery of vaccines administered orally and intramuscularly. Bioadhesive systems based on polysaccharide microparticles have been reported to significantly enhance the systemic absorption of conventional drugs and polypeptides across the nasal mucosa, even when devoid of absorption enhancing agents. A major area of application of microparticles is as dry powder inhalations formulations for asthma and for deep-lung delivery of various agents. It has also been reported that, particles reaching the lungs are phagocytosed rapidly by alveolar Macrophages. Although phagocytosis and sequestration of inhaled powders may be a problem for drug delivery to other cells comprising lung tissue, it is an advantage for chemotherapy of tuberculosis. Phagocytosed microparticles potentially can deliver larger amounts of drug to the cytosol than oral doses. It is also opined strongly that, microparticles have the potential for lowering dose frequency and magnitude, which is especially advantageous for maintaining drug concentrations and improving patient compliance. This is the main reason this dosage form is an attractive pulmonary drug delivery system. (Le Corre, et al., 2010).

3.7.2 Starch microcapsules

Microencapsulation is the process of enclosing a substance inside a membrane to form a microcapsule. it provides a simple and cost-effective way to enclose bioactive materials within a semi-permeable polymeric membrane. Both synthetic/semi-synthetic polymers and natural polymers have been extensively utilized and investigated as the preparation materials of microcapsules. Although the synthetic polymers display chemical stability, their unsatisfactory biocompatibility still limits their potential clinical applications. Because the natural polymers always show low/non toxicity, low immunogenicity and thereafter good biocompatibility, they have been the preferred polymers used in microencapsulation systems. Among the natural polymers, alginate is one of the most common materials used to form microcapsules, however, starch derivatives are now gaining attention. For instance starch nasal bioadhesive microspheres with significantly extended half-life have been reported for several therapeutic agents including insulin. Improved bioavailability of Gentamycin-encapsulated starch microspheres as well as magnetic starch microspheres for parenteral administration of magnetic iron oxides to enhance contrast in magnetic resonance imaging has been reported. (Le Corre, et al., 2010).

3.7.3 Starch nanoparticles

Nanoparticles are solid or colloidal particles consisting of macromolecular substances that vary in size from 10-1000 nm. The drug may be dissolved, entrapped, adsorbed, attached or encapsulated into the nanoparticle matrix. The matrix may be biodegradable materials such as polymers or proteins or biodegradable/biocompatible/bioasborbable materials such as starch. Depending on the method of preparation, nanoparticles can be obtained with different physicochemical, technical or mechanical properties as well as modulated release characteristics for the immobilized bioactive or therapeutic agents. (Le Corre, et al., 2010).

4. Application of modified starches in drug delivery

Native starch irrespective of their source are undesirable for many applications, because of their inability to withstand processing conditions such as extreme temperature, diverse pH, high shear rate, and freeze thaw variation. To overcome this, modifications are usually done to enhance or repress the inherent property of these native starches or to impact new properties to meet the requirements for specific applications. The process of starch modification involves the destructurisation of the semi-crystalline starch granules and the effective dispersion of the component polymers. In this way, the reactive sites (hydroxyl groups) of the amylopectin polymers become accessible to electrophilic reactants (Rajan, et al., 2008). Common modes of modifications useful in pharmaceuticals are chemical, physical and enzymatic with, a much development already seen in chemical modification. Starch modification through chemical derivation such as etherification, esterification, cross-linking, and grafting when used as carrier for controlled release of drugs and other bioactive agents. It has been shown that, chemically modified starches have more reactive sites to carry biologically active compounds, they become more effective biocompatible carriers and can easily be metabolized in the human body (Prochaska, et al., 2009; Simi and Emilia Abraham, 2007).

4.1 Chemical modification of starch

There are a number of chemical modifications made to starch to produce many different functional characteristics. The chemical reactivity of starch is controlled by the reactivity of its glucose residues. Modification is generally achieved through etherification, esterification, crosslinking, oxidation, cationization and grafting of starch. However, because of the dearth of new methods in chemical modifications, there has been a trend to combine different kinds of chemical treatments to create new kinds of modifications. The chemical and functional properties achieved following chemical modification of starch, depends largely on the botanical or biological source of the starch,, reaction conditions (reactant concentration, reaction time, pH and the presence of catalyst), type of substituent, extent of substitution (degree of substitution, or molar substitution), and the distribution of the substituent in the starch molecule (Singh, et al., 2007). Chemical modification involves the introduction of functional groups into the starch molecule, resulting in markedly altered physico-chemical properties. Such modification of native granular starches profoundly alters their gelatinization, pasting and retrogradation behavior (Choi and Kerr, 2003; Kim, et al., 1993) (Perera, et al.) and (Liu, et al., 1999) (Seow and Thevamalar, 1993). The rate and efficiency of the chemical modification process depends on the reagent type, botanical origin of the starch and on the size and structure of its granules (Huber and BeMiller, 2001).This also includes the surface structure of the starch granules, which encompasses the outer and inner surface, depending on the pores and channels (Juszczak, 2003).

4.1.1 Carboxymethylated starch

Starches can have a hydrogen replaced by something else, such as a carboxymethyl group, making carboxymethyl starch (CMS). Adding bulky functional groups like carboxymethyl and carboxyethyl groups reduces the tendency of the starch to recrystallize and makes the starch less prone to damage by heat and bacteria. Carboxymethyl starch is synthesized by reacting starch with monochloroacetic acid or its sodium salt after activation of the polymer with aqueous NaOH in a slurry of an aqueous organic solvent, mostly an alcohol. The total degree of substitution (DS), that is the average number of functional groups introduced in the polymer, mainly determines the properties of the carboxymethylated products (Heinze, 2005). The functionalization influences the properties of the starch. For example, CMS have been shown to absorb an amount of water 23 times its initial weight. This high swelling capacity combined with a high rate of water permeation is said to be responsible for a high rate of tablet disintegration and drug release from CMS based tablets. CMS has also been reported to be capable of preventing the detrimental influence of hydrophobic lubricants (such as magnesium stearate) on the disintegration time of tablets or capsules.. Some of the recent use of carboxymethylated starch in pharmaceuticals are summarised in Table 1.

Study Title	Methodology	Drug used	Summary	References
Synthesis and in vitro evaluation of carboxymethyl starch-chitosan nanoparticles as drug delivery system to the colon.	Complex coacervation process.	5-aminosalicylic acid	Chitosan-carboxymethyl starch nanoparticles developed based on the modulation of ratio show promise as a system for controlled delivery of drugs to the colon.	(Saboktakin, et al., 2010)
Carboxymethyl-Starch Excipients for Gastrointestinal Stable Oral Protein Formulations Containing Protease Inhibitors.	Monochloroacetic acid Semisynthesis	Protease inhibitors	The feasibility to deliver stable bioactive proteins into intestinal environment using CMS was demonsttrated. In addition to its protective role, the CMS allows the release of the bioactive agents in less than 6 h, fitting well with the upper intestinal transit. Such stability is needed for formulation of oral vaccines with specific antigens.	(Patrick De Koninck, 2010)
Novel Polymeric Biomaterial Based on Carboxymethyl starch and its Application in Controlled Drug Release.	Blender, mixing and sieving	Acetylsalicylic acid	Matrix (CMS) releases the enclosed drug at a much faster rate in neutral and alkaline pH than in acidic pH, thus holding the promise of being developed into a vehicle for targeted drug delivery (oral route) to the lower gastrointestinal tract.	(Sen and Pal, 2009b)
Carboxymethyl high amylose starch as excipient for controlled drug release: Mechanistic study and the influence of degree of substitution.	Non aqueous medium synthesis, protanated using acid treatment.	Acetaminophen	The results showed that, CMHAS(Na) with DS between 0.1 and 0.2 can be used as sustained release excipients, while those with high DS, between 0.9 and 1.2, are better as delayed release excipient.	(Lemieux, et al., 2009)
High-amylose sodium carboxymethyl starch matrices for oral, sustained drug-release: Formulation aspects and in vitro drug-release evaluation.	Spray Drying	Acetaminophen	The results proved that the new Spray Drying process developed for High-amylose sodium carboxymethyl starch (HASCA) manufacture is suitable for obtaining similar-quality (HASCA) in terms of drug release and compression performances.	(Brouillet, et al., 2008)
Carboxylated high amylose starch as pharmaceutical excipients and formulation Structural insights of pancreatic enzymes.	Carboxylation and protonation	Pancreatic enzymes	High loading capacity (up to 70–80% enzymes) was obtained. An advantage of these formulations is that gastroprotection is afforded by the carboxylated matrices (carboxylic groups), without enteric coating.	(Massicotte, et al., 2008)

High-amylose carboxymethyl starch matrices for oral sustained drug-release: In vitro and in vivo evaluation.	Blending of HASCA with NaCl	Acetaminophen	In vitro drug-release from an optimized HASCA formulation was not affected by either acidic pH value or acidic medium residence time. Compressed blend of HASCA with an optimized quantity of sodium chloride provided a pharmaceutical sustained-release tablet with improved integrity for oral administration. In vivo studies demonstrated extended drug absorption.	(Nabais, et al., 2007)
Physicochemical and Pharmaceutical Properties of Carboxymethyl Rice Starches Modified from Native Starches with Different Amylose Content.	Monochloroacetic acid Reaction	Material science	Carboxymethyl rice starches (CMRS) can function as tablet binder in the wet granulation of both water-soluble and water-insoluble diluents. The tablets compressed from these granules showed good hardness with fewer capping problems compared with those prepared using the pregelatinized native rice starch as a binder.	(Kittipongpatana, et al., 2007)
An Aqueous Film-coating Formulation based on Sodium Carboxymethyl Mungbean Starch	Petri dish method	Material scienc	Carboxymethyl mungbean starch (SCMMSs) exhibited the ability to form a clear, thin film with greater flexibility and strength than that of the native starch. This study reports the potential of SCMMS as tablet film coating agent	(Kittipongpatana, et al., 2006)

Table 1. Use of carboxymethylated starch in drug delivery

4.1.2 Acetylated starch

Acetylated starch has also been known for more than a century. Starches can be esterified by modifications with an acid. When starch reacts with an acid, it loses a hydroxyl group, and the acid loses hydrogen. An ester is the result of this reaction. Acetylation of cassava starch has been reported to impart two very important pharmaceutical characters to it; increased swelling power (Rutenberg, 1984) and enhanced water solubility of the starch granules (Aziz, 2004). Starch acetates and other esters can be made very efficiently on a micro scale without addition of catalyst or water simply by heating dry starch with acetic acid and anhydride at 180°C for 2-10 min (Shogren, 2003). At this temperature, starch will melt in acetic acid (Shogren, 2000)and thus, a homogeneous acetylation would be expected to occur. Using acetic acid, starch acetates are formed, which are used as film-forming polymers for pharmaceutical products. A much recent Scandium triflate catalyzed acetylation of starch at low to moderate temperatures is reported by (Shogren, 2008). Generally, starch acetates have a lower tendency to create gels than unmodified starch. Acetylated starches are distinguishable through high levels of shear strength. They are particularly stable against heat and acids and are equally reported to form flexible, water-soluble films. Some of the recent uses of acetylated starch in pharmaceuticals are summarized in Table 2.

Study Title	Methodology	Drug used	Summary	References
An Oral Colon-Targeting Controlled Release System Based on Resistant Starch Acetate: Synthetization, Characterization, and Preparation of Film-Coating Pellets.	Acetic anhydride synthesis	5-ASA, BSA, HGF, and insulin	The study suggests that an oral colon-targeting controlled release system based on resistant starch acetate (RSA) as a film-coating material has an excellent colon-targeting release performance and the universality for a wide range of bioactive components.	(Pu, 2011)
Preparation and Properties of Starch Acetate Fibers for Potential Tissue Engineering Applications	Extruded as fibers onto a rotating drum with variable speed using a syringe and needle	Tissue engineering scaffolds	The starch acetate fibers support the adhesion of fibroblasts demonstrating that the fibers would be suitable for tissue engineering and other medical applications. They possess better mechanical properties and water stability than most polysaccharide-based biomaterials and protein fibers used in tissue engineering.	(Narendra Reddy, 2009)
Acetylated starch-based biodegradable materials with potential biomedical applications as drug delivery systems.	Acetyl esterification	Bovine serum albumin	Drug release studies show that the starch acetate coated tablets could deliver the drug to the colon suggesting that it can be a potential drug delivery carrier for colon-targeting.	(Chen, et al., 2007)
Optimization and characterization of controlled release multi-particulate beads coated with starch acetate	Organic synthesis	Dyphylline	Starch acetate-coated beads provided controlled release of dyphylline.	(Nutan, et al., 2005)
Drug release from starch-acetate microparticles and films with and without incorporated alpha-amylose.	Acetyl Esterification /aqueous synthesis	Timolol calcein and bovine serum albumin	This study demonstrates the achievement of slow release of different molecular weight model drugs from the starch acetate microparticles and films as compared to fast release from the native starch preparations.	(Tuovinen, et al., 2004a)
Starch acetate microparticles for drug delivery into retinal pigment epithelium-in vitro study.	Modified water-in-oil-in-water double-emulsion technique.	Calcein	The study indicates that the natural enzyme-sensitive starch acetate is suitable for drug delivery into retinal pigment epithelium (RPE). The starch acetate microparticles were easily taken up by cultured human RPE cells without significant toxicity.	(Tuovinen, et al., 2004b)

Starch acetate as a tablet matrix for sustained drug release	Monolithic matrix system	Propranolol hydrochloride	The drug release was considerably slower from sodium acetate tablet matrices. Also, a decrease in starch acetate concentration increased the drug release rate. Crack formation increased area available for Fickian diffusion, which caused slow attenuation of drug release rate.	(Pohja, et al., 2004)
Drug release from starch-acetate films.	Monolithic matrix system	BSA, FITC-dextran timolol and sotalol–HCl	The results showed that acetylation of potato starch can substantially retard drug release thus allowing sustained drug delivery. The drug release profiles may be controlled by the degree of substitution.	(Tuovinen, et al., 2003)
Acetylation enhances the tabletting properties of starch	Acetic Anhydride synthesis	materials science	The acetate moiety, perhaps in combination with existing hydroxyl groups, was a very effective bond-forming substituent. The formation of strong molecular bonds increased, leading to a very firm and intact tablet structure.	(Raatikainen, et al., 2002)

Table 2. Some acetylated starches and their application in drug delivery.

4.1.3 Hydroxypropylated starch

Hydroxypropyl groups introduced into starch chains are said to be capable of disrupting inter- and intra-molecular hydrogen bonds, thereby weakening the granular structure of starch leading to an increase in motional freedom of starch chains in amorphous regions (Choi and Kerr, 2003; Seow and Thevamalar, 1993; Wootton and Manatsathit, 1983). Such chemical modification involving the introduction of hydrophilic groups into starch molecules improves the solubility of starch and the functional properties of starch pastes, such as its shelf life, freeze/thaw stability, cold storage stability, cold water swelling, and yields reduced gelatinization temperature, as well as retarded retrogradation. Owing to these properties, hydroxypropylated starches is gaining interest in medicine.

Study Title	Methodology	Drug used	Summary	References
Hydroxypropylated starches of varying amylose contents as sustained release matrices in tablets	Monolithic matrix tablet formulation	Propranolol hydrochloride	Hydroxypropylation improved the sustained release ability of amylose-containing starch matrices, and conferred additional resistance to the hydrolytic action of pancreatin under simulated gastrointestinal conditions.	(Onofre and Wang, 2010)

Table 3. Hydroxyl-propylated starch in drug delivery.

4.1.4 Succinylated starch

Modification of starch by Succinylation has also been found to modify its physicochemical properties, thereby widening its applications in food and non-food industries like pharmaceuticals, paper and textile industries. Modification of native starch to its succinate derivatives reduces its gelatinisation temperature and the retrogradation, improves the freeze-thaw stability as well as the stability in acidic and salt containing medium (Trubiano Paolo, 1997; Trubiano, 1987; Tukomane and Varavinit, 2008). Generally, succinylated starch can be prepared by treating starches with different alkenyl succinic anhydride, for example, dodecenyl succinic anhydride, octadecenyl succinic anhydride or octenyl succinic anhydride. The incorporation of bulky octadecenyl succinic anhydride grouping to hydrophilic starch molecules has been found to confer surface active properties to the modified starch (Trubiano Paolo, 1997). Unlike typical surfactants, octadecenyl succinic anhydride starch, forms strong films at the oil–water interface giving emulsions that are resistant to reagglomeration. A recent application of succinylated starch in pharmaceuticals are summarized in Table 4.

Study Title	Methodology	Drug used	Summary	References
Preparation and characterisation of octenyl succinate starch as a delivery carrier for bioactive food components	Pyridine-catalyzed esterification	Bovine serum albumin/ASA	Ocetyl succcinate starch was found to be a potential carrier for colon targeted drug delivery	(Wang, et al., 2011)

Table 4. A recent application of succinylated starch in drug delivery

4.1.5 Phosphorylated starch

Phosphorylation was the earliest method of starch modification. The reaction gives rise to either monostarch phosphate or distarch phosphate (cross-linked derivative), depending upon the reactants and subsequent reaction conditions. Phosphate cross-linked starches show resistance to high temperature, low pH, high shear, and leads to increased stability of the swollen starch granule. The presence of a phosphate group in starch increases the hydration capacity of starch pastes after gelatinization and results in the correlation of the starch phosphate content to starch paste peak viscosity , prevents crystallization and gel-forming capacity (Nutan, et al., 2005). These new properties conferred on starch by phosphorylation, makes them useful as disintegrants in solid dosage formulations and as matrixing agents. Interestingly, it has been documented that, the only naturally occurring covalent modification of starch is phosphorylation. Traditionally, starch phosphorylation is carried out by the reaction of starch dispersion in water with reagents like mono- or di sodium orthophosphates, sodium hexametaphosphate, sodium tripolyphosphate (STPP), sodium trimetaphosphate (STMP) or phosphorus oxychloride. Alternative synthetic methods such as extrusion cooking, microwave irradiation and vacuum heating have been reported (A. N. Jyothi, 2008; Sitohy and Ramadan, 2001). Some of the recent uses of phosphorylated starch in pharmaceuticals are summarized in Table 5.

Study Title	Methodology	Drug used	Summary	References
Starch Phosphate: A Novel Pharmaceutical Excipient For Tablet Formulation.	Phosphorylation using mono sodium phosphate dehydrate	Ziprasidone	At low concentration, starch phosphate proved to be a better disintegrant than native starch in tablet formulation.	(N.L Prasanthi, 2010)
Starch phosphates prepared by reactive extrusion as a sustained release agent.	Reactive extrusion	Metoprolol tartrate	Starch phosphate prepared by reactive extrusion produced stronger hydrogels with sustained release properties as compared with native starch.	(O'Brien, et al., 2009)

Table 5. Use of phosphorylated starch in Pharmaceuticals.

4.1.6 Co-polymerized starch

Chemical modification of natural polymers by grafting has received considerable attention in recent years because of the wide variety of monomers available. Graft copolymerization is considered to be one of the routes used to gain combinatorial and new properties of natural and synthetic polymers. In graft copolymerization the guest monomer benefits the host polymer with some novel and desired properties in which the resultant copolymer gains characteristic properties and applications (Fares, et al., 2003). As a rule, graft copolymerization produces derivatives of significantly increased molecular weight. Starch grafting usually entails etherification, acetylation, or esterification of the starch with vinyl monomers to introduce a reaction site for further formation of a copolymeric chain. Such a chain would typically consist of either identical or different vinyl monomers (block polymers), or it may be grafted onto another polymer altogether. Graft copolymers find application in the design of various stimuli-responsive controlled release systems such as transdermal films, buccal tablets, matrix tablets, microsphers/hydrogel bead system and nanoparticulate system (Sabyasachi Maiti, 2010). Some of the recent uses of graft co-polymerized starch in pharmaceuticals is summarized in Table 6.

Study Title	Methodology	Drug used	Summary	References
Grafted Starch-Encapsulated Hemoglobin (GSEHb) Artificial Red Blood Cells Substitutes.	Self –assembly	Artificial RBCs	Satisfactory artificial RBCs with good oxygen carrying capacity obtained.	(Xu, et al., 2011)
Preparation and using of acrylamide grafted starch as polymer drug carrier.	Chemical grafting of acrylamide on the starch polymer	Ceftriaxone Sodium	Acrylamide grafted starch showed higher uptake of water compared with native starch. The starch exhibited good controlled release properties.	(Al-Karawi and Al-Daraji, 2010)
Graft copolymers of ethyl methacrylate on waxy maize starch derivatives as novel excipients for matrix tablets: Physicochemical and technological characterisation	Free radical polymerization and alternatively dried in a vacuum oven (OD) or freeze-dried (FD)	Material science	The copolymers could be used as excipients in matrix tablets and have potentials for use as controlled release materials.	(Marinich, et al., 2009)
Physical blends of starch graft copolymers as matrices for colon targeting drug delivery systems.	Synthesis by Cerium iv ion method	Theophylline, procaine and bovineserum	The physical blend offered good controlled release of drugs, as well as of proteins and presented suitable properties for use as hydrophilic matrices for colon-specific drug delivery.	(Silva, et al., 2009)
Microwave initiated synthesis of polyacrylamide grafted carboxymethylstarch (CMS-g-PAM): application as a novel matrix for sustained drug release.	Microwave initiated synthesis	5-amino salicylic acid	'In vitro' release of a model drug (5-amino salicylic acid) from CMS-g-PAM matrix showed a sustained drug release. In this matrix, the rate of release of the enclosed drug could be precisely programmed simply by adjustment of percentage grafting during synthesis.	(Sen and Pal, 2009a)
Starch–Acrylics Graft Copolymers and Blends: Synthesis, Characterization, and Applications as Matrix for Drug Delivery	Polymerization reaction	Paracetamol	The graft copolymers, provided a pH sensitive matrix system for site-specific drug delivery. The authors concluded that graft copolymers may be a useful tool to overcome the harsh environment of the stomach and can possibly be used in future as excipient for colon-targeted drug delivery.	(Shaikh and Lonikar, 2009)

Characterization and in vitro evaluation of starch based hydrogels as carriers for colon specific drug delivery systems	γ-rays induced polymerization and crosslinking	Ketoprofen	Hydrogels prepared showed pH responsive property; preventing drug release pH 1, but released it at pH 7.	(El-Hag Ali and AlArifi, 2009)
Hydrophobic grafted and crosslinked starch nano particles for drug delivery.	Grafted using potassium persulphate as catalyst	Indomethacin	Fatty acid grafted starch nano particle with high swelling power was obtained and was found to be a good vehicle for oral controlled drug delivery	(Simi and Emilia Abraham, 2007)
Bioadhesive grafted starch copolymers as platforms for peroral drug delivery: a study of theophylline release.	Radiation of starch and acrylic acid mixtures with ⁶⁰Co	Theophylline	Results show that,the release of theophylline from the graft copolymer tablets was practically independent of the pH of the dissolution medium and the type of starch used for grafting. Incorporation of divalent cations into the graft copolymers led to a significant decrease in swelling and retardation of drug release.	(Geresh, et al., 2004)
Ethyl Methacrylate grafted on two starches as polymeric matrices for drug delivery	Ceric ion redox initiation method	Theophylline and procaine hydrochloride	The HS-EMA and S-EMA were found to be efficient matrices for insoluble drugs	(Echeverria, et al., 2005)

Table 6. Use of graft co-polymerized starch in Pharmaceuticals.

4.2 Physical modification of starch

Physical modification of starch is mainly applied to change the granular structure and convert native starch into cold water-soluble starch or small-crystallite starch. The major methods used in the preparation of cold water-soluble starches involve instantaneous cooking–drying of starch suspensions on heated rolls (drum-drying), puffing, continuous cooking–puffing–extruding, and spray-drying (Jarowrenko, 1986). A method for preparing granular cold water-soluble starches by injection and nozzle-spray drying was described by (Pitchon & Joseph 1981). Among the physical processes applied to starch modification, high pressure treatment of starch is considered an example of 'minimal processing'(Stute, et al., 1996). A process of iterated syneresis applied to the modification of potato, tapioca, corn and wheat starches resulted in a new type of physically modified starches (Lewandowicz and Soral-Smietana, 2004). Some of the recent uses of physically modified starch in pharmaceuticals are summarized in Table 7.

Study Title	Methodology	Drug used	Summary	References
Microwave Assisted Modification of Arrowroot Starch for Pharmaceutical Matrix Tablets	Microwave Assisted Modification	Theophylline	The modified arrowroot starch, demonstrated promising properties as hydrophilic matrix excipients for sustained release tablets	(Pornsak Sriamornsak 2010)
Effect of heat moisture treatment on the functional and tabletting properties of corn starch Gelatinized/freeze-dried starch as excipient in sustained release tablets.	Heat moisture treatment	Material science	Heat moisture treatment (HMT) of corn starch could be useful when fast disintegration, lower swelling power, and lower crushing strength are desired in tablet.	(Iromidayo Olu-Owolabi.B, 2010)
Pregelatinized glutinous rice starch as a sustained release agent for tablet preparations	Heat treatment and spray drying	Propranolol HCl	In this study glutinous rice starch slurry was physically modified by heat and then dried by spray drying. The tablet containing pregelatinized glutinous starch and propranolol HCl were prepared by wet granulation method. The mechanisms of drug release from the matrices were anomalous (non-Fickian) diffusion in both hydrochloric buffer (pH 1.2) and phosphate buffer media (pH 6.8).	(Peerapattana, et al.)
A Novel Pregelatinized Starch as a Sustained-Release Matrix Excipient.	Controlled thermal pregelatinization And spray-drying	Ethenzamide acetaminophen and sodium salicylic acid	The study shows that highly functional pregelatinized starch is a new matrix excipient that can control drug-release profiles from first- to zero-order sustained release and enables drug release independent of drug solubility and external conditions.	(Masaaki Endo, 2009)
Effects of Drying Process for Amorphous Waxy Maize Starch on Theophylline Release from Starch-Based Tablets.	Oven drying and freeze drying procedure	Theophylline	The drying method was found to affect the morphology and drug release profiles of the compressed tablets.	(Yoon, et al., 2007)

Gelatinized/freeze-dried starch as excipient in sustained release tablets.	Gelatinization and freeze-drying	Material science	A new technique for the production of cold water-swellable starch using gelatinization and freeze-drying processes was obtained and the matrices containing different modified starch -hydroxypropyl methylcellulose mixtures possess good sustained release properties.	(Sánchez, et al., 1995)
Modified starches as hydrophilic matrices for controlled oral delivery III. Evaluation of sustained-release theophylline formulations based on thermal modified starch matrices in dogs	Thermal Modification	Theophylline	Several thermally modified starch matrices evaluated in dogs demonstrated good sustained-release performance	(Herman and Remon, 1990)
Modified starches as hydrophilic matrices for controlled oral delivery. II. In vitro drug release evaluation of thermally modified starches.	Thermal Modification	Theophylline	Thermally modified starches containing a low amount of amylose (25% and lower) revealed promising properties as directly compressible tabletting excipients for sustained release purposes.	(Herman and Remon, 1989)

Table 7. Some of the recent uses of physically modified starch in medicine.

Study Title	Methodology	Drug used	Summary	References
Enzymatic modification of cassava starch by fungal lipase.	Esterification using fungal lipase	Material science	Esterification of starch using fungal lipase with long chain fatty acids like palmitic acid gives thermoplastic starch which has got wide use in plastic industry, pharmaceutical industries, and in biomedical applications such as materials for bone fixation and replacements, carriers for controlled release of drugs and other bioactive agents. Unlike chemical esterification, enzymatic esterification is ecofriendly and avoids the use of nasty solvents.	(Rajan, et al., 2008)
Enzyme-Catalyzed Regioselective Modification of Starch Nanoparticles	Etherification of starch nanoparticles catalyzed by Candida antartica Lipase B	Material science	Selective etherification of starch nanoparticles catalyzed by Candida antartica Lipase B (CAL-B) in its immobilized (Novozyme 435) and frees (SP-525) forms. The starch nanoparticles were made accessible for acylation reactions by formation of Aerosol-OT (AOT, bis[2-ethylhexyl]sodium sulfosuccinate) stabilized microemulsions. The close proximity of the lipase and substrates promotes the modification reactions.	(Chakraborty, et al., 2005)

Table 8. Some enzymatically modified starches and potential medical/pharmaceutical applications

4.3 Enzymatic modification of starch

An alternative to obtaining modified starch is by using various enzymes. These include enzymes occurring in plants, e.g pullulanase and isoamylase groups. Pullulanase is a 1,6-α-glucosidase, which statistically impacts the linear α-glucan, a pullulan which releases maltotriose oligomers. This enzyme also hydrolyses α-1,6-glycoside bonds in amylopectin and dextrines when their side-chains include at least two α-1,4-glycoside bonds. Isoamylase is an enzyme which totally hydrolises α-1,6-glycoside bonds in amylopectin, glycogen, and some branched maltodextrins and oligosaccharides, but is characterised by low activity in relation to pullulan (Norman 1981). In a study (Kim and Robyt, 1999) starch granules was modified in situ by using a reaction system in which glucoamylase reacts inside starch granules to give conversions of 10-50% D-glucose inside the granule. Enzymatic modification of starch still needs to be explored and studied. Some of the recent uses of enzymatically modified starch in pharmaceuticals is summarized in Table 8.

Study Title	Methodology	Drug used	Summary	References
Evaluation of glutinous rice starch based matrix microbeads using scanning electron microscopy	Micro orifice ionotropic-getation method	Material science	Microbeads were prepared by ionotropic-getation method using glutinous rice starch from Assam Bora rice, and sodium alginate backbone with different crosslinking agents. Photomicrographs provides information on the surface texture, size, mechanistic properties, suitability of drying condition and mechanism of drug release from the prepared micro devices.	(Nikhil K Sachan, 2010)
Development of porous HAp and β-TCP scaffolds by starch consolidation with foaming method and drug-chitosan bilayered scaffold based drug delivery system.	Starch consolidation with foaming method	Ceftriaxone	This study confirmed the ability of starch consolidation scaffolds to release drugs suitable for treating osteomyelitis.	(Kundu, et al.)
Preparation of starch-based scaffolds for tissue engineering by supercritical immersion precipitation.	Supercritical immersion precipitation	Scaffolds	Highly porous and interconnected scaffolds were obtained in this report which made use of polymeric blend of starch and poly (L-lactic acid) for tissue engineering purposes. Good porosity is critical in scaffolds technology	(Duarte, et al., 2009)
Porous scaffold of gelatin-starch with nanohydroxyapatite composite processed via novel microwave vaccum drying.	Microwave vaccum drying	Hydroxyapatite	Gelatin blended with starch results in scaffold composites with enhanced mechanical properties. A gelatin–starch blend reinforced with HA nanocrystals (nHA) gave biocompatible composites with enhanced mechanical properties.	(Sundaram, et al., 2008)

Novel Starch-Based Scaffolds for Bone Tissue Engineering: Cytotoxicity, Cell Culture, and Protein Expression	Melt-based technology	Scaffolds	A study on 50/50 (wt%) blend of corn starch/ethylene-vinyl alcohol (SEVA-C) led these authors to conclude that, starch-based scaffolds should be considered as an alternative for bone tissue-engineering applications in the near future.	(Salgado, et al., 2004)
Microwave Processing of Starch-Based Porous Structures for Tissue Engineering Scaffolds	Microwave Processing	Scaffolds	This study reports the suitability of different types of starch-based polymers; Potato, sweet potato, corn starch, and nonisolated amaranth and quinoa starch in preparing porous structures.	(Torres, et al., 2007)
Scaffold development using 3D printing with a starch- based polymer.	Rapid prototyping	Scaffolds	In this study, a unique blend of starch-based polymer powders (corn starch, dextran and gelatin) was developed for the 3DP process. Cylindrical scaffolds of five different designs were fabricated and found to possess enhanced mechanical and chemical properties.	(Lam, et al., 2002)
New partially degradable and bioactive acrylic bone cements based on starch blends and ceramic fillers	Free radical polymerization of methyl methacrylate and acrylic acid	Acrylic bone cements	This work reports the development of new partially biodegradable acrylic bone cements based on corn starch/cellulose acetate blends (SCA).	(Espigares, et al., 2002)
Porous starch-based drug delivery systems processed by a microwave route	Microwave baking method	Non-steroid anti-inflammatory agent	A new simple processing route to produce starch-based porous materials was developed for the delivery of non-steroid anti-inflammatory agents.	(Malafaya, 2001)

Table 9. Other starch derivatives and starch scaffolds with potential medical/pharmaceutical applications

5. Conclusions

It is obvious that starch has moved from its traditional role as food to being an indispensable medicine. The wide use of starch in the medicine is based on its adhesive, thickening, gelling, swelling and film-forming properties as well as its ready availability, low cost and controlled quality. From the foregoing, to think that starch is still ordinary inert excipients is to be oblivious of the influence this important biopolymer plays in therapeutic outcome of bioactive moieties. Starch has proven to be the formulator's "friend" in that, it can be utilized in the preparation of various drug delivery systems with the potential to achieve the formulator's desire for target or protected delivery of bioactive agents. It is important to note that apart from the low cost of starch, it is also relatively pure and does not need intensive purification procedures like other naturally occurring biopolymers, such as celluloses and gums. A major limitation to starch use appears to be its higher sensitivity to the acid attack; however, modification has been proved to impart acid-resistance to the product. It is important to optimize the process of transition of starch granules from its native micro- to the artificial submicron levels in greater detail and also pay greater attention to its toxicological profiles especially when it is desired to be used at nanoscale. Although starch is generally regarded as safe, its derivatives and in fact at submicron levels it may pose some safety challenges especially as carriers in drug delivery systems. It is possible to conclude that, although starch is food, it is also medicine.

6. References

A. N. Jyothi, M.S.S., S. N. Moorthy, J. Sreekumar and K. N. Rajasekharan, (2008). 'Microwave-assisted Synthesis of Cassava Starch Phosphates and their Characterization'. *Journal of Root Crops*, 34 (1):34-42.

Al-Karawi, A.J.M. and Al-Daraji, A.H.R., (2010). 'Preparation and using of acrylamide grafted starch as polymer drug carrier'. *Carbohydrate Polymers*, 79 (3):769-774.

Aziz, A., R. Daik, M.A. Ghani, N.I.N. Daud and B.M. Yamin, , (2004). 'Hydroxypropylation and acetylation of sago starch'. *Malaysian J. Chem.*, 6 (48-54).

Balmayor, E., Tuzlakoglu, K., Marques, A., Azevedo, H. and Reis, R., (2008). 'A novel enzymatically-mediated drug delivery carrier for bone tissue engineering applications: combining biodegradable starch-based microparticles and differentiation agents'. *Journal of Materials Science: Materials in Medicine*, 19 (4):1617-1623.

Boesel, L.F., Mano, J.F. and Reis, R.L., (2004). 'Optimization of the formulation and mechanical properties of starch based partially degradable bone cements'. *Journal of Materials Science: Materials in Medicine*, 15 (1):73-83.

Brouillet, F., Bataille, B. and Cartilier, L., (2008). 'High-amylose sodium carboxymethyl starch matrices for oral, sustained drug-release: Formulation aspects and in vitro drug-release evaluation'. *International Journal of Pharmaceutics*, 356 (1-2):52-60.

Chakraborty, S., Sahoo, B., Teraoka, I., Miller Lisa, M. and Gross Richard, A., (2005). 'Enzyme-Catalyzed Regioselective Modification of Starch Nanoparticles'. *Polymer Biocatalysis and Biomaterials*: American Chemical Society, 246-265.

Chen, L., Li, X., Li, L. and Guo, S., (2007). 'Acetylated starch-based biodegradable materials with potential biomedical applications as drug delivery systems'. *Current Applied Physics*, 7 (Supplement 1):e90-e93.

Choi, S.G. and Kerr, W.L., (2003). 'Water mobility and textural properties of native and hydroxypropylated wheat starch gels'. *Carbohydrate Polymers*, 51 (1):1-8.

Cristina Freire, A., Fertig, C.C., Podczeck, F., Veiga, F. and Sousa, J., (2009). 'Starch-based coatings for colon-specific drug delivery. Part I: The influence of heat treatment on the physico-chemical properties of high amylose maize starches'. *European Journal of Pharmaceutics and Biopharmaceutics*, 72 (3):574-586.

Duarte, A.R.C., Mano, J.F. and Reis, R.L., (2009). 'Preparation of starch-based scaffolds for tissue engineering by supercritical immersion precipitation'. *The Journal of Supercritical Fluids*, 49 (2):279-285.

Echeverria, I., Silva, I., Goñi, I. and Gurruchaga, M., (2005). 'Ethyl methacrylate grafted on two starches as polymeric matrices for drug delivery'. *Journal of Applied Polymer Science*, 96 (2):523-536.

Efentakis, M., Pagoni, I., Vlachou, M. and Avgoustakis, K., (2007). 'Dimensional changes, gel layer evolution and drug release studies in hydrophilic matrices loaded with drugs of different solubility'. *International Journal of Pharmaceutics*, 339 (1-2):66-75.

El-Hag Ali, A. and AlArifi, A., (2009). 'Characterization and in vitro evaluation of starch based hydrogels as carriers for colon specific drug delivery systems'. *Carbohydrate Polymers*, 78 (4):725-730.

Espigares, I., Elvira, C., Mano, J.F., Vázquez, B., San Román, J. and Reis, R.L., (2002). 'New partially degradable and bioactive acrylic bone cements based on starch blends and ceramic fillers'. *Biomaterials*, 23 (8):1883-1895.

Fares, M.M., El-faqeeh, A.S. and Osman, M.E., (2003). 'Graft Copolymerization onto Starch– I. Synthesis and Optimization of Starch Grafted with N-tert-Butylacrylamide Copolymer and its Hydrogels'. *Journal of Polymer Research*, 10 (2):119-125.

Freire, C., Podczeck, F., Veiga, F. and Sousa, J., (2009). 'Starch-based coatings for colon-specific delivery. Part II: Physicochemical properties and in vitro drug release from high amylose maize starch films'. *European Journal of Pharmaceutics and Biopharmaceutics*, 72 (3):587-594.

Geresh, S., Gdalevsky, G.Y., Gilboa, I., Voorspoels, J., Remon, J.P. and Kost, J., (2004). 'Bioadhesive grafted starch copolymers as platforms for peroral drug delivery: a study of theophylline release'. *Journal of Controlled Release*, 94 (2-3):391-399.

Gomes, M.E., Sikavitsas, V.I., Behravesh, E., Reis, R.L. and Mikos, A.G., (2003). 'Effect of flow perfusion on the osteogenic differentiation of bone marrow stromal cells cultured on starch-based three-dimensional scaffolds'. *Journal of Biomedical Materials Research Part A*, 67A (1):87-95.

Heinze, T., (2005). 'CARBOXYMETHYL ETHERS OF CELLULOSE AND STARCH – A REVIEW'. *Chemistry of plant raw materials.*, 3:13-29.

Herman, J. and Remon, J.P., (1989). 'Modified starches as hydrophilic matrices for controlled oral delivery. II. In vitro drug release evaluation of thermally modified starches'. *International Journal of Pharmaceutics*, 56 (1):65-70.

Herman, J. and Remon, J.P., (1990). 'Modified starches as hydrophilic matrices for controlled oral delivery III. Evaluation of sustained-release theophylline formulations based on thermal modified starch matrices in dogs'. *International Journal of Pharmaceutics*, 63 (3):201-205.

Huber, K.C. and BeMiller, J.N., (2001). 'Location of Sites of Reaction Within Starch Granules1'. *Cereal Chemistry*, 78 (2):173-180.

Iromidayo Olu-Owolabi.B, A.A.T.K.A.O., (2010). 'Effect of heat moisture treatment on the functional and tabletting properties of corn starch'. *African Journal of Pharmacy and Pharmacology*, 4 (7):498-510.

Jansook, P. and Loftsson, T., (2009). 'CDs as solubilizers: Effects of excipients and competing drugs'. *International Journal of Pharmaceutics*, 379 (1):32-40.

Jarowrenko, W., (1986). 'Pregelatinised starches'. *In O. B. Wurzburg (Ed.), Modified starches: Properties and uses.Boca Raton, FL: CRC Press*:71.

Juszczak, (2003). 'Surface of triticale starch granules—NC-AFM observations'. *Electronic Journal of Polish Agricultural Universities, Food Science and Technology*, 6.

Killen, B.U. and Corrigan, O.I., (2006). 'Effect of soluble filler on drug release from stearic acid based compacts'. *International Journal of Pharmaceutics*, 316 (1-2):47-51.

Kim, H.R., Muhrbeck, P. and Eliasson, A.-C., (1993). 'Changes in rheological properties of hydroxypropyl potato starch pastes during freeze−thaw treatments. III. Effect of cooking conditions and concentration of the starch paste'. *Journal of the Science of Food and Agriculture*, 61 (1):109-116.

Kim, Y.-K. and Robyt, J.F., (1999). 'Enzyme modification of starch granules: in situ reaction of glucoamylase to give complete retention of -glucose inside the granule'. *Carbohydrate Research*, 318 (1-4):129-134.

Kittipongpatana, O.S., Chaichanasak, N., Kanchongkittipoan, S., Panturat, A., Taekanmark, T. and Kittipongpatana, N., (2006). 'An Aqueous Film-coating Formulation based on Sodium Carboxymethyl Mungbean Starch'. *Starch - Stärke*, 58 (11):587-589.

Kittipongpatana, O.S., Chaitep, W., Kittipongpatana, N., Laenger, R. and Sriroth, K., (2007). 'Physicochemical and Pharmaceutical Properties of Carboxymethyl Rice Starches Modified from Native Starches with Different Amylose Content'. *Cereal Chemistry*, 84 (4):331-336.

Kundu, B., Lemos, A., Soundrapandian, C., Sen, P., Datta, S., Ferreira, J. and Basu, D., 'Development of porous HAp and β-TCP scaffolds by starch consolidation with foaming method and drug-chitosan bilayered scaffold based drug delivery system'. *Journal of Materials Science: Materials in Medicine*, 21 (11):2955-2969.

Lam, C.X.F., Mo, X.M., Teoh, S.H. and Hutmacher, D.W., (2002). 'Scaffold development using 3D printing with a starch-based polymer'. *Materials Science and Engineering: C*, 20 (1-2):49-56.

Langoth, N., Kalbe, J. and Bernkop-Schnürch, A., (2003). 'Development of buccal drug delivery systems based on a thiolated polymer'. *International Journal of Pharmaceutics*, 252 (1-2):141-148.

Le Corre, D., Bras, J., Dufresne, A., (2010). 'Starch Nanoparticles: A Review'. *Biomacromolecules 11, 1139–1153*.

Le Corre, D.b., Bras, J. and Dufresne, A., (2010). 'Starch Nanoparticles: A Review'. *Biomacromolecules*, 11 (5):1139-1153.

Lemieux, M., Gosselin, P. and Mateescu, M.A., (2009). 'Carboxymethyl high amylose starch as excipient for controlled drug release: Mechanistic study and the influence of degree of substitution'. *International Journal of Pharmaceutics*, 382 (1-2):172-182.

Lewandowicz, G. and Soral-Smietana, M., (2004). 'Starch modification by iterated syneresis'. *Carbohydrate Polymers*, 56 (4):403-413.

Li, S., Lin, S., Daggy, B.P., Mirchandani, H.L. and Chien, Y.W., (2003). 'Effect of HPMC and Carbopol on the release and floating properties of Gastric Floating Drug Delivery System using factorial design'. *International Journal of Pharmaceutics*, 253 (1-2):13-22.

Liu, H., Ramsden, L. and Corke, H., (1999). 'Physical properties and enzymatic digestibility of hydroxypropylated ae, wx, and normal maize starch'. *Carbohydrate Polymers*, 40 (3):175-182.

Malafaya, P.B.E., C.; Gallardo, A.; San Román, J.; Reis, R.L., (2001). 'Porous starch-based drug delivery systems processed by a microwave route'. *Journal of Biomaterials Science, Polymer Edition*, 12:1227-1241.

Manek, R.V., Kunle, O.O., Emeje, M.O., Builders, P., Rao, G.V.R., Lopez, G.P. and Kolling, W.M., (2005). 'Physical, Thermal and Sorption Profile of Starch Obtained from Tacca leontopetaloides'. *Starch - Stärke*, 57 (2):55-61.

Marinich, J.A., Ferrero, C. and Jiménez-Castellanos, M.R., (2009). 'Graft copolymers of ethyl methacrylate on waxy maize starch derivatives as novel excipients for matrix tablets: Physicochemical and technological characterisation'. *European Journal of Pharmaceutics and Biopharmaceutics*, 72 (1):138-147.

Masaaki Endo, K.O., Yoshihito Yaginuma, (2009). 'A Novel Pregelatinized Starch as a Sustained-Release Matrix Excipient'. *Pharmaceutical Technology*.

Mason, W.R., (2009). ' Starch Use in Foods. In Starch: Chemistry and Technology, Third Edition; chapter 20'.746 - 772

Massicotte, L.P., Baille, W.E. and Mateescu, M.A., (2008). 'Carboxylated high amylose starch as pharmaceutical excipients: Structural insights and formulation of pancreatic enzymes'. *International Journal of Pharmaceutics*, 356 (1-2):212-223.

Michailova, V., Titeva, S., Kotsilkova, R., Krusteva, E. and Minkov, E., (2001). 'Influence of hydrogel structure on the processes of water penetration and drug release from mixed hydroxypropylmethyl cellulose/thermally pregelatinized waxy maize starch hydrophilic matrices'. *International Journal of Pharmaceutics*, 222 (1):7-17.

Munday, D.L. and Cox, P.J., (2000). 'Compressed xanthan and karaya gum matrices: hydration, erosion and drug release mechanisms'. *International Journal of Pharmaceutics*, 203 (1-2):179-192.

N.L Prasanthi, N.R.a.R., (2010). *Starch Phosphate: A Novel Pharmaceutical Excipient For Tablet Formulation*.

Nabais, T., Brouillet, F., Kyriacos, S., Mroueh, M., Amores da Silva, P., Bataille, B., Chebli, C. and Cartilier, L., (2007). 'High-amylose carboxymethyl starch matrices for oral sustained drug-release: In vitro and in vivo evaluation'. *European Journal of Pharmaceutics and Biopharmaceutics*, 65 (3):371-378.

Narendra Reddy, Y.Y., (2009). 'Preparation and Properties of Starch Acetate Fibers for Potential Tissue Engineering Applications'. *Biotechnology and Biengineering*, 103 (5).

Nikhil K Sachan, S.K.G.a.A.B., (2010). 'Evaluation of glutinous rice starch based matrix microbeads using scanning electron microscopy'. *Journal of Chemical and Pharmaceutical Research*, 2 (3):433-452.

Norman 1981, B.E., 'New developments in starch syrup technology'. *w: Enzymes and Food Processing, G.G. Birch, N. Blakebrough, K.J. Parker eds., Applied Science Publishers, London,*:15-50.

Nutan, M.T.H., Soliman, M.S., Taha, E.I. and Khan, M.A., (2005). 'Optimization and characterization of controlled release multi-particulate beads coated with starch acetate'. *International Journal of Pharmaceutics*, 294 (1-2):89-101.

Nykänen, P., Lempää, S., Aaltonen, M.L., Jürjenson, H., Veski, P. and Marvola, M., (2001). 'Citric acid as excipient in multiple-unit enteric-coated tablets for targeting drugs on the colon'. *International Journal of Pharmaceutics*, 229 (1-2):155-162.

O'Brien, S., Wang, Y.-J., Vervaet, C. and Remon, J.P., (2009). 'Starch phosphates prepared by reactive extrusion as a sustained release agent'. *Carbohydrate Polymers*, 76 (4):557-566.

Onofre, F.O. and Wang, Y.J., (2010). 'Hydroxypropylated starches of varying amylose contents as sustained release matrices in tablets'. *International Journal of Pharmaceutics*, 385 (1-2):104-112.

Park, K., (2007). 'Nanotechnology: What it can do for drug delivery'. *Journal of Controlled Release*, 120 (1-2):1-3.

Patrick De Koninck, D.A., Francine Hamel, Fathey Sarhan and Mircea Alexandru Mateescu, (2010). 'Carboxymethyl-Starch Excipients for Gastrointestinal Stable Oral Protein Formulations Containing Protease Inhibitors'. *Journal of Pharmacy and Pharmaceutical Sciences*, 13 (1):78-92.

Peerapattana, J., Phuvarit, P., Srijesdaruk, V., Preechagoon, D. and Tattawasart, A., 'Pregelatinized glutinous rice starch as a sustained release agent for tablet preparations'. *Carbohydrate Polymers*, 80 (2):453-459.

Perera, C., Hoover, R. and Martin, A.M., 'The effect of hydroxypropylation on the structure and physicochemical properties of native, defatted and heat-moisture treated potato starches'. *Food Research International*, 30 (3-4):235-247.

Pitchon. E, O.R.J.D., & Joseph T. H (1981). 'Process for cooking or gelatinizing materials'. *US Patent 4280 851*.

Pohja, S., Suihko, E., Vidgren, M., Paronen, P. and Ketolainen, J., (2004). 'Starch acetate as a tablet matrix for sustained drug release'. *Journal of Controlled Release*, 94 (2-3):293-302.

Pornsak Sriamornsak , M.J., Suchada Piriyaprasarth, (2010). 'Microwave-Assisted Modification of Arrowroot Starch for Pharmaceutical Matrix Tablets'. *Advanced Materials Research*, 93-94:358-361.

Prochaska, K., Konowal, E., Sulej-Chojnacka, J. and Lewandowicz, G., (2009). 'Physicochemical properties of cross-linked and acetylated starches and products of their hydrolysis in continuous recycle membrane reactor'. *Colloids and Surfaces B: Biointerfaces*, 74 (1):238-243.

Pu, H.C., Ling Li, Xiaoxi Xie, Fengwei Yu, Long Li, Lin, (2011). 'An Oral Colon-Targeting Controlled Release System Based on Resistant Starch Acetate: Synthetization, Characterization, and Preparation of Film-Coating Pellets'. *Journal of Agricultural and Food Chemistry*, 59 (10):5738-5745.

Raatikainen, P., Korhonen, O., Peltonen, S. and Paronen, P., (2002). 'Acetylation Enhances the Tabletting Properties of Starch'. *Drug Development and Industrial Pharmacy*, 28 (2):165-175.

Rahmouni, M, Lenaerts, V, Leroux and J, C., (2003). *Drug permeation through a swollen cross-linked amylose starch membrane*. Paris, FRANCE: Editions de santé.

Rajan, A., Sudha, J.D. and Abraham, T.E., (2008). 'Enzymatic modification of cassava starch by fungal lipase'. *Industrial Crops and Products*, 27 (1):50-59.

Reis, A.V., Guilherme, M.R., Moia, T.A., Mattoso, L.H.C., Muniz, E.C. and Tambourgi, E.B., (2008). 'Synthesis and characterization of a starch-modified hydrogel as potential carrier for drug delivery system'. *Journal of Polymer Science Part A: Polymer Chemistry*, 46 (7):2567-2574.

Robertson, M.I., (1999). 'Regulatory issues with excipients'. *International Journal of Pharmaceutics*, 187 (2):273-276.

Rutenberg, M.W.a.D.S., Whistler, R.L., J.N. BeMiller and E.F. Paschall (Eds.), (1984). 'Starch Derivatives: Production and Uses. In: Starch: Chemistry and Technology, '. *Academic Press, New York,*:312-388.

Saboktakin, M.R., Tabatabaie, R.M., Maharramov, A. and Ramazanov, M.A., (2010). 'Synthesis and in vitro evaluation of carboxymethyl starch-chitosan nanoparticles as drug delivery system to the colon'. *International Journal of Biological Macromolecules,* 48 (3):381-385.

Sabyasachi Maiti, S.R., Biswanath Sa, (2010). 'Polysaccharide-Based Graft Copolymers in Controlled Drug Delivery'. *International Journal of PharmTech Research,* 2 (2):1350-1358.

Sajilata, M.G., Singhal, R.S. and Kulkarni, P.R., (2006). 'Resistant Starch–A Review'. *Comprehensive Reviews in Food Science and Food Safety,* 5 (1):1-17.

Salgado, A.J., Coutinho, O.P. and Reis, R.L., (2004). 'Novel Starch-Based Scaffolds for Bone Tissue Engineering: Cytotoxicity, Cell Culture, and Protein Expression'. *Tissue Engineering,* 10 (3-4):465-474.

Sánchez, L., Torrado, S. and Lastres, J., (1995). 'Gelatinized/freeze-dried starch as excipient in sustained release tablets'. *International Journal of Pharmaceutics,* 115 (2):201-208.

Sen, G. and Pal, S., (2009a). 'Microwave initiated synthesis of polyacrylamide grafted carboxymethylstarch (CMS-g-PAM): Application as a novel matrix for sustained drug release'. *International Journal of Biological Macromolecules,* 45 (1):48-55.

Sen, G. and Pal, S., (2009b). 'A novel polymeric biomaterial based on carboxymethylstarch and its application in controlled drug release'. *Journal of Applied Polymer Science,* 114 (5):2798-2805.

Seow, C.C. and Thevamalar, K., (1993). 'Internal Plasticization of Granular Rice Starch by Hydroxypropylation: Effects on Phase Transitions Associated with Gelatinization'. *Starch - Stärke,* 45 (3):85-88.

Serrero, A.l., Trombotto, S.p., Cassagnau, P., Bayon, Y., Gravagna, P., Montanari, S. and David, L., 'Polysaccharide Gels Based on Chitosan and Modified Starch: Structural Characterization and Linear Viscoelastic Behavior'. *Biomacromolecules,* 11 (6):1534-1543.

Shaikh, M.M. and Lonikar, S.V., (2009). 'Starch–acrylics graft copolymers and blends: Synthesis, characterization, and applications as matrix for drug delivery'. *Journal of Applied Polymer Science,* 114 (5):2893-2900.

Shangraw, R.F., (1992). 'International harmonization of compendia standards for pharmaceutical excipients'. *D.J.A. Crommelin, K.Midha (Eds.), Topics in Pharmaceutical Sciences, MSP, Stuttgart, Germany*:205-223.

Shogren, R., (2008). 'Scandium triflate catalyzed acetylation of starch at low to moderate temperatures'. *Carbohydrate Polymers,* 72 (3):439-443.

Shogren, R.L., (2000). 'Modification of maize starch by thermal processing in glacial acetic acid'. *Carbohydrate Polymers,* 43 (4):309-315.

Shogren, R.L., (2003). 'Rapid preparation of starch esters by high temperature/pressure reaction'. *Carbohydrate Polymers,* 52 (3):319-326.

Silva, I., Gurruchaga, M. and Goñi, I., (2009). 'Physical blends of starch graft copolymers as matrices for colon targeting drug delivery systems'. *Carbohydrate Polymers,* 76 (4):593-601.

Simi, C. and Emilia Abraham, T., (2007). 'Hydrophobic grafted and cross-linked starch nanoparticles for drug delivery'. *Bioprocess and Biosystems Engineering*, 30 (3):173-180.

Singh, J., Kaur, L. and McCarthy, O.J., (2007). 'Factors influencing the physico-chemical, morphological, thermal and rheological properties of some chemically modified starches for food applications--A review'. *Food Hydrocolloids*, 21 (1):1-22.

Sitohy, M.Z. and Ramadan, M.F., (2001). 'Degradability of Different Phosphorylated Starches and Thermoplastic Films Prepared from Corn Starch Phosphomonoesters'. *Starch - Stärke*, 53 (7):317-322.

Stute, R., Heilbronn, Klingler, R.W., Boguslawski, S., Eshtiaghi, M.N. and Knorr, D., (1996). 'Effects of High Pressures Treatment on Starches'. *Starch - Stärke*, 48 (11-12):399-408.

Sundaram, J., Durance, T.D. and Wang, R., (2008). 'Porous scaffold of gelatin-starch with nanohydroxyapatite composite processed via novel microwave vacuum drying'. *Acta Biomaterialia*, 4 (4):932-942.

Te Wierik, G.H.P., Eissens, A.C., Bergsma, J., Arends-Scholte, A.W. and Bolhuis, G.K., (1997). 'A new generation starch product as excipient in pharmaceutical tablets: III. Parameters affecting controlled drug release from tablets based on high surface area retrograded pregelatinized potato starch'. *International Journal of Pharmaceutics*, 157 (2):181-187.

Torres, F.G., Boccaccini, A.R. and Troncoso, O.P., (2007). 'Microwave processing of starch-based porous structures for tissue engineering scaffolds'. *Journal of Applied Polymer Science*, 103 (2):1332-1339.

Trubiano Paolo, C., (1997). 'The Role of Specialty Food Starches in Flavor Encapsulation'. *Flavor Technology*: American Chemical Society, 244-253.

Trubiano, P.C., (1987). ' Succinate and substituted succinate derivatives of starch. ' *Modified starches: Properties and uses, CRC Press, Boca Raton, Florida,* In: Wurzburg, O.B., Editor, 1987:131-148.

Tukomane, T. and Varavinit, S., (2008). 'Influence of Octenyl Succinate Rice Starch on Rheological Properties of Gelatinized Rice Starch before and after Retrogradation'. *Starch - Stärke*, 60 (6):298-304.

Tuovinen, L., Peltonen, S. and Järvinen, K., (2003). 'Drug release from starch-acetate films'. *Journal of Controlled Release*, 91 (3):345-354.

Tuovinen, L., Peltonen, S., Liikola, M., Hotakainen, M., Lahtela-Kakkonen, M., Poso, A. and Järvinen, K., (2004a). 'Drug release from starch-acetate microparticles and films with and without incorporated [alpha]-amylase'. *Biomaterials*, 25 (18):4355-4362.

Tuovinen, L., Ruhanen, E., Kinnarinen, T., Rönkkö, S., Pelkonen, J., Urtti, A., Peltonen, S. and Järvinen, K., (2004b). 'Starch acetate microparticles for drug delivery into retinal pigment epithelium--in vitro study'. *Journal of Controlled Release*, 98 (3):407-413.

Wang, X., Li, X., Chen, L., Xie, F., Yu, L. and Li, B., (2011). 'Preparation and characterisation of octenyl succinate starch as a delivery carrier for bioactive food components'. *Food Chemistry*, 126 (3):1218-1225.

Williams, H.D., Ward, R., Culy, A., Hardy, I.J. and Melia, C.D.2011, 'Designing HPMC matrices with improved resistance to dissolved sugar'. *International Journal of Pharmaceutics*, 401 (1-2):51-59.

Wootton, M. and Manatsathit, A., (1983). 'The Influence of Molar Substitution on the Water Binding Capacity of Hydroxypropyl Maize Starches'. *Starch - Stärke*, 35 (3):92-94.

Xu, R., Feng, X., Xie, X., Xu, H., Wu, D. and Xu, L., (2011). 'Grafted Starch-Encapsulated Hemoglobin (GSEHb) Artificial Red Blood Cells Substitutes'. *Biomacromolecules*: null-null.

Yoon, H.-S., Kweon, D.-K. and Lim, S.-T., (2007). 'Effects of drying process for amorphous waxy maize starch on theophylline release from starch-based tablets'. *Journal of Applied Polymer Science*, 105 (4):1908-1913.

4

Trends in Functional Food Against Obesity

José C.E. Serrano, Anna Cassanyé and Manuel Portero-Otín
Department of Experimental Medicine, University of Lleida
Spain

1. Introduction

The increasingly accepted notion of the relationship between diet and health has opened new perspectives on the effects of food ingredients on physiological functions and health. Among the nutritional complications, increased incidence of obesity and its associated medical complications is creating a pressure from consumers towards the food industry which may provide an opportunity for the development of functional foods designed for the prevention and/or treatment of these pathologies.

Obesity is a multifactor disease where several factors may influence its onset, which includes the contributions of inherited, metabolic, behavioural, environmental, cultural, and socioeconomic factors as it is shown in Figure 1. Most of these factors may play together in different grades of contribution, which may differ between patients, and may influence treatment objectives in each individual.

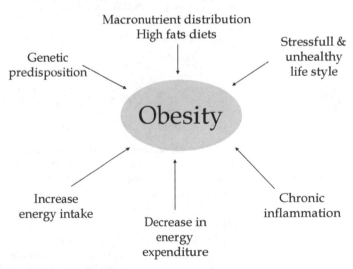

Fig. 1. Obesity, a multifactorial disease.

Moreover, overweight and obesity may raise the risk of other related pathologies like high blood pressure, high blood cholesterol, heart disease, stroke, diabetes, certain types of cancer, arthritis, and breathing problems. As weight increases, so does the prevalence of health risks. The health outcomes related to these diseases, however, may be improved through weight loss or, at a minimum, no further weight gain. The main goal of any nutritional intervention is to individually determine the principal factors that may contribute with individual obesity predisposition and find specific tools to counteract each factor. Food Industry may play an important role providing enough tools, functional foods, for the prevention and treatment of obesity

In simplified terms, overweight and obesity can be defined as an imbalance where the amount of energy intake exceeds the amount of energy expended. Treatment and prevention of obesity requires changes in one or two of the components of this simplified equation. In this sense, the development of functional foods should be aimed to decrease the amount of energy intake (by lowering the energy density of foods or reducing the food intake) or increasing caloric expenditure through the stimulation of thermogenesis and/or modifying the distribution and use of nutrients as energy fuel between tissues, discouraging fat deposition. A summary of possible strategic features for the development of functional foods against obesity is shown in Figure 2.

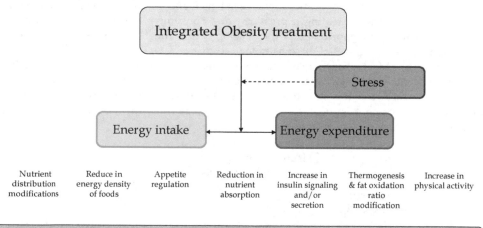

Fig. 2. Main strategic features for the development of functional foods against obesity.

Since several decades ago, in the fight against obesity, food manufactures had offer, a variety of food products named in the beginnings as "dietetic products" based principally in substitutions of sugars and fat by non-nutritive sweeteners and fat replacers respectively. Nowadays, the research and development of new products, should offer the market, food products indicated especially for obese people that besides their low caloric content, can offers the possibility to influence the energy metabolism as well as in the physiological sensation of satiety. Currently, there is a wide variety of products in the market with a low energy density, while the supply of products with bioactive ingredients that decrease appetite, increase caloric expenditure and/or affect the distribution of body fat is scare and in some cases of doubtful effectiveness.

In this context, the new European regulation regarding food labelling, may encourage the food industry to carry out more investment in research and in the determination of the effectiveness of the functional products launched to the market at different levels (biochemical, molecular, genomic and psychological). In this way, the confidence and scientifically contrasted effectiveness of these products as preventive and therapy tools against obesity, may contribute to reduce the incidence of obesity in the whole population.

The first step in the development of functional foods is the identification of functional factor, condition or compound that produces a specific effect, which is effective as an adjunct in the treatment of obesity. The new European regulation demands that the effectiveness of such functional foods should be properly established with sufficient scientific evidence, including intervention studies in human populations. It is also desirable to establish the possible interactions of the functional ingredient within the body at different levels (genomic, molecular, cellular and psychological). On the other hand, is very important and necessary to investigate the functional ingredients incorporated into food as such, taking into account possible interactions between "functional ingredient" and other food matrix components, its dose, culinary preparation processes and the usual form of consumption. The key points in the evaluation of functional foods would be the safety and efficacy, thereby avoiding misleading advertising to the consumer.

In this sense, it is necessary to establish specific biomarkers (e.g. body mass index, blood cholesterol levels, percentage of body fat) the effectiveness of consumption of functional foods designed against obesity. However, there is still no consensus on the specific relevance and applicability of each of these biomarkers in the context of obesity, so that there is unanimity on tests of functional assessment for all food companies to launch a new functional food to the market.

The objective of this chapter is to describe possible guidelines for the development of functional foods based on the scientific evidence of the actions of several bioactive compounds and nutritional/technological modifications of foods to be used for the prevention and/or treatment of obesity. It describes possible actions that could be undertaken from different levels, starting with the technological modification of food with the aim to produce a satiety feeling towards the incorporation of functional ingredients that may modify energy intake and expenditure.

2. Energy balance, factors that influence energy intake & expenditure

As mentioned before, the strategy for functional food development should be based in the reduction of energy intake and/or in the increase in energy expenditure. The reduction in energy intake can be obtained by increasing satiety feeling either by the activation of satiety centres or by the modification of hunger feeling delaying it's unset. Optimally, a good satiety functional food must be satisfying and have a reduce energy content. The increase in energy expenditure can be regulated by the modification in the metabolic rate via an increase in thermogenesis or by the control in hormonal energetic metabolisms like insulin sensitivity. Additionally it should be also taken into consideration other factors like inflammation, psychological and physiological stress and life styles that may influence the response and adaptation to energy intake and expenditure.

This section includes a brief description of physiological mechanisms that may influence energy balance and possible strategies for counteracting the effect of each mechanism in obesity, as a tool for the manufacture of functional foods.

2.1 Satiety control, a tool for energy intake reduction

Human behaviour towards food can be defined as a physiological and psychological process that may be influenced by genetic and environmental factors in which the individual is involved. The physiological regulation of the act of eating (hunger and satiety sensations) is a complex interaction between peripheral signals and central nervous system interpretation of these signals, to which must be added physio-psychological variables, such as differences in taste perception and the strictly psychological variables likely influenced by the individual's surrounding environment.

From a physiological point of view the satiety and hunger regulation has been described using two paradigms: the glucostatic hypothesis (Mayer & Thomas, 1967) and the lipostatic model (Kennedy, 1953). The glucostatic hypothesis is based on the assumption that small changes in plasma glucose levels induced signal initiation and termination of eating. However, this model does not take into account how the body regulates the long-term storage and use of energy. The lipostatic model hypothesizes that there are peripheral signals that gives information about the amount of fat or stored energy and therefore the amount of energy needed to maintain a good energy balance. This hypothesis has been supported by the discovery of leptin, an adipokine that is released by adipose tissue in proportion to its fat content. However, since there are no significant fluctuations throughout the day on the composition of body fat and thus leptin, this model may not explain the dynamic behaviour and varying feelings of satiety-hungry induced throughout the day.

The interpretation of these signals is done by the central nervous systems. Recently it has been reported that short- and long-term satiety and hunger feelings may be regulated by several neural circuits at the ventromedial, dorsomedial and paraventricular hypothalamic nuclei for satiety sensations and at the lateral hypothalamus for hunger sensations. Although the hypothalamus is an important centre in the energy balance regulation, there are other brain regions such as the medulla oblongata and cortical and striatal structures, essential for the eating behaviour modulation. For example, some neural circuits of the medulla oblongata seem to have an important role in autonomic eating regulation, limiting the quantity of ingested food through the satiety responses regulation. Whereas, other parts of the brain, like the nucleus accumbens and ventral tegmental area, where dopamine, opioids and cannabinoids signals are integrated, regulated the motivation to eat, the rewards and the acts before eating. In this context, although hunger is connected to the biological needs, there are also psychological factors involved in the food intake regulation. Learning and emotions play a powerful role in determining what to eat, when to eat, and even how much to eat. In this context, the psychological desire of eating and its complicated mechanisms of influence in satiety interpretation difficult the design of functional foods for this purpose.

Although, there are many peripheral signals that can contribute to feeding behaviour and body weight regulation and can be modified by food and food ingredients. It is important to recognize that short-term and long-term food intake and energy balance are regulated through distinct, but interacting, mechanisms. Figure 3, shows a brief review of nowadays known possible satiety signals that may influence eating behaviour which included, beside short and long-term signals, individual social behaviour and other metabolism compounds that may influence satiety feeling.

Short-term regulation of food intake results from an integrated response from neural and humoral signals that originate mainly at the brain, gastrointestinal tract and adipose tissue. Ingested food evokes satiety in the gastrointestinal tract primarily by two distinct ways, i.e. by mechanical stimulation and therefore stimulation of the nerve endings; and by the release of satiety peptides. The scheme is more complicated as both ways seems to be intimately related, since many of the intestinal peptides released may inhibit also gastric emptying thus enhancing gastric mechanoreceptor stimulation.

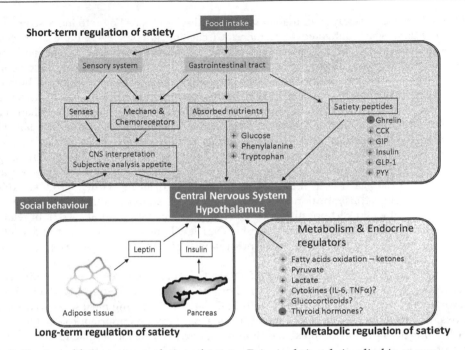

Fig. 3. Short- and long-term regulation of satiety. Principal signals implied in energy regulation and satiety.

The postprandial satiety consequences of food intake are determined both by the specific chemical composition and the characteristics physical properties of the food. Accordingly, different foods, despite their equal energy content, can differ in their capacity to affect postprandial metabolism, especially secretion of gastrointestinal peptides, thereby regulating energy homeostasis. A classical example is fibre were several differences in chemical structures and characteristic physical properties can be observed. For example, bulk/volume, viscosity, water-holding capacity, adsorption/binding, or fermentability may determine the subsequent physiological behaviour of fibre eventhough it is ingested in the same quantity. Table 1 and below is included a brief description of the principal satiety signals derived from the gastrointestinal tract and its possible effects in food intake behaviour.

Peptide	Organ of synthesis	Receptor related with satiety signals	Effects on food intake
CCK	Proximal intestine I-cells	CCK1R	Decrease
GLP-1	Distal intestine L-cells	GLP1R	Decrease
PYY$_{3-36}$	Distal intestine L-cells	Y2R	Decrease
PP	Pancreatic F cells	Y4R, Y5R	Decrease
Amylin	Pancreatic β-cells	CTRs, RAMPs	Decrease
Gastric leptin	Stomach P-cells	Leptin receptor	Decrease
Ghrelin	Gastric X/A-cells	Ghrelin receptor	Increase

Table 1. Gastrointestinal satiety peptides that may regulate food intake.

Ghrelin. Ghrelin is the only mammalian substance that has been shown to increase appetite and food intake when delivered to humans. Circulating ghrelin levels typically rise just before and fall shortly after a meal, thus playing a role in meal time hunger and meal initiation. The postprandial ghrelin response is affected principally by the caloric content of meals. Thus, high energy-rich meals suppress ghrelin more than lower ones. In humans, across the range of intakes of 220 to 1000 kcal, the lowest point of postprandial ghrelin was found to decrease by about 2.4% for every 100 kcal increase of energy intake (Callahan et al., 2004). Recent findings suggest that postprandial suppression of ghrelin is not mediated by nutrients in the stomach or duodenum but rather from post-ingestive increases in lower intestinal osmolarity (via enteric nervous signaling) as well as from insulin surges (Cummings, 2006). In contrast, the short-term parenteral administration of glucose and insulin in physiological doses may not suppress ghrelin levels (Gruendel et al., 2007). Moreover, ghrelin concentration is not affected by stomach distension, since the administration of water did not influence its concentrations (Shiiya et al., 2002). In relation to the possible modification of its plasmatic leveles, it is known that increased fibre content of the meal has shown both to decrease postprandial ghrelin concentrations as well as to inhibit the decrease.

Cholecystokinin (CCK). The inhibitory effect of CCK on food intake has been confirmed in numerous species, including humans. It is however short-lived, lasting less than 30 min. Therefore, CCK may inhibit food intake within the meal by reducing meal size and duration but does not affect the onset of a next meal. Thus it may have an important role in the causal chain leading to satiation or meal termination. Gastric distension augments the anoretic effects of CCK in humans. However, other mechanisms including activation of duodenal chemosensitive fibres and activation of CCK receptor 1 in the pyloric sphincter thay may slow down gastric emptying may be implicated.

Glucose-dependent insulinotropic polypeptide (GIP). GIP is released in response to the presence of nutrients in the intestinal lumen. The major stimuli for GIP release are dietary fat and carbohydrates. Protein seems to have no effect, although some evidence exists indicating that the intraduodenal administration of aminoacids can stimulate GIP release.

Glucagon-like peptide 1 (GLP-1). GLP-1 is an incretin hormone released in response to food intake. GLP-1 is typically very low in the fasting state, but rise quickly after food intake, especially after carbohydrate intake. The rise of GLP-1 has been correlated with increased satiety and less hunger. GLP-1 is thought to play an important role in the "ileal break", a mechanism that regulates the flow of nutrients from the stomach into the small intestine. It is also suggested that portal GLP-1 might influence the production of ghrelin (Lippl et al., 2004) and the increase in β-cell mass in the pancreas, thus improving the insulin production.

Peptide tyrosine-tyrosine (PYY). PYY is a member of the pancreatic polypeptide fold family including neuropeptide Y (NPY) and pancreatic polypeptide (PP). PYY mediates ileal and colonic breaks, mechanisms that ultimately slow gastric emptying and promote digestive activities to increase nutrient absorption. Plasma concentration of PYY increases after meals consistent with a meal-related signal of energy homeostasis. Nutrients stimulate PYY release within 30 minutes of ingestion, reaching usually a maximum within 60 min. The release is directly proportional to caloric intake; however meal composition may affect postprandial PYY release. In humans, infusion of PYY_{3-36} (active form) comparable to those after a meal result in decreased energy intake at subsequent meals compared with a control group (Batterham et al., 2002)

Amylin. Amylin is co-secreted together with insulin from pancreatic β-cells. It is a 37 amino acid peptide with anorexigenic effects that have shown to reduce meal size as well as the number of meals. The inhibitory effect of amylin on food intake is thought to be due to the inhibition of gastric emptying.

Pancreatic polypeptide (PP). PP is secreted by F-cells on the endocrine pancreas comprising approximately less than 5% of islet volume. The main function of PP is thought to be the inhibition of exocrine pancreas. Its secretion is controlled by the parasympathetic nervous system and shows a biphasic manner in proportion to food intake.

Leptin. Leptin is a peptide hormone which is released from white adipose tissue and acts in the hypothalamus to promote weight loss, both by reducing appetite and food intake and by increasing energy expenditure. Circulating leptin concentrations are highly positively correlated with body mass index. Despite this strong association, leptin levels shows large individual variation for a given degree of adiposity, indicating the likely effect of variables other than adipose mass, such as genetic and environmental factors. Additionally, food consumption stimulates leptin secretion after a meal and high carbohydrate meals results in greater leptin responses. Although leptin does not seem to play an important role in the short-term regulation of food intake, when subjects are in energy balance, plasma leptin is negatively correlated with appetite and food intake when energy balance is disturbed. Leptin therefore seems to have a role in the regulation of food intake when energy stores changes.

Insulin. Insulin is the major endocrine and metabolic polypeptide hormone secreted by β-cells of the endocrine pancreas and one of the key adiposity signals in the brain influencing energy homeostasis. Plasma insulin concentrations are in direct proportions to changes in adipose mass. Insulin concentrations are increased at positive energy balance and decreased at the times of negative energy balance. Additionally, plasma insulin concentrations are largely determined by peripheral insulin sensitivity which is related to the amount and distribution of body fat in insulin-resistant patients. Insulin thus provides information to the central nervous system about the size and distribution of the adipose mass to regulate metabolic homeostasis.

Besides the described satiety peptides and hormones, other absorbed food derived compounds, metabolites and hormones may also serve as satiety signals for the central nervous system. For example, aminoacids such as phenylalanine and tryptophan that are precursors to monoamine neurotransmitters suppress food intake in humans. The ratio of plasma tryptophan to other amino acids may influence brain serotonin levels, which are known to have inhibitory influence on food intake. Some authors have suggested that the oral administration of 5-hydroxytryptophan may reduce food intake, as well as a reduction in carbohydrate intake and a higher satiety feeling in obese subjects (Cangiano et al., 1992).

Other metabolisms by-products like ketones (from fatty acids), which are metabolic substrate for the central nervous system, are known to inhibit feeding. Moreover, lactate and pyruvate have been reported to induce satiety effects in animal models (Nagase et al., 1996). And finally, endocrine regulators of food intake like cytokines (IL-6 and TNFα), glucocorticoids and thyroid hormones have also been described to regulate satiety and hunger feelings. However the mechanisms of how they may influence feeding behaviour, possibly related with energy expenditure administration, are not well understood.

Food formulation for the induction of the release of satiety peptides or the decrease of ghrelin could be basically obtained by the incorporation of functional ingredients that may increase satiety peptides release in the gut, as well as by rheological modifications with the objective of producing sensory stimulation in the first phases of digestion and also by a higher gastric distension and emptying rate time. The possible strategies are described below.

2.1.1 Energy reduced satisfying foods, combined strategy to reduce energy intake

The combination of energy reduced foods with functional ingredients that stimulate satiety peptides release may be an interesting and ambitious approach. It should be taken in consideration that the release of satiety peptides is normally proportional to the energy intake and energy density of food. In this sense, it seems difficult to design energy reduced food with also the ability to stimulate satiety. However, some approximations to this goal can be developed.

There are some evidences that macronutrients may regulate the secretion of satiety peptides. For example, fats and proteins may increase the release of CCK and reduce ghrelin concentrations by its effect in gastric emptying delay. Other authors suggest also some effects in GLP-1, GIP and PYY, however the exact mechanisms of action is still unknown.

In the other hand, carbohydrates, especially viscous and fermentable fibre (with a reduce energy content), have been widely investigated because of its satiety effects in humans. Figure 4 briefly schematises the possible mechanisms of action of fibres in the induction of satiety.

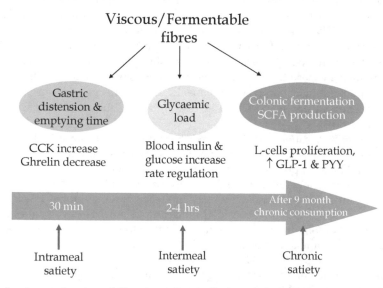

Fig. 4. Mechanisms of action of fibre in satiety induction. SCFA: short-chain fatty acids.

Viscous fibres have been related to satiety because of its effects in gastric distension and emptying rates which may increase CCK and reduce ghrelin secretion. Later, in the small intestine, viscous fibre may regulate bolus transit time, which may affect the total glycemic load, regulating glucose and insulin levels, hence inducing direct satiety stimulation in central nervous system satiety centre. And finally, fibre colonic fermentation by products like short-chain fatty acids are recognized to induce L-cells proliferation in the colon, which are the main cells where GLP-1 and PYY is produced. Thus, chronic fibre intake may chronically increase the amount of these satiety peptides in plasma. Notwithstanding, several authors suggest that the amount of fibre that may exert satiety feeling in humans should be higher than 8 g/day, which difficult its incorporation in technologically modified foods. Moreover, it should be taken into consideration that viscous fibre may produce acute satiety feeling (intrameal and intermeal satiety); whereas fermentable fibres may produce chronic satiety feeling, however after a chronic consumption higher than 9 months.

Other bioactive compounds like polyphenols have been described also to induce a higher rate of secretion of satiety peptides like GLP-1 possibly because of its effects in the inhibition of glucose uptake by enterocytes (McCarty, 2005). In this context, the addition of antioxidant dietary fibre (fibres with associated antioxidant compounds) to food formulation may help to increase satiety feeling of foods.

In brief, it can be said that glucose and fat are the main inductors of satiety peptides. However, it is widely accepted that the increase in glucose and fat uptake are strongly related with obesity. The next step should be the reduction of energy density of formulated foods.

Currently there are a variety of food ingredients that can be used with the purpose to replace fat and sugar in foods formulations (Table 2 & 3) with the intention to reduce the total calorie intake. For example, the average consumption of simple sugars in the Spanish diet by 2005 (MAPA, 2005) was around 113 g/person/day an estimated 18% of total calories (2424 kcal/person/day), exceeding by 8% the recommended dietary intake of simple sugars. Dairy products, beverages, cakes, pastries, etc.., target products for the replacement of simple sugars with non-nutritive sweeteners, contributed to 68% and 62% of the total intake of simple sugars and sucrose in the Spanish diet respectively (Figure 5). Its replacement with non-nutritive sweeteners could result in a reduction in energy intake around 310 kcal/day. However, the abuse in the intake of non-nutritive sweeteners is also criticized by the scientific community, since some of them are banned in other countries, such as the use of cyclamate by the Food and Drug Administration in the United States.

Generic name (commercial name)	Sweetener capacity (compare with sucrose)
Potassium acesulfame (Sunnett)	180-200
Alitame	2000
Aspartame (Equal, Nutrasweet)	180
Cyclamates	30-50
Glycyrrhizin	30-50
Lo han guo	30
Neotame	8000-13000
Perillartine	2000
Saccharine (Sweet'n low)	300
Stevioside	40-300
Sucralose (Splenda)	600

Table 2. Available non-nutritive sweeteners in the market.

Generic name	Commercial name	Caloric value
Based on proteins		
Whey protein	Simplesse, Dairy-Lo	1-2 kcal/g
Based on carbohydrates		
Cellulose	Avicel cellulose gel, Methocel, Solka-Floc	0 kcal/g
Dextrins	Amylum, N-Oil	4 kcal/g
Gums	Kelcogel, Keltrol	0 kcal/g
Inulin	Raftiline, Fruitafit, Fibruline	1-1.2 kcal/g
Maltodextrins	Crystal Lean, Lorelite, Lycadex, Maltrin	4 kcal/g
Polydextrose	Litesse, Sta-Lite	1 kcal/g
Polyols	Several brands	1.6-3.0 kcal/g
Fat analogous		
Olestra	Olean	0 kcal/g
Sorbestrin		1.5 kcal/g
Salatrim	Benefat	5 kcal/g
Emulsifiers	Dur-Lo	9 kcal/g

Table 3. Fat replaces available in the market.

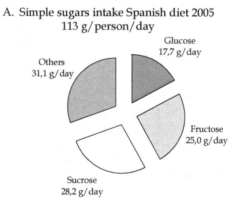

A. Simple sugars intake Spanish diet 2005
113 g/person/day

Glucose
17,7 g/day

Others
31,1 g/day

Fructose
25,0 g/day

Sucrose
28,2 g/day

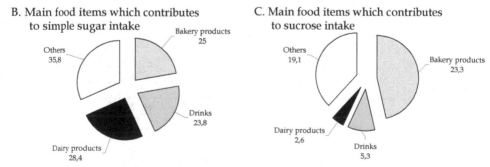

B. Main food items which contributes to simple sugar intake

Others 35,8
Bakery products 25
Drinks 23,8
Dairy products 28,4

C. Main food items which contributes to sucrose intake

Others 19,1
Bakery products 23,3
Dairy products 2,6
Drinks 5,3

Fig. 5. Simple sugars intake and principal food items that contributes to its intake in the Spanish diet.

With respect to fat substitutes, several fat replacers have been developed to assimilate the sensory properties of fat. It can be divided basically into 3 groups; 1) based on proteins, used mainly in dairy products; 2) based on carbohydrates, used in dressings, meat products, etc. and 3) based on modified fat, that may contains organoleptic characteristics similar to fats but with a lower caloric content. Among them, Olestra approved in 1996 by the United States to replace fats and oils, is a sucrose polyester containing between 6 and 8 fatty acids per molecule, that has organoleptic properties similar to those of typical fat, but without the capacity of being hydrolyzed by lipases; and Salatrim, a mixture of long- (mainly stearic acid) and short-chain fatty acids (acetic, propionic and butyric) esterified with glycerol that greatly reduce the caloric intake per gram of fat. The main drawback of the use of fat-based substitutes is the possible decrease in the absorption of fat soluble vitamins, which must be taken into account in the formulation or modification of food that are sources of fat-soluble vitamins in the diet.

Naturally, foods with lower energy density and satiety properties, such as fruits and vegetables, compared to energy-dense foods, are low in fat and high in water and/or dietary fibre content, since these compounds may add weight and volume to foods without increasing its caloric content. Therefore, increasing the water content through wetting agents and/or increasing the content of dietary fibre could be a possible approach.

Possibly the best option is to reduce the energy density of foods by the incorporation of fibre (with an average energy value of 1-2 kcal/g) to the formulations, which also includes the possibility to increase the water content of food products. Physiologically, fibre may decrease the absorption of macronutrients like fat and carbohydrates reducing its energy value. On the other hand, some fibres obtained from fruits and vegetables may also provide functional antioxidant compounds valuables for obesity treatments.

It should be noted, however, that an increased intake of foods with low energy density is not enough to lose weight, except when they move to higher-energy density. The choice of food can often be influenced also by their quantity, volume or weight, in this sense changes in the appearance of food can be another simple strategy to develop other types of functional foods.

2.1.2 Rheological modifications of foods to induce satiety

The main objective in developing this type of functional foods is to induce satiety by modifying the perception of consumers toward food. That is, sensory (sight, smell and taste mainly) stimulations that able to production of an "apparent" fullness feeling in the early stages of feeding. For example, some studies suggest that the volume of food ingested psychologically affect hunger, satiety and the amount of food that individuals want to eat (Rolls et al., 2000). In this sense, an example could be the development of products with a high volume and low density in both weight and energy (for example by cereal extrusion or air emulsions desserts like "mousse").

Satiety may also be induced through the stimulation of retro-nasal aroma by food (Ruijschop & Burgering, 2007). There are some indications that not all types of foods produce the same quality (flavour) or quantity (intensity) of sensory stimulation. The physical structure largely seems to be responsible for stimulating aroma. For example, solids tend to produce a greater sense of satiety than liquid foods, probably because of the increased contact time of food in the oral cavity and therefore the greater sensory stimulation. It is also suggested the addition of encapsulated flavours that may extend specialized sensory flavour in the mouth in foods with a low energy density.

Rheological modifications for satiety peptide release are mainly based in the induction of a higher gastric distension and a reduction in gastric emptying which may promote the release of CCK and the decrease of ghrelin levels. Viscous fibres are the main nutrient associated with the ability to slow gastric emptying as discussed above.

2.1.3 Digestive enzyme inhibition and gut energy absorption inhibition

Another possible strategy is to limit the absorption of nutrients in the intestinal tract, by limiting the action of digestive enzymes and/or interacting with them to physically interrupt its absorption in the intestinal tract. The most common example is the use of fibre, which modulates the intestinal transit time resulting in increased satiety, reduced physical accessibility of nutrients for being absorbed and a reduced the calorie intake of the formulation.

An interesting example of functional fibre is chitosan (product obtained from chitin, located in the shells of shellfish), which is a positively charged polymer that could bind to negatively charged fat molecules in the intestinal lumen (mainly free fatty acids) therefore inhibiting its absorption. On the other hand there are other bioactive compounds with the ability to inhibit the activity of digestive enzymes, the most common digestive enzyme inhibitor found in foods are condensed tannins, which have the ability to precipitate proteins (including enzymes), reducing its action.

In connection with the absorption of carbohydrates, some researchers have shown that certain polyphenols (for example from tea) have the ability to inhibit *in vitro* the translocation of glucose transporter, GLUT2, in the intestinal epithelial cells, thus inhibiting the absorption of glucose (Kwon et al., 2007). This same effect has been demonstrated *in vivo* by oral sucrose tolerance test in the presence of epigallocatechin gallate, observing a decrease in blood glucose values (Serrano et al., 2009).

2.2 Thermogenesis and possible modifications in energy expenditure

Small differences in energy expenditure might have long-term effects on body weight. The human energy budget is usually divided into four major components (Figure 6), which, together, constitute the total energy expenditure: 1) basal metabolic rate; 2) diet-induced thermogenesis; 3) physical activity and 4) adaptative thermogenesis.

One of the suggested metabolic factors involved in the development of obesity is adaptive thermogenesis. It is defined as the regulated production of heat in response to environmental temperature or diet. It can be seen as a mechanism for dissipation, in a regulated manner, of the food energy as heat instead of its accumulation in fat. Skeletal muscle is potentially one of the largest contributors to adaptative thermogeneis in humans. For example, an adrenaline infusion may stimulate muscles to consume over a 90% more oxygen, thus increasing energy expenditure.

The possible mechanism of action of the increase in heat production may be due to the higher stimulation/synthesis of uncoupling proteins at the mitochondria. The uncoupling protein can dissipate a proportion of metabolic energy through uncoupled metabolism that creates heat rather than the generation of ATP. Consequently, these have been the topic of extensive research as possible targets to fight against obesity. The favoured route has been to develop specific β3-adrenergic compounds that might stimulate UCP1 without the undesirable β1 and β2 effects on heart rate and blood pressure.

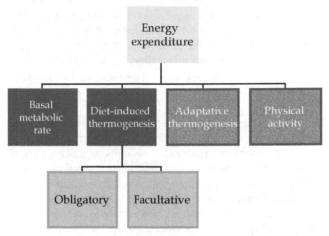

Fig. 6. Principal components of energy expenditure in humans.

In relation to nutrition, it has been shown that carbohydrates induced an increased adrenaline concentration, resulting in increased muscle thermogenesis (Astrup et al., 1986). Other authors have reported a variety of compounds that can alter energy expenditure. For example, in rodents, acute treatment with retinoic acid increases the thermogenic capacity in brown adipose tissue, which further in time may induce a significant decrease in body weight and adiposity (Bonet et al., 2000). Green tea extracts have been also described to stimulate thermogenesis in brown adipose tissue, mainly due to the interaction between its high content of catechins and caffeine that may stimulate the noradrenaline released by the sympathetic nervous system. Overall, the action of caffeine and catechins may prolong the stimulatory effects of noradrenaline on energy metabolism and lipid.

In another context, it is interesting to observe that after overfeeding, the same amount of excess energy intake and nutrients does not always invoke the same body weight gain in all people because of differences in diet-induced thermogenesis (energy expenditure in response to food intake).

Diet-induced thermogenesis can be divided in two categories: obligatory and facultative thermogenesis. The obligatory part consists of all process related to the digestion, absorption and processing of food. Stimulation of adenosine triphosphate (ATP) hydrolysis during intestinal absorption, initial metabolic steps and nutrient storage, are responsible for this food thermic effect. For example, measured thermic effects of nutrient digestion are 0-3% for fat, 5-10% for carbohydrates and 20-30% for proteins. While, the facultative component enables wasting of energy after a high caloric meal and prevents the storage of energy.

In relation to facultative thermogenesis, it has been observed that the thermic effect of food is reduced in obese and insulin-resistant patients, possibly because of the effect of the autonomic nervous system activation by certain nutrients and the insulin secretion stimulation. Insulin resistances may induce a decrease turnover of ATP by muscle tissue, therefore reducing its metabolic rate. It has also been suggested that insulin, via unidentified receptors, most probably located in the central nervous system, may stimulate muscle sympathetic nerve activity and facultative thermogenesis, and therefore its probably reduce in obese insulin resistance patients. Functional foods against insulin resistance may also help to reduce obesity by increasing insulin sensibility thus optimizing the use of energy. This aspect is described deeply in the next section.

And finally, protein turnover defined as degradation of proteins into amino acids and resynthesis of new proteins could be responsible for large part of the energy expenditure, around 15-20% of basal metabolic rate. Most tissues exhibit protein turnover, specifically the skeletal muscle tissue, liver, skin and small intestine. In humans, it has been shown that after carbohydrate overfeeding, protein turnover is increased by 12%, may be because of the reduction in protein intake.

2.3 Insulin metabolism and the regulation of adiposity

The causal links between obesity and insulin resistance are complex and controversial. Weight gain from overfeeding induces insulin resistance, whereas weight loss by caloric restriction reverse insulin resistance. The main concern should be to reduce the insulin secretion, associated to an increase in fat deposition in the adipocyte; as well as to increase its sensibility which is associated to an increase in metabolic rate.

For food formulation it is important to know that increased plasma glucose levels increase the secretion of insulin from pancreatic β-cells, although other substrates such as free fatty acids, ketone bodies, and certain amino acids can also directly stimulate insulin's release or augment glucose's ability to trigger insulin release.

It has been suggested that a low-glycemic diet can help control obesity because of the ability to reduce the rate of glucose uptake in the small intestine and therefore the insulin secretion, as well as to increase satiety value of food and appetite regulation (Brand-Miller et al., 2002; Roberts, 2003). While, high glycemic index foods, by its higher stimulation in insulin secretion, may promote a higher postprandial oxidation of glucose at the expense of fat oxidation, which can lead to increased body weight gain.

The glycemic index of a food is directly proportional to the degree of intestinal absorption of carbohydrates, in this sense the incorporation of factors that replace simple carbohydrates or decrease the absorption of carbohydrates such as fibre, fat or by inhibiting the action of digestive enzymes can reduce the glycemic index of food.

However, not all strategies to reduce the glycemic index of foods are widely accepted by the scientific community. For example, replacing simple sugars like sucrose with fructose, as a sweetener with low glycemic index, has been associated with increased body adiposity (Bray et al., 2004). After the ingestion of fructose, insulin is not increased, leptin is reduced, and ghrelin is not inhibited. Because these hormones play important roles in regulating food intake, the combined effects of excessive fructose intake could result in a lower induction of satiety and increase in total intake. Moreover, these foods should be low in saturated fatty acids, especially *trans* fatty acids, which are associated with a higher induction in adiposity. In this context, the addition of conjugated linoleic acid appears to have an effect based on the inhibition of the activity of lipoprotein lipase, which may reduce the uptake of lipids by the adipocyte.

3. Functional diets based on traditional foods

The use of appropriate combinations of nutrients that affect different processes of energy intake and expenditure could be the best strategy to tackle obesity control, usually a multi-causal problem. A higher effectiveness can be obtained from the combination of functional ingredients that inhibit the appetite sensations, the bioavailability of macronutrients and induce a thermogenic response to an individual. An interesting approach could be the formulation of "Ready Meal" type product.

A ready meal is a type of convenience food that consists of a pre-packaged meal that needs little preparation. In 2009 the global market for ready meals was worth $71.6bn. In the Western Europe and the US the ready meals markets are fairly mature and well-established with moderate growth rates. The only pre-requisite for such diets is that they should be nutritionally appropriate and balanced with a low caloric content that allows consumers to replace any meal time with some of these diets on the market.

4. Conclusions

To create and consume a satisfying diets and foods with high intrameal and intermeal satiety feeling that may help to fight against obesity, ideally should consist of low-energy dense foods with high palatability; however, such foods do not commonly exist. Moreover, individual genetic variation may influence the effectiveness of such preparations against obesity.

Notwithstanding, the diversity of strategies to combat the obesity problem makes possible to develop a variety of food products that may satisfy individual requirements. However, before its market launch, the metabolic effects should be widely contrasted as well as it should be defined the concrete profile of the people who can benefit from the consumption of the formulated functional food.

5. References

Astrup, A., Bulow, J., Christensen, N.J., Madsen, J. & Quaade, F. (1986). Facultative thermogenesis induced by carbohydrate: a skeletal muscle component mediated by epinephrine. American Journal of Physiology, Vol. 250, pp. E226-E229.

Batterham, R.L., Cowley, M.A., Small, C.J., Herzog, H., Cohen, M.A. & Deakin, C.L. (2002). Gut hormone PYY(3-36) physiologically inhibits food intake. Nature, Vol. 418, pp. 415-421.

Bonet, M.L., Oliver, J., Pico, C., Felipe, F., Ribot, J., Cinit, S. & Palou, A. (2000). Opposite effects of feeding a vitamin A-deficient diet and retinoic acid treatment on brown adipose tissue uncoupling protein 1 (UCP-1), UCP2 and leptin expression. Journal of Endocrinology, Vol. 166, pp. 511-517.

Brand-Miller, J.C., Holt, S.H., Pawlak, D.B. & McMillan J. (2002). Glycemic index and obesity. American Journal of Clinical Nutrition, Vol. 76, pp. 281S-285S.

Bray, G.A., Nielsen, S.J. & Popkin, B.M. (2004). Consumption of high fructose corn syrup in beverages may play a role in the epidemic of obesity. American Journal of Clinical Nutrition, Vol. 79, pp. 537-543.

Callahan, H.S., Cummings, D.E., Pepe, M.S., Breen, P.A., Matthys, C.C. & Weigle, D.S. (2004). Postprandial suppression of plasma ghrelin level is proportional to ingested caloric load but does not predict intermeal interval in humans. Journal of Clinical Endocrinology & Metabolism, Vol. 89, pp. 1319-1324.

Cangiano, C., Ceci, F., Cascino, A., Del Ben, M. & Laviano, A. (1992). Eating behaviour and adherence of dietary prescriptions in obese adults subjects with 5-hydroxytryptophan. American Journal of Clinical Nutrition, Vol. 56, pp. 863-867.

Cummings, D.E. (2006). Ghrelin and the short- and long-term regulation of appetite and body weight. Physiology & Behaviour, Vol. 89, pp. 71-84.

Gruendel, S., Otto, B., Garcia, A.L., Wagner, K., Mueller, C., Weickert, M.O., Heldwein, W. & Koebnick, C. (2007). Carob pulp preparation rich in insoluble dietary fibre and polyphenols increases plasma glucose and serum insulin response in combination with a glucose load in humans. British Journal of Nutrition, Vol. 98, pp. 101-105.

Kennedy, G.C. (1953). The role of depot fat in the hypothalamic control of food intake in the rat. Proceedings of the Royal Society of London B, Biological Sciences, Vol. 140, pp. 578-596.

Kwon, O., Eck, P., Chen, S., Corpe, C.P., Lee, J.H., Kruhlak, M. & Levine, M. (2007). Inhibition of the intestinal glucose transporter GLUT 2 by flavonoids. FASEB Journal, Vol. 21, pp. 366-377.

Lippl, F., Kircher, F., Erdmann, J., Allescher, H.D., Schusdziarra V. (2004). Effect of GIP, GLP-1, insulin and gastrin on ghrelin release in the isolated rat stomach. Regulatory Peptides, Vol. 119, pp. 93-98.

MAPA Ministerio de Agricultura Pesca y Alimentación. (2006). La Alimentación en España. Secretaría General de Agricultura y Alimentación, Madrid, Spain.

Mayer, J. & Thomas, D.W. (1967). Regulation of food intake and obesity. Science, Vol. 156, pp. 328.

McCarty, M.F. (2005). A chlorogenic acid-induced increase in GLP-1 production may mediate the impact of heavy coffee consumption on diabetes risk. Medical Hypotheses, Vol. 64, pp. 848-853.

Nagase, H., Bray, G.A. & York, D.A. (1996). Effects of pyruvate and lactate on food intake in rat strains sensitive and resistant to dietary obesity. Physiology & Behaviour, Vol. 59, pp. 555-560.

Roberts, S.B. (2003). Glycemic index and satiety. Nutrition in Clinical Care, Vol. 6, pp. 20-26.

Rolls, B.J., Bell, E. A. & Waugh, B.A. (2000). Increasing the volume of a food by incorporating air affects satiety in men. American Journal of Clinical Nutrition, Vol. 72, pp. 361-368.

Ruijschop, R.M. & Burgering, M.J.M. (2007). Aroma induced satiation possibilities to manage weight through aroma in food products. Agro Food Industry Hi-Tech, Vol. 18, pp. 37-39.

Serrano, J., Boada, J., Gonzalo, H., Bellmunt, M.J., Delgado, M.A., Espinel, A.E., Pamplona, R. & Portero-Otin, M. (2009). The mechanism of action of polyhenols supplementation in oxidative stress may not be related to their antioxidant properties. Acta Physiologica, Vol. 195, pp. 61.

Shiiya, T., Nakazato, M., Mizuta, M., Date, Y., Mondal, M.S., Tanaka, M., Nozoe, S., Hosoda, H., Kangawa, K. & Matsukura, S. (2002). Plasma ghrelin levels in lean and obese humans and the effect of glucose on ghrelin secretion. Journal of Clinical Endocrinology Metabolism, Vol. 87, pp. 240-244.

5

Improving Nutrition Through the Design of Food Matrices

Rommy N. Zúñiga[1] and Elizabeth Troncoso[2]
[1]Center for Research and Development CIEN Austral, Puerto Montt
[2]Pontificia Universidad Católica de Chile, Santiago,
Chile

1. Introduction

Increasing epidemiological evidence has linked the prevalence of diseases, such as obesity, cardiovascular disease, hypertension, type II diabetes mellitus, and even cancer, to dietary factors. Besides, overweight and obesity are major risk factors to the prevalence of the mentioned diet-related diseases. The obesity epidemic around the world has been attributed to energy imbalance, mainly because of increased food consumption and/or sedentary lifestyle. Changes in eating behavior and the massive amount of high-calorie foods readily available has allowed obesity to reach epidemic proportions around the world. The logical strategy to attack the obesity problem is lowering the total energy intake along with a reduction in fat and sugar/digestible carbohydrate intake, which can have a substantial impact on body weight (American Dietetic Association [ADA], 2005). In past years, the food industry developed "light" products by diminishing or replacing the amount of fat and/or sugar from high-calorie products. However, the replacement of fat and sugar decreases the palatability of foods and for consumers foods must be simultaneously safe, healthy, delicious and convenient (German & Watzke, 2004). Hence, the challenge for the food industry is much more complex than simply providing healthy foods; to provide healthy and delicious foods is the real challenge.

The control of digestion and the release of nutrients from the food matrix is an alternative approach to attack the obesity problem and other diet-related diseases. The concept of a food matrix points to the fact that nutrients are contained into a larger continuous medium that may be of cellular origin (*i.e.*, fruits and vegetables) or a structure produced by processing, where nutrients interact at different length scales with the components and structures of the medium (Aguilera & Stanley, 1999). Research on human digestion has often been undertaken with a view to changing the rates of digestion and delivery sites of macronutrients that might affect satiety and thus caloric intake. Several aspects of human eating behavior and food digestion may be relevant for identifying effective measures to treat or prevent diet-related diseases. Although most of people in developed and developing countries eat an unbalanced diet, an increasing part of the consumers are progressively more aware of the relationship between diet and health. Thus, the demand for functional food products that address specific health benefits is growing steadily (Palzer, 2009). The food industry is currently responding by reformulating its products, especially looking at salt, sugar and fat content with a particular emphasis on healthier fat compositions (Lundin et al., 2008). New designed foods with lower amounts of fat, controlled release of bioactives, in-body self-assembly structures or slowly digestible starches are already being developed by the food industry. The structure of these products must be modified accordingly to equalize the physical (*e.g.*, rheology) and sensory (*e.g.*, taste and release of aromas) properties of the original food. Therefore, designed food structures are required to diminish the sensorial impacts of such product modifications.

The food industry has assembled considerable information about composition, biochemistry, structure, and physical properties of foods. The principles of process engineering of biomaterials and the fundamental role of food structure in understanding the behavior of foods during processing are well established (Aguilera, 2005; German & Watzke, 2004). This knowledge is the basis for the food industry to formulate, process, preserve and distribute foods. Although the total amount of a nutrient in natural and formulated foods may be obtained from composition tables, its bioavailability (*i.e.*, the rate and extent to which a nutrient contained in a food is absorbed and become available at the site of action) depend on many factors, for instance, food microstructure, processing conditions, presence of other components, among others. However, there is still a lack of information about the performance of foods inside our body, which has limited the capacity of the industry to create products with tailored nutritional properties. Therefore, this chapter is an attempt to relate how the changes induced in food matrices affect their physicochemical properties and macronutrient (protein, fat, and carbohydrate) bioavailability, improving the nutritional performance of tailored foods. This chapter initially reviews the current situation about consumer trends and health. Next, the structures of the main macronutrients are briefly revised and the steps of the digestion process are explained. Finally, examples of structured foods to improve nutrition are presented and discussed.

2. Why improving the nutritional performance of foods? Past and present situation

Health and wellbeing are the major drivers for the food industry today. Scientific evidence that the quantity, composition and microstructure of the food ingested affects health is growing steadily (Norton et al., 2007; Parada & Aguilera, 2007). Although food digestibility and nutrient bioavailability had not been taken into account in the food design until now; a better understanding of the relationship between food properties, digestion and absorption would help in the rational design of foods with enhanced nutritional properties. This section deals with modifications in consumer perception about foods and the increasing need for foods for health and wellbeing.

2.1 Changes in human eating behavior

As a consequence of changes in lifestyle, diet is increasingly affecting health. A fast-paced lifestyle has left less time to cook, thus the consumption of fast food meals (mostly cheap foods available in large quantities) is augmenting constantly. In developed and developing countries families are eating out of home more often, portion sizes of foods consumed are getting larger at the same time body weights of people continue to increase. Taking as example the United States, the incidence of obesity increased continuously over the past decades (Fig. 1). Over-consumption of high–energy-dense (kcal/g) foods and beverages, and the increased portion sizes, may contribute to positive energy balance and lead to increasing incidence and prevalence of overweight and obesity. High-fat and sugar foods are problematic for the regulation of energy intake because they are high in energy density and very palatable. In addition, a decrease in physical activity due to the increasingly sedentary nature of many forms of work, changing modes of transportation, and increasing urbanization has contributed to the energy imbalance.

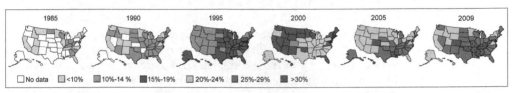

Fig. 1. Evolution of percentage of United States population by state, with a body mass index above 30, regarded as obese. Source: Centers for Disease Control and Prevention.

Nowadays, eating is driven by necessity and by pleasure. From the beginning of the humanity foods had been consumed to obtain energy and essential nutrients for living, but in modern societies foods are also consumed to have the pleasuring feeling of food flavors and textures into the mouth. Therefore, we eat because we have to and because we like to. As societies increase their purchase level, they begin to recognize the pleasurable aspects of eating, as comfort or reward and to satisfy, delight or stimulate the senses. The modern food industry has recognized this opportunity and much effort has been investing in the development of foods with enhanced sensory properties, transforming raw materials in palatable structures to gain acceptability from the consumer. Currently, we are surrounded by a huge variety of tempting high-calorie foods and cutting out or drastically restricting this kind of foods is simply not sustainable, probably because tastier foods are more rewarding and initiates some kind of seeking behavior.

The worldwide increase in the above mentioned diet-related problems are likely to change eating habits, processing technologies, and products. Hence, the food chain is facing a major challenge and now it is the consumer who indicate to producers what they want to eat. This global tendency is reshaping the industry into one that provides, in addition to safe and high-quality foods, products that contribute to the health and wellness of consumers. On this regard, the major food companies have understood that health and wellbeing are the major drivers for the current food industry, which is well represented in their logos, "*Good food, good life*" from Nestlé, "*Feel good, look good and get more out life*" from Unilever and "*The Beverage Institute for health and wellness*" from The Coca-Cola Company. The above are some examples of the messages that these companies want to give to the current health-conscious consumer.

2.2 Food-related diseases focused on obesity

Obesity, a preventable disease, with its co-morbidities such as the metabolic syndrome and cardiovascular diseases, are the major medical problems of the last decades. The health consequences and compromised quality of life associated with obesity provide major incentives to reduce the continuing obesity epidemic. The large numbers of children entering adulthood overweight together with weight gain in adulthood, produce an enormous burden in terms of human suffering, lost productivity and health care expenditures. The solution to overweight and obesity problem is body weight maintenance after body weight loss. This seems simple, but the required conditions are difficult to achieve for many individuals. Fortunately, the evidence has shown that many of the health risks associated with obesity can be reversed with weight loss.

Once considered a problem only in developed countries, obesity is now dramatically increasing in developing countries. According to the World Health Organization (WHO, 2011), some important key facts about obesity are:

* Worldwide obesity has more than doubled since 1980.
* 1.5 billion adults were overweight in 2008.
* 43 million children under the age of five were overweight in 2010.
* 2.8 million adults die each year as a result of being overweight or obese.
* 44% of the diabetes burden, 23% of the ischemic heart disease burden and between 7% and 41% of certain cancer burdens are attributable to overweight and obesity.
* 2.3 billion adults will be overweight and more than 700 million obese by 2015.

According to Wansink (2007), over 85% of the population with weight problems has consumed an average excess of only 25 kcal/day over a prolonged period of time. A sustained reduction in daily calorie consumption could prevent or reduce this long term weight gain in a large proportion of the population, and thus to reduce the incidence of obesity and associated health problems. Fatty foods have high-energy density and palatability but exert a relatively weak effect on satiation (compared calorie per calorie with protein and carbohydrate foods), which may encourage calorie over-consumption (Marciani et al., 2007). Lowering energy density of foods can decrease energy intake independent of macronutrient content and palatability.

The human mechanism of appetite regulation is highly complex involving neurophysiological interactions between the gut and the brain. The stomach is able to signal by distension of its wall how full it is and, therefore, how much more should be eaten, and duodenal receptors sensitive to the nutrient content of the chyme (*i.e.*, semifluid mass of partly digested food expelled by the stomach into the duodenum) also signal satiety through the secretion of gut hormones. Besides, neurological pathways including the hypothalamus, where two major neuronal populations stimulate or inhibit food intake, and the brain steam are involved (Marciani et al., 2001a; Suzuki et al., 2010). Understanding the neurophysiological mechanism of appetite regulation could help to food technologists in developing more satiating foods.

The energy density of foods has been demonstrated to have a robust and substantial effect on both satiety and satiation. Satiety refers to the effects of a food or meal after eating has ended, whereas satiation (sensation of fullness) refers to the process involved in the termination of a meal (ADA, 2005). Satiation has been found to be independent of the administered macronutrient (fat, protein or carbohydrate) for isocaloric liquid foods, but it was linearly related to the meal volumes, suggesting that stomach distension is a key factor in the sensation of fullness (Goetze et al., 2007).

Enhancing the satiety level of foods while keeping a low energy density may restrict the daily food intake and the desire of overeating, which it can be a strategy for preventing over-consumption and energy imbalances. Reduced-energy foods should preserve the sensory properties of the original foods to play a potential role in helping against obesity. Designed foods with tailored mechanical properties of the material, low caloric density and designed flavor properties may help in developing new foods for a sustainable reduction in energy intake, thus helping us of fighting against obesity.

2.3 Evidence relating the impact of food structure on nutrition

Physicochemical and sensory properties of manufactured foods depend largely on the food matrix structure. Currently, there is an emerging interest in the impact of food structure on digestion behavior and its relationship to human nutrition (Lundin et al., 2008). New interest has arisen regarding the function that food structure may play once foods are inside the body and, consequently, in our nutrition, health and wellness. Attention is further supported by the increased belief that foods and not nutrients are the fundamental unit of nutrition (Jacobs & Tapsell, 2007). The last assumption is based on recent scientific data demonstrating that in the case of certain nutrients the state of the food matrix of natural or processed foods may favor or hinder their nutritional response *in vivo*.

In recent years, there has been an upsurge in efforts to understand how food structure influences the rates of macronutrients digestion. This research is being undertaken with a view to developing novel foods that regulate calorie intake, provide increased satiety responses, provide controlled lipid digestion and/or deliver bioactive molecules (Singh & Sarkar, 2011). Foods are consumed to maintain human biological processes and the food matrix influences these processes (Jacobs & Tapsell, 2007). It has been shown that disruption of the natural matrix may influence the release, transformation and subsequent absorption of certain nutrients in the digestive tract (Parada & Aguilera, 2007).

Food processing modifies physical and chemical properties of food and thus may influence the release and uptake of nutrients from the food matrix. In this complex scenario, food scientists have proposed to develop novel foods to control the impact of physical properties and food microstructure on the digestion behavior and its relationship to human nutrition, because in many cases the interactions between individual macronutrients control the rate of digestive processes, conditioning the absorption of nutrients. Thus, the release of nutrients from the food matrix as well as the interactions between food components and restructuring phenomena during transit in the digestive system becomes far more important than the original contents of nutrients (Troncoso & Aguilera, 2009).

2.4 Modifying the food matrix to minimize or maximize nutrients bioavailability

According to Aguilera (2005) *"the creation of new products to satisfy expanding consumer's demands during this century will be based largely on interventions at the microscopic level"*. Thus, to make this next generation of designed foods, a combination of understanding of material chemistry and material science is needed, together with an understanding of how the processing affects food structure, chemistry and attractiveness (Norton et al., 2006).

The food industry is extremely innovative in terms of new products, but is highly traditional in term of processes. For a particular product type, a limited range of unit operations have been employed for some considerable time and the same process lines are used to make a range of different product structures. The current challenge is to develop novel functional structures through innovative processing or new units operations. It is conceivable that enhanced nutritional properties of foods may be achieved by proper assembly of hierarchical structures from the microscopic level up to the macroscale (*i.e.*, bottom up approach). These new fabrication techniques will require understanding and precise control of assembly processes at all scales.

Nutrient bioavailability is gaining considerable attention in food technology. While one of the ongoing concerns of the food industry has always been to produce and provide the consumer with safe foods, the nutritional and caloric composition is now becoming equally important. For this reason, during the last years considerable research has been conducted to modify food matrices in basis on their physicochemical properties and effects on food digestion. For instance, the direct effect of physical properties (*e.g.*, microstructure, particle size and physical state) of foods has been evaluated by its influence on nutrient absorption. So, different intentional modifications could be induced by food technologists to design and fabricate foods with controlled bioavailability.

Foods can be viewed as delivery systems of macro and micronutrients to improve nutrition. Delivery systems to encapsulate, protect and deliver bioactive components are widely used by the pharmaceutical industry to carry these active agents to specific locations within the gastrointestinal tract (GIT) and release them at controlled rates. Using this knowledge the food industry, through the design of food matrices, is developing similar systems to encapsulate, protect and deliver food components. However, the effectiveness of the encapsulation process relies on the preservation of the bioavailability of the encapsulated component and the release of it in the correct portion within the GIT. As mentioned before, little it is known about the influence of food structure and breakdown on nutrient release in the GIT. This point is the primary importance because only what is released can be bioavailable for absorption (Parada & Aguilera, 2007).

Structuring matrices for nutrient delivery is a subject of enormous interest and several structuring food biopolymers are under study. To develop structured foods and develop a strategy for controlled release of food nutrients at desired sites in the GIT, it is essential to understand the kinetics of food disintegration and predict its digestion and subsequent metabolism. Biochemical, physiological, and physicochemical parameters that influence these processes need to be understood. This knowledge will benefit the food-processing industry in developing proper food structures for health purposes. The possibility of predicting the release of nutrients from food matrices under simulated GIT conditions is the upmost relevance in order to be able to define relationships between food matrix-nutrient, as well as for looking at the interactions of ingredients with the enzymes involved in the digestive process. Although *in vivo* methods provide direct data of bioavailability, ethical restrictions and complex protocols limit this type of studies when humans are used in biological research (Parada & Aguilera, 2007). Therefore, the need for validated *in vitro* methods is urgent in order to evaluate and compare the effect of the microstructure over the amount and the rate of nutrients release in the GIT.

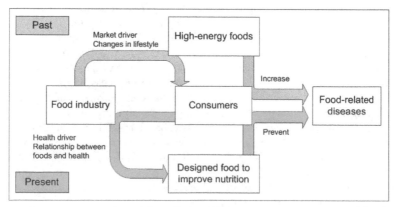

Fig. 2. Past and present scenarios of the relationship between consumers and food industry.

Summarizing, the terms "quality" and "health" are two main drivers of the modern food industry and the food microstructure influences both. Figure 2 gathers the principal subjects discussed in this section. To accomplish the ultimate goal of designing healthy foods with enhanced quality, knowledge of the overall food structure including the main building blocks of foods (proteins, carbohydrates and lipids) and a vast comprehension of the human digestion process are urgently needed to properly address the design of the future foods, which it will be discussed in the next sections.

3. Understanding the main food components and their structures

Foods consist of a complex group of components that differ in chemical composition and physical structure. Foods are largely composed by polymers (*i.e.*, proteins, carbohydrates, and lipids), micronutrients and water. Altogether, these components are arranged in different food matrices as the result of the multiple interactions that polymers can display under different conditions in an aqueous medium, as well as of the abundance of each respective substance. Proteins, carbohydrates and lipids are simple macromolecules made up of even simpler repeating units that play an important role in human nutrition. Each of these polymers has well-defined structures in the foods that contain them. This section discusses about the structure of proteins, carbohydrates, and lipids in foods and its impact on nutrient bioavailability.

3.1 Proteins

Protein is the most effective macronutrient found in foods providing satiety, on an energy equivalent basis, and protein hydrolysates comprising short chain peptides have been shown to induce release of satiety hormones, such as cholecystokinin (CCK), as part of this process (Lundin et al., 2008; Mackie & Macierzanka, 2010).

Proteins are certainly the most complex macromolecule encountered in foods. This complexity is evident in their secondary structure, which is intrinsically related to the sequence of amino acids (AAs) along the backbone. It is commonly recognized that 20 AAs form the building blocks of most proteins, being linked by peptide (amide) bonds. As a consequence, various proteins with different AA sequences will also differ in structure, modifying their physical properties.

Four levels of hierarchical organization are used to describe protein structure. The primary structure of a protein refers to the order of AA sequence in the polypeptide chain. The secondary structure describes the regular local conformations of the polypeptide backbone. The description of the overall three-dimensional folding of the protein is referred to as its tertiary structure. It involves the folding pattern of the polypeptide backbone including the secondary structures, arrangements of motifs into domains, and conformations of the side chains. Ultimately, the quaternary structure refers to the fourth-dimensional level of structure of protein complexes that may arise from the association of identical or heterogeneous polypeptide chains (Li-Chan, 2004). Changes in the secondary, tertiary, and quaternary structures without cleavage of backbone peptide bonds constitute "denaturation". In practical terms, denaturation means the unfolding of the native structure (Aguilera & Stanley, 1999). Protein denaturation usually has a negative connotation, since it is associated with loss of functionality in foods. However, it is often a prerequisite for improved digestibility, biological availability, and improved performance (*e.g.*, foam formation, emulsification or gelation). Denaturation not only affects the physical state of proteins but also their susceptibility to pepsin and trypsin/chymotrypsin mixture during digestion (Troncoso & Aguilera, 2009).

The most important factor in proteolysis is the accessibility of the enzyme to substrate and this involves both exposure of the cleavage site and local flexibility of the molecule and its side chains (Mackie & Macierzanka, 2010). In addition, hydrolysis of food proteins by digestive enzymes generates bioactive peptides that may exert a number of physiological effects *in vivo*, for example on the gastrointestinal, cardiovascular, endocrine, immune and nervous systems. On the other hand, some proteolysis-resistant proteins may produce allergenic reactions. Different factors regulate the allergenic potential (*i.e.*, defined as immunoglobulin E-mediated hypersensitivity reactions) of protein foods, such as type and composition of AAs, physicochemical characteristics of proteins, and relative abundance of the allergen in the food. It is known that β-lactoglobulin (β-lg), a recognized allergenic protein, is hardly hydrolyzed by enzymes such as pepsin and chymotrypsin, but not by trypsin. The relative resistance to proteolysis is generally explained by the compact globular tertiary structure of the protein at low pH (<pH 3), which protects most of the peptide bonds susceptible to enzyme action. Physical and/or chemical denaturation of β-lg generally leads to a higher rate of hydrolysis by gastrointestinal enzymes. Stănciuc et al. (2008) showed that thermal treatment caused partial unfolding of β-lg and consequently an increased rate of hydrolysis by both trypsin and chymotrypsin at 37 °C. This phenomenon was attributed to the accessibility of the specific peptide bonds to the enzymes being enhanced. After heating at 78°C, a decrease in hydrolysis after prolonged heating time was thought to be due to aggregation phenomena, which could have hidden some susceptible bonds. High pressure treatment (HP) of β-lg between 400 and 800 MPa promoted digestion by pepsin, because HP treatment caused significant unfolding of the protein molecule (Chicón et al., 2008; Zeece et al., 2008).

Aggregation (cross-linking) of proteins results in the formation of high-molecular-weight polymers. These modifications are expected to affect the kinetics of protein digestion due to the reduced availability of reactive sites to proteolysis. Depending on the pH, temperature and presence of ions, proteins can be transformed into fibrils, spherical micro-gels, micro-particles, fractal aggregates and gels (Schmitt et al., 2007). Such controlled aggregation allows the management of molecular interactions to achieve the desired protein structures. Enzymatically cross-linked β-casein decreased their digestibility in comparison with native β-casein (Monogioudi et al., 2011). Heat treatment of milk (sterilization) increases protein resistance to the *in vitro* digestion in comparison with unheated milk, probably because of the structural changes caused by denaturation and aggregation of whey proteins and/or casein with whey proteins through disulfide bounds (Almaas et al., 2006; Dupont et al., 2010). In addition, structuring β-lg into fibrils (formed by heating solutions at 80°C and pH 2.0 for 20 h) induces a more complete digestion by pepsin (Bateman et al., 2010).

Hence, by using carefully controlled processing parameters (mainly temperature, pH, ionic strength, and pressure), protein structures can be designed to release selected bioactive peptides, decrease allergenicity of protein and also induce a higher sensation of satiety during protein digestion.

3.2 Lipids

It is usually considered that, in the normal diet, some 25 to 30% of the total calories are conveniently supplied as lipid. Certain lipids are needed for good health (*e.g.*, essential fatty acids and fat-soluble vitamins). Nevertheless, the over-consumption of some dietary lipids (*e.g.*, cholesterol, saturated fats, and *trans*-fatty acids) increases the prevalence of some public health problems, including cardiovascular disease and obesity (Simopoulus, 1999). In this scenario, an improved knowledge of lipid structure could address the rational design of dietary lipids to enhance or retard lipid digestion, aspect having strong nutritional and health impact.

Lipids are a chemically heterogeneous group of compounds characterized by their insolubility in water, but soluble in organic solvents. In general, lipids are conformed by fats and oils. The basic unit of natural fats and oils is the triglyceride (TG) molecule; however, others molecules, such as monoglycerides (MGs), diglycerides (DGs), and phospholipids (PLs), can be structuring dietary lipids too. The TG molecule consists of a glycerol backbone to which are acylated three fatty acids (FAs). Two FAs are at the ends of the glycerol molecule (sn-1 and sn-3 positions) pointing in one direction, while the third FA molecule in the middle (sn-2 position) is pointing in the opposite direction. The dietary FAs molecules that compose the TG structure may vary in the number of carbon atoms and the presence of double bounds from saturated FA to unsaturated FA. The type and position of the FA chains on the glycerol backbone affect the structure of the TG molecule and, consequently, its bioavailability. TGs containing short- and medium-chain FAs have higher rate of FA release during lipolysis than for long-chain FAs, and is faster for FAs at the sn-1 and sn-3 positions than in sn-2 position (Fave et al., 2004). On the other hand, a typical TG molecule can exhibit three physical states: crystal, bulk, and interface. These physical states will determine the properties and characteristics of the physical state of the lipid phases in foods, which vary from liquid to crystalline phases. The lipid phases in most foods are usually liquid at body temperature (~37°C), but in some foods they may be either fully or partially crystalline. The crystallinity of the lipid phase may alter the ability of enzymes to digest the emulsified lipids. For example, the release rate of a lipophilic drug can be decreased with increasing crystallinity degree of lipid droplets (Olbrich et al., 2002). Hence, according to the above mentioned aspects, composition, structural organization and physical state of lipid phases will control lipid digestion. In addition, lipids have the property of forming, in the presence of water, different self-assembled structures. These mesophase structures include monolayers, micelles, reverse micelles, bilayers, and hexagonal phases. The aqueous medium behavior of lipids is determined by intrinsic parameters such as lipid polarity, length and branching of acylated fatty acids, presence of double bounds, among others (Ulrich, 2002). Mesophase structures of lipids have different sizes and configurations and, for this reason, they are finding many applications in food, pharmaceutical, and cosmetics industries. Nowadays, bilayer vesicles (*i.e.*, often called liposomes) have received widespread attention because of their ability to entrap functional components. This characteristic allows use of these structures as drug-delivery vehicles for controlling the release of incorporated agents.

The lipid molecules in foods may be organized into a number of different forms, including as bulk, structural, emulsified, colloidal, or interfacial structures (McClements, 2005). However, invariably, all ingested lipids ends up as an emulsion either by gastric emulsification or prior to ingestion during manufacturing process. The structure of these emulsions is determined by the nature of the lipid phase, the aqueous phase and the interface (Lundin et al., 2008). Particularly the properties of the interface modulate lipid digestion. The interface of emulsified lipids is determined by the physicochemical properties of lipid, such as lipid droplet size (which determines the interfacial area of lipid droplets), structure of lipid droplet, and the molecular structure of the TGs that constitute the lipid droplet (Armand et al., 1999). Several works have investigated the impact of emulsion structure on lipid digestion. For example, it has been demonstrated that a lower initial fat droplet size facilitates fat digestion by lipase (Armand et al., 1999). Additionally, the composition of the emulsion interface can limit the lipase activity. Interfaces composed by phospholipid limit fat digestion in the absence of bile salts because there are few possible points where lipase can access the emulsified substrate (Wickham et al., 1998).

In principle, rational design of lipid structures may be a useful tool for food processors to control lipid digestibility and bioavailability. Nevertheless, there is clearly a need for further research to establish the key physicochemical factors that impact the performance of food lipids within the GIT.

3.3 Starch

Carbohydrates constitute the most heterogeneous group among the major food elements, ranging widely in size, shape, and function. Polysaccharides such as starch, cellulose, hemicellulose, pectic substances and plant gums provide textural attributes such as crispness, hardness, and mouthfeel to many foods. Many of them can form gels that will constitute their structure and also enhance viscosity of solutions owing to their high molecular weight (Aguilera & Stanley, 1999).

Although carbohydrates are not essential in the diet, they generally make up ~40–45% of the total daily caloric intake of humans, with plant starches generally comprising 50–60% of the carbohydrate calories consumed (Goodman, 2010). Hence, starch becomes the most important source of energy in the human diet. Due to its functionality, starch is used in a wide range of foods for a variety of purposes including thickening, gelling, adding stability and replacing expensive ingredients. There are a number of different structural scales to be considered in starch, ranging through the distribution of individual starch branches, through the overall branching structure of the starch molecules in a granule, to the macroscopic structure of the grain (Dona et al., 2010).

Starch is formed by two polymers of glucose, amylose and amylopectin. Starch is a plant storage polysaccharide synthetized in the form of granules (normally 1-100 μm in diameter) in which molecules are organized into a radially anisotropic, semi-crystalline unit. Starch molecule is composed of the straight-chain glucose polymer amylose (with α-1,4 glycosidic linkages) and the branched glucose polymer amylopectin (with α-1,6 glycosidic bonds) (Pérez et al., 2009). Normal starches, such as maize, rice, wheat and potato, contain 70-80% amylopectin and 20–30% amylose (Jane, 2009). Amylopectin, the major component in starch, strongly influences its physicochemical and functional properties.

There are several levels of structural complexity in starch granules. The first level is the "cluster arrangement", in which most starch granules are made up of alternating amorphous and crystalline shells. This structural periodicity is due to regions of ordered, tightly packed parallel glucan chains (crystalline zones) alternate with less ordered regions corresponding to branch points (amorphous zones). Thus, the starch granule appears to be formed by alternating concentric rings (growth rings).

Starch granules are insoluble in cold water and are densely packed in its native form. In the raw form, the native granule is generally indigestible for humans due to this semi-crystalline structure. The addition of water and the application of heat to native granules is essential to transform them into foods with pleasing textural attributes. Gelatinization and retrogradation are the main microstructural changes related to starch digestion. Gelatinization is the collapse (disruption) of the molecular order within the starch granule manifested in irreversible changes in properties such as granular swelling, native crystallite melting and starch solubilization (Mason, 2009). When granules of starch are heated in the presence of water, the amorphous regions that pervade the whole granule swell (up to 50 times), forming a continuous gel phase. As the temperature exceeds a value between 50 and 80°C (depending on the crystallinity degree), the crystalline structure of the matrix is broken down. As gelatinization proceeds, the granule network is destroyed and amylose diffuses into the aqueous medium, increasing its viscosity. Further heating and/or shear disrupts the granule and a starch paste consisting of a continuous phase of amylose and amylopectin is formed (Aguilera & Stanley, 1999; Jane, 2009). The total amount of bioaccessible starch is the principal factor affected by the gelatinization process, thus a gelatinized granule is more digestible than a raw granule because the digestive enzymes can attack more easily the active sites. In fully isolated amylose or amylopectin molecules the digestion rate would be basically the same and occur relatively fast (less than 10 min). Then, the change in the total amount of digestible starch in a food can be explained because in most foods starch is present as a combination of raw or partly gelatinized granules (more resistant to digestion) and gelatinized granules (more digestible).

Retrogradation is the phenomenon occurring when the molecules comprising gelatinized starch begin to re-associate in an ordered structure (Mason, 2009). During aging, starch molecules can re-associate into crystalline segments (retrograde), losing water from the structure and undergoing an incomplete re-crystallization. Amylose and amylopectin have different behaviors, for example, amylose has a high tendency to retrograde and produce tough gels, whereas amylopectin, in an aqueous dispersion, is more stable and produces soft gels (Jane, 2009). Re-crystallization can lower the digestibility of starch because the resulting structure is less accessible to enzymes (Parada & Aguilera, 2011a).

Starch retrogradation and water-limited gelatinization should present an opportunity to redesign starchy foods aiming at reducing the glycemic response (GR), defined as the concentration of glucose in the blood after ingestion. An understanding of the internal organization of starch granules is crucial for food scientists and engineers to comprehend the functionalities and improve the nutritional properties of designed starch products.

4. Digestion: A step-by-step process controlled by different food structures

The manner in which a food is structured impacts its breakdown during digestion and consequently certain nutrional functions such as nutrient release and bioaccessibility (Parada & Aguilera, 2007). This requires understanding what happens inside the gut when a food is ingested and how the GIT behaves as the interface between the body and the food supply (Norton et al., 2007). The GIT is a highly specialized system that allows humans to consume diverse food matrices in discrete meals to meet nutrient needs. The main organs of the GIT include the mouth, the stomach and the small intestine. Figure 3 summarizes some components and phenomena occurring in the GIT during food digestion.

The digestive system is connected to the vascular, lymphatic and nervous systems to facilitate regulation of the digestive response, delivery of absorbed compounds to organs in the body and the regulation of food intake (Schneeman, 2002). In the mouth the food is broken down by the chewing action, mixed with saliva, and undergoes a temperature change as heat flow occurs (heating or cooling to body temperature). The foods are processed to the stage at which a bolus can be formed and swallowed, passing through the esophagus to the stomach (Lucas et al., 2002). The motility of the stomach (*i.e.*, with a maximum contraction force ranging from 0.1 to 1 N) continues the process of mixing food with the digestive secretions (Marciani et al., 2001b), now including gastric juice which contains acid and some digestive enzymes (*e.g.*, gastric lipase and pepsin), inducing further reduction of particle size, remixing and phase separation (*e.g.*, oil and aqueous phases). Typically, there is an appreciable increase in the pH of the stomach contents after ingestion of a food, followed by a gradual decrease to around pH 2 over the next hour or so. After being ingested a food component may remain in the stomach for a period ranging from a few minutes to a few hours depending on its quantity, physical state, structure and location (Weisbrodt, 2001). In the stomach occurs a complex dynamic step affecting the kinetics and pattern of subsequent digestion in the small intestine. The action of the stomach continues to break down the food into smaller particles prior to passage to the small intestine (Hoebler et al., 2002). Once the chyme is in the small intestine, peristaltic motor activity propels it along the length of the intestine and segmentation allows mixing with digestive juices, which include pancreatic enzymes, bile salts (BSs) and sloughed intestinal cells. Digestion of macronutrients, which began in the mouth, continues in the small intestine primarily through the action of enzymes. Each of the macronutrients has a unique set of enzymes that break the macromolecules into sub-units that can be taken up by the absorptive systems in the intestinal cells, allowing the subsequent transport of nutrients to the systemic circulation (Salminem et al., 1998). In this section we give a brief overview of the basic physiological events that occur during the digestion of the main macronutrients (proteins, lipids and carbohydrates) found in natural and processed foods.

ORGANS	COMPONENTS	CONDITIONS	PROCESSES
Mouth	Amylase	pH 5-7	Mixing/Dilution
	(salivary α-amylase)	Mechanical forces	Matrix disruption
	Lipase	5-60s	Structural changes
	(lingual lipase)		Phase transitions
	Mucin, Salts		Digestion
Stomach	Proteases	pH 1-3	Mixing/Dilution
	(pepsin/pepsinogen)	Gastric grinding (0.1-1 N)	Emulsification
	Lipases	30 min-4 hours	Droplet breakup/Coalescence
	(gastric lipase)		Molecular interactions
	HCl, Salts		Competitive absorption
			Precipitation
			Gelification
			Digestion
Small intestine	Proteases	pH 6-7.5	Mixing/Dilution
	(trypsin, chymotripsin,	Mixing	Emulsification
Gall bladder	elastase, carboxypeptidase)	1-2 hours	Droplet breakup/Coalescence
	Lipases		Molecular interactions
Pancreas	(pancreatic lipase, pancreatic		Competitive absorption
	lipase-related proteins, carboxyl		Surface denaturation
	ester lipase, phospholipase A2)		Micellization
	Amylase		Digestion
	(pancreatic α-amylase)		Transport
	Salts, Bile, Phospholipids		Absorption

Fig. 3. Summary of the major components and processes involved in the digestion of foods.

4.1 Protein digestion

Proteins typically make up about 10% of caloric intake in a normal diet, being a dietary component essential for nutritional homeostasis in humans. In general, ingested protein undergoes a complex series of degradative processes promoted by hydrolytic enzymes (*i.e.*, proteases) originating in the stomach, pancreas, and small intestine. The product of this proteolytic activity is a mixture of AAs and small peptides that are absorbed by the small intestinal enterocytes (Erickson & Sim, 1990). Consequently the nutritional value of protein, also known as protein quality, is related to its AA content, to its digestibility (which can be regulated by the food processing) and to the subsequent physiological utilization of specific AAs after digestion and absorption (Friedman, 1996).

Protein digestion begins in the stomach by the action of gastric proteases. When the bolus enters the gastric lumen is not only exposed to hydrochloric acid and salts but also to different pepsins, the major gastric proteases. An acidic milieu is required for the proteolytic activity of pepsins, with an optimum activity between pH 1.8 and 3.2 (Untersmayr & Jensen-Jarolim, 2008). The action of the gastric proteases results in a mixture of large polypeptides, smaller oligopeptides, and some free AAs. These hydrolytic products regulate diverse gastric functions that are under hormonal control, such as secretion of acid and pepsinogen and rate of gastric emptying (Erickson & Sim, 1990). Subsequently, the remaining proteins and polypeptides present in the chyme are released into the small intestine, where they are exposed to a variety of proteases and peptidases (see Figure 3) synthesized and released by the pancreas (i.e., the major source of proteases in the digestive system), and by the specific peptidases of the brush border of the intestinal mucosa (Whitcomb & Lowe, 2007). The single AAs, di-, and tri-peptides resulting from the intestinal digestion are taken up by enterocytes (where small peptides of up to 3 AAs are split into AAs by cytosolic peptidases) and then are used as nutrients for the human body (Untersmayr & Jensen-Jarolim, 2008).

Nowadays, the fate of dietary proteins during gastrointestinal digestion has become of particular interest because of the potential role that digestion may play in determining the allergenic potential of foods. In this context, the fundamental role of the stomach as the primary organ of protein digestion is very well recognized, leading to the classification of proteins as digestion-resistant or digestion-labile proteins (Moreno, 2007). Resistance of proteins to pepsin digestion has been proposed as a marker for potential allergenicity because it does appear to be a characteristic shared by many food allergens (e.g., β-lg, α-lactalbumin and casein in cow's milk, parvalbumin in fish, ovomucoid and ovalbumin in egg, tropomyosin in shellfish and seafood, and Ara h 1 and Ara h 3 in peanut) (Mills & Breiteneder, 2005). Hence, there are groups of proteins, such as storage proteins or structural proteins, more stable to proteolysis in the GIT. Consequently, it has been postulated that for a food protein sensitizing an individual, it must have properties which preserve its structure from degradation in the GIT (such as resistance to low pH, bile salts, and proteolysis), thus allowing enough allergen to survive in a sufficiently intact or immunologically active form to be taken up by the gut and sensitize the mucosal immune system (Moreno, 2007). In this scenario, further investigation is needed to reveal the mechanisms controlling the allergenic potential of foods, as well as the influence of the food matrix on the gastrointestinal digestion and absorption of protein allergens.

4.2 Lipid digestion

Lipids are a major source of calories (9 kcal/g) in our daily diets and food emulsions (e.g., mayonnaise, sauces, dressings) a major carrier of fat calories. The composition, structure and properties of fatty foods change appreciably during digestion. The structural organization and properties of the lipids within the bolus depend on their initial structural organization within the food, as well as the duration and intensity of the mastication process. In most cases, the lipids in the bolus are present as oil droplets stabilized at their interfaces by particles, proteins or surfactants, which may vary in size from around a micrometer (for some food emulsions) to more than a millimeter (for some bulk fats). These emulsified oil droplets may have been present in the original food or they may have been formed within the mouth due to breakdown of a bulk fat phase and the interaction with proteins (Hernell et al., 1990). In general, there is still a relatively poor understanding of the physicochemical and structural changes that occur within the mouth when fatty foods are consumed, although considerable progress has been made for some lipid foods such as emulsions and gels (Malone et al., 2003; Vingerhoeds et al., 2005).

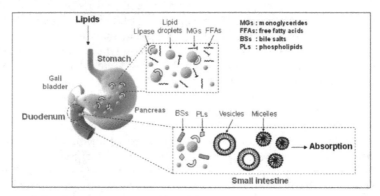

Fig. 4. Scheme of the digestive process of lipid foods.

After the food is swallowed it rapidly passes down the esophagus and enters the stomach, where it is mixed with acidic digestive juices containing gastric enzymes, minerals and various surface active compounds, and is also subjected to mechanical agitation due to movements of the stomach. During gastric digestion occurs breakdown of the food matrix structure, lipid droplet coalescence/disruption processes, and changes in the interfacial composition of the lipid phase due to adsorption/desorption of surface active substances (*e.g.*, proteins) to the surfaces of lipid droplets. Emulsion stability in the acidic gastric environment can readily be manipulated by altering the nature and state of the emulsifier agents, as revealed by non-invasive MRI studies in humans (Marciani et al., 2004). To be effective, the gastric lipase has to reach the surface of the lipid droplets in order to hydrolyze the emulsified TGs to DGs, MGs, and free fatty acids (FFAs). Typically, lipid hydrolysis stops when 10-30% of the FFAs have been released from the TGs (Armand, 2007). The emulsified lipids within the chyme are transferred from the stomach to the duodenum where they are mixed with sodium bicarbonate, BSs, phospholipids (PLs) and enzymes. The sodium bicarbonate secreted into the small intestine causes the pH to increase from highly acidic in the stomach to neutrality (pH 6.0-7.5) in the duodenum, where the pancreatic enzymes work most efficiently (Bauer et al., 2005). BSs and PLs are surface-active agents that can facilitate emulsification of the lipids by adsorbing to the droplet surfaces (Porter et al., 2007). The combined action of BSs and pancreatic juice brings about a marked change in the physicochemical form of the luminal lipid emulsion. In addition, the chyme is subjected to shear flow patterns in the small intestine that promote mixing and further emulsification. The lipid hydrolysis continues within the duodenum through the actions of lipases originating from the pancreas (Armand, 2007). A pancreatic lipase/colipase complex has to bind to the surface of the lipid droplet to hydrolyze the TGs to DGs, MGs, and FFAs, hence any compound covering the surface must be previously liberated (Jurado et al., 2006). Finally, lipid digestion continues in the small intestine with desorption and dispersion of insoluble lipid into an absorbable form. The digested lipids are solubilized in the lumen of the small intestine into at least two types of nanostructures: bile salt micelles and unilamellar vesicles. These assemblies are subsequently absorbed into the enterocyte's brush border membrane lining the surface of the small intestine. Thus, the absorption of digested fat and fat-soluble molecules that occurs in the small intestine is usually >90% (relatively high efficiency)

(German & Dillard, 2006). Figure 4 shows a highly schematic diagram of lipid digestion in the GIT where it is possible to observe that the dynamics of lipid digestion leads to the breakdown of complex structures, which disassemble during transit in the GIT.

4.3 Carbohydrate digestion

Starch, sucrose and lactose are the most important digestible carbohydrates in the human diet. However, from a nutritional point of view, the starch is the main carbohydrate derived from foods contributing significantly to the exogenous supply of glucose and total food energy intake (Slaughter et al., 2001). Starch is stored in plants as partially crystalline granules that contain two distinct polysaccharide fractions - amylose and amylopectin (*i.e.*, both polysaccharides composed of glucose molecules linked by digestible glycoside bonds). Upon cooking in the presence of water starch granules swell and partly disintegrate, facilitating the action of degrading enzymes that progressively transform them into maltose and glucose. The main characteristic of starch, compared with simple carbohydrates, is its slow digestion in the small intestine, which it produces a moderate GR. Additionally, food microstructure also affects the kinetics of starch hydrolysis and, consequently, the GR of starchy products, which it can be considered a reflection of the final nutritive effect of the food (Parada & Aguilera, 2007). In this way, the effect of the microstructure on digestibility of starchy foods may be manifested in two ways: the degree of starch gelatinization and the effect of the food matrix (non-starchy). These microstructural factors change from one food to another, modifying the digestibility of starch and its glycemic response (Fernandes et al., 2005). In humans there are several enzymes that hydrolyze starch. In the mouth, saliva contains α-amylase, enzyme that hydrolyses accessible starch, but since this starch remains in the mouth for a short period, the level of digestion is small. Once the food is partly digested in the mouth it is transported to the stomach, where there is no starch digestion but breakdown of food matrix will facilitate the access of hydrolytic enzymes to active sites for starch degradation in the small intestine. In the small intestine, the food receives pancreatic juice that contains pancreatic α-amylase, which hydrolyzes glycoside bonds producing glucose (absorbed in the intestinal epithelium), oligosaccharides and dextrins (Tester et al., 2004). However, it is known that the extent of starch digestion within the small intestine is variable and that a substantial amount of starch escapes digestion in the small intestine and enters the colon. The reasons for the incomplete digestion of starch may be separated into intrinsic factors (*i.e.*, physical form, presence of amylose-lipid complexes, and α-amylase inhibitors) and extrinsic factors (*e.g.*, chewing and transit through the bowel) (Cummings & Englyst, 1995).

Commonly, the GR is a way to know the bioavailability of starch and the nutritive effect of starchy foods. This metabolic response after food ingestion is not the same for different foods with the same amount of starch (Englyst & Englyst, 2005) and, principally, food processing affects this response. As an example, for many years it has been suggested that starch gelatinization affects the glycemic response (Björck et al., 1994).

In summary, food microstructure affects the nutritive value or quality of food products. However, the health benefit of different nutrients after absorption is not only conditioned by the food matrix, but also by the regular and harmonic functioning of the GIT. As mentioned before, designed food matrices can help in the prevention of diet-related diseases through through the controlled released of macronutrients or bioactive compounds. The next section deals with the latest research aiming to improve the nutritional performance of foods.

5. Structuring food matrices to improve bioavailability

The bioavailability of macro and micronutrients may be either increased or decreased by manipulating the microstructure of the foods that contain them. The food manufacturing processes affect the resulting structures and properties such as appearance, texture, taste, etc. However, food is ultimately going to be eaten. The ingestion of food leads to its breakdown and deconstruction through the different processes involved in the digestive stage (Lundin et al, 2008). Hence, food structures can be designed to control not only material properties before ingestion but also to control digestive breakdown rate and extent in the GIT. This section deals with the main aspects of designing protein, lipid and starch matrices in order to facilitate the rational design and fabrication of functional foods for improved health and wellness.

5.1 Protein systems

The food industry has widely used proteins in food formulations because possess unique functional properties. In addition to their contribution to the nutritional properties of foods through provision of AAs (that are essential for human growth and maintenance), proteins impart the structural basis for various functional properties of foods, such as water binding, gelling, foaming, and emulsifying. In particular, their ability to form gels and emulsions allow them to be an ideal material for controlling the rate of digestion and releasing of bioactive compounds at specific sites in the GIT.

Protein gels are undoubtedly the most convenient and widely used matrix in food applications. Food gels can be defined as three-dimensional continuous polymeric network holding large quantities of aqueous solution that shows mechanical rigidity during the observation time (Aguilera & Stanley, 1999). Gelation of food proteins and particularly of globular proteins (*e.g.*, egg white, soybean, and whey proteins) has received much attention lately among food scientists because they are a useful way to modulate texture and sensory perception of foods. Various processing routes, such as control of ionic strength, change in pH, heating, high pressure and enzymes are used to produce gels with different microstructural properties, which are strongly related to their aggregate molecular structure.

Two types of globular protein gels can be described depending on their microstructure: (i) fine-stranded gels, composed of more or less flexible linear strands making up a regular network characterized by its elastic behavior and high resistance to rupture, and (ii) particulate gels, composed of large and almost spherical aggregates, characterized by their lower elastic behavior and lower rupture resistance. Fine-stranded gels are created by linear aggregation of structural units maintained by hydrophobic interactions, whereas the particulate gels are produced by random aggregation of structural units essentially controlled by van der Waals interactions.

Protein gels can behave as pH-sensitive matrices through the presence of acidic (*e.g.*, carboxylic) or basic (*e.g.*, ammonium) groups in proteins chains, which either accept or release protons in response to changes in pH of the medium. This behavior could strongly influence the rate of molecule release by gels exposed to either gastric (low pH) or intestinal medium (neutral pH) by decreasing polypeptide chain interaction and thereby uptaking water inside the network and allowing the diffusion of molecules outward by osmotic pressure.

Gels of diverse mechanical and microstructural properties can be formed by controlling the assembly of protein molecular chains. Among the factors that modulate the release of protein molecules from gels, factors that promote gel breakdown play a major role.

Remondetto et al. (2004) showed that the release of nitrogen from cold-induced β-lg gels, as an index of matrix degradation, was slightly higher for particulate than for filamentous gels. In addition, the release of iron from these matrices was not correlated with the food matrix degradation at GIT conditions, suggesting that gel microstructure and not proteolysis influences the release of iron. In the case of particulate gels, the release of iron was lower due to its location inside aggregates that associate to form networks; in contrast iron is located at the outer surface of the linear aggregates of filamentous gels which it facilitates its release. Both types of gels may protect iron from gastric environment, but filamentous gels would release a larger amount of iron in the small intestine favoring the increase in iron absorption. In agreement with previous results, Maltais et al., (2009) found that fine-stranded cold-induced soy protein isolate (SPI) gels degraded more slowly than particulate gels, probably because their lower porosity slows down the digestion. Besides, the release of riboflavin was delayed from fine-stranded gels, because of the lower rate of protein digestion. From this study it was demonstrated that fine-stranded and particulate SPI protein gels were able to protect riboflavin from gastric conditions and release it under intestinal conditions. SPI films (*i.e.*, thin dehydrated gels) degraded more slowly as cross-linking density increases. Protein matrix undergoes a so-called first-order degradation during gastric and intestinal enzyme digestion. The behavior of the films was attributed to a more rigid structure and greater entanglement of the polypeptide chains in the SPI films due to increased cross-linking, leading to decreased penetration of digestive enzymes into the network (Chen et al., 2008).

Protein gel micro-beads have been recently used to encapsulate microorganisms with probiotic activity. The main challenge of this technique is to improve the survival of probiotics in the human digestive environment. Hébrard et al., (2006) produced whey protein isolate (WPI) beads containing yeasts from cold-induced gelation. Beads were resistant to acidification and pepsin attack during simulated human gastric digestion. Only about of 2% of initially entrapped yeasts were recovered in the gastric medium. In addition, WPI micro-beads were stable after incubation in simulated gastric juice in the presence of pepsin, allowing a targeted disintegration of protein matrices under physiological intestinal conditions (Doherty et al., 2011). Both results suggest that WPI beads might cross the gastric barrier and can deliver probiotics in the small intestine.

In conclusion, microstructure of protein gels largely affects the digestion of proteins and the release of bioactive compounds. If a digestive enzyme needs to degrade the gel matrix, then a large porosity and a and a high-pore interconnectivity would increase the rate of digestion. In addition, to design efficient systems for specific intestinal absorption of bioactive compounds or probiotics, gels should be gastroresistant. Hence, controlling microstructure and tailoring the porous structure of protein gels, it would give more opportunities for protection or release of a nutrient or a physiologically bioactive component in the GIT.

5.2 Lipid matrices

Many food scientists are currently focusing their efforts on developing novel lipid structures to decrease or increase the bioavailability of food lipids. In this context, much attention has been paid to the formulation of emulsion-based food systems to encapsulate, protect and release lipophilic constituents. A variety of emulsion-based delivery systems are already available (Figure 5), including conventional emulsions, nanoemulsions, solid lipid particles, multiple emulsions, and multilayer emulsions. Nevertheless, the structural design of these systems is still very far from an exact science due to their compositional and structural complexity. Further research is still needed to achieve a detailed understanding of the molecular characteristics, structural organization, physicochemical properties, and functional performance of these delivery systems (McClements et al., 2009).

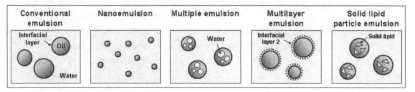

Fig. 5. Examples of different emulsions-based delivery systems

Conventional emulsions consist of oil droplets (~µm) dispersed in an aqueous medium, with the oil droplets being surrounded by emulsifier molecules. Oil-in-water (O/W) emulsions contain a non-polar region (the oil phase), a polar region (the aqueous phase), and an amphiphilic region (the interfacial layer). Then, within O/W emulsions it is possible to incorporate functional agents that are polar, non-polar, and amphiphilic (McClements, 2005). Nevertheless, conventional emulsions have some limitations (*e.g.*, physical instability) that promote the development of more sophisticated structured systems. Like conventional emulsions, nanoemulsions consist of small oil droplets (radius < 100 nm) dispersed within a continuous phase, with each droplet being surrounded by a protective coating of emulsifier molecules. However, nanoemulsions do have a number of advantages over conventional emulsions for certain applications due to their relatively small particle size: (i) they scatter light weakly and so tend to be transparent; (ii) they have high physical stability; (iii) they have unique rheological characteristics; and, (iv) they can greatly increase the bioavailability of encapsulated lipophilic components (Mason, et al., 2006). Additionally, the very small oil droplet size in nanoemulsions increases the digestion rate and the total amount of FFAs released during digestion in comparison with conventional emulsions (Li & McClements, 2010). This last characteristic may be of interest for the realization of lipid food for humans with disorders that prevent efficient digestion or absorption of lipids (*e.g.*, cystic fibrosis or pancreatitis) (Fave et al., 2004).

Multiple emulsions are systems in which dispersed droplets contain smaller droplets inside (Figure 5). Particularly, water-oil-water (W/O/W) emulsions consist of small water droplets contained within larger oil droplets that are dispersed in an aqueous continuous phase. Within these systems, functional components can potentially be located in a number of different ways. Water soluble components can be incorporated into the inner or outer water phase, while oil soluble components can be incorporated into the oil phase. Potential advantages of the multiple emulsions as delivery systems over conventional emulsions are: (i) functional ingredients could be trapped inside the inner water droplets and released at a controlled rate in the GIT; (ii) functional ingredients could be protected from chemical degradation; and (iii) reduction of the overall fat content of food products by loading the oil phase with water droplets (McClements et al., 2009).

The quality and functional performance of conventional O/W emulsions can be improved by the formation of multilayered interfaces using the layer-by-layer electrostatic deposition technique (Guzey & McClements, 2006). Multilayered emulsion delivery systems (Figure 5) may have a number of advantages over conventional single-layered emulsions: (i) improved physical stability to environmental stresses; (ii) greater control over the release rate of functional agents due to the ability to manipulate the thickness and permeability of the laminated interfacial coating; and (iii) ability to trigger release of functional agents in response to specific changes of environmental conditions in the GIT, such as dilution and/or pH (McClements et al., 2009).

Finally, it is possible to control the physical location of a lipophilic component and to slow down molecular diffusion processes using conventional emulsions by crystallizing the lipid phase. These systems are known as solid lipid particle emulsions, which consist of emulsifier coated (partially) solid lipid particles dispersed in an aqueous continuous phase (Figure 5) (Videira et al., 2002). By controlling the morphology of the crystalline lipid matrix it is possible to obtain more precise control over the release kinetics of functional compounds. Additionally, both lipophilic and hydrophilic bioactives can be incorporated within the same system using solid lipid particle emulsions (McClements et al., 2009). Hence, there are a large number of different ways that can be used by food manufacturers to embed lipophilic compounds within food matrices with different degradation rates. However, the selection of a particular matrix is a matter of functional preference.

5.3 Starch

Elevated glucose levels and high postprandial blood glucose cause a metabolic stress concentrations associated with increased risk of diseases, such as type II diabetes, cardiovascular disease and obesity (Kim et al., 2008; Parada & Aguilera, 2011a). The slow digestion of starch, in comparison with simple carbohydrates (*e.g.*, glucose, fructose), involves a gradual release of glucose to the bloodstream, thus producing a low GR.

Different starchy foods (pasta, baked foods, among others) elicit different GRs. The extent of starch digestion within the small intestine is variable, depending on food physical form, and a substantial amount of undigested starch enters the colon, where may be fermented by bacteria or simply appear in faeces (Troncoso & Aguilera, 2009).

The starch in its native state is resistant to enzymatic digestion and the availability of starch chains to the digestive enzymes is increased as gelatinization progresses. Gelatinization breaks the weak hydrogen bonds between polymer chains, so the exposed polar groups can further interact with water molecules. Due to this biophysical phenomenon, the granule structure changes from a crystalline to a disordered structure that is more easily accessible for the digestive enzymes (Parada & Aguilera, 2011a). Digestibility is strongly influenced by shearing and heating of starch samples. Both heat and shear during the preparation of starch suspensions alter the progression towards gelatinization, which increases the availability of polysaccharide starch chains to digestive enzymes, thus affecting the rates of hydrolysis (Dona et al., 2010).

The degree of gelatinization of starch granules and the recrystallization of the released polymers is of vital importance in the breakdown of starch molecules into sugars during digestion, because both phenomena influence the susceptibility to enzymatic degradation. Hence, a nomenclature has emerged describing the susceptibility of starch to digestion by enzymes in the small intestine: rapidly digestible starch (RDS), slowly digestible starch (SDS) that has slow but complete digestion, while resistant starch (RS) is resistant to digestion (Parada & Aguilera, 2011a).

RS obtained through appropriate physical or chemical modification can withstand the environmental changes in the upper GIT and can be rapidly degraded by the enzymes produced by the colonic microbiota. The incomplete digestion of starch is related to the matrix surrounding starch, the nature and physicochemical properties of the starch *per se* at the granule and molecular levels (*e.g.* granule size and amylose/amylopectin ratio), and the presence of other dietary components (*e.g.* sugar, dietary fiber and lipid) (Troncoso & Aguilera, 2009).

Starch granules in a real food are not isolated but exist within a three-dimensional matrix structure formed by proteins or other biopolymeric materials (such as fiber), which affects both the degree of gelatinization during processing and the digestion of starch (Giacco et al., 2001). If a food matrix is relatively impervious to enzymes and the granules are not properly exposed to their action, the digestion could be limited. The "encapsulation" of starch granules by a protein network is an important factor in explaining the slow degradation of starch by α-amylase in cooked pasta, the microstructure of the protein matrix as well as the physical state of starch (degree of gelatinization and retrogradation, amylose-to-amylopectin ratio, etc.) may explain the differences in the *in vivo* and *in vitro* enzymatic susceptibility of starch (Petitot et al., 2009).

The presence of a structured and continuous protein network is an important factor in explaining the slow degradation of starch in pasta. It had been shown that the degree of gelatinization and the interaction starch-gluten are a key factor during digestion of starch. Over-mixing could disrupt the gluten matrix producing an augmented rate of digestion; whereas higher heating temperatures produced a more compact structure (higher denaturation of gluten matrix) delaying digestion (Parada & Aguilera, 2011b). In agreement with these results, Kim et al. (2008) found that the disruption of the starch-coating protein matrix could be responsible for the increase in starch digestibility in pasta.

The degree of gelatinization and the gluten matrix conformation affect the digestibility of starch; the degree of gelatinization probably predominates in the extent of starch digestion, while the state of the gluten matrix is more related to the rate of digestion of starch.

Starch retrogradation and water-limited gelatinization should present an opportunity to redesign starchy products aiming at reducing the GR. Moreover, understanding the mechanisms by which the protein matrix slows down starch digestion could help to a rational design of new products with controlled starch digestion.

6. Conclusion

Recent knowledge supports the hypothesis that, beyond considering nutritional composition, food matrix microstructure may modulate various physiological functions in the human body and play detrimental or beneficial roles in some diet-related diseases. Advanced concepts in physical chemistry and material science provide a convenient and powerful framework for developing an understanding of the interactions and assemblies of food components into microstructure, which influence food macrostructure and functional properties. In turn, medical research coupled with food engineering approaches allow to obtain a deep comprehension of how foods behave during the digestive process. Advances in technologies for producing food matrices with tailored properties make possible the production of foods that have potential impact on the human health. This presents a challenge for the scientific community and the food industry due to the necessity of considering relationships between targets for functional food science and gastrointestinal behavior.

7. Acknowledgment

Author Zúñiga acknowledge to CIEN Austral for the financial support of this chapter.

8. References

Aguilera, J.M. (2005). Why food microstructure? *Journal of Food Engineering*, Vol. 67, No. 1-2, (March 2005), pp. 3-11, ISSN 0260-8774

Aguilera, J.M & Stanley, D.W. (1999). *Microstructural Principles of Food Processing and Engineering 2nd Edition*, Aspen Publishers Inc., ISBN 978-083-4212-56-0, Gaitherburg, MA, USA

Almaas, H.; Cases, A-L.; Devold, T.G.; Holm, H.; Langsrud, T.; Aabakken, L.; Aadnoey, T. & Vegarud, G.E. (2006). In Vitro Digestion of Bovine and Caprine Milk by Human Gastric and Duodenal Enzymes. *International Dairy Journal*, Vol. 16, No. 9, (September 2006), pp. 961-968, ISSN 0958-6946

American Dietetic Association. (2005). Position of the American Dietetic Association: Fat Replacers. *Journal of the American Dietetic Association*, Vol. 105, No. 2, (February 2005), pp. 266-275, ISSN 0002-8223

Armand, M. (2007). Lipases and Lipolysis in the Human Digestive Tract: Where Do We Stand? *Current Opinion in Clinical Nutrition and Metabolic Care*, Vol. 10, No. 2, (March 2007), pp. 156-164, ISSN 1473-6519

Armand, M. ; Pasquier, B. ; André, M. ; Borel, P. ; Senft, M. ; Peyrot, J. ; Salducci, J. ; Portugal, H. ; Jaussan, V. & Lairon, D. (1999). Digestion and Absorption of 2 Fat Emulsions with Different Droplet Sizes in the Human Digestive Tract. *American Journal of Clinical Nutrition*, Vol. 70, No. 6, (December 1999), pp. 1096-1106, ISSN 1938-3207

Bateman, L.; Ye, A. & Singh, H. (2010). In Vitro Digestion of β-Lactoglobulin Fibrils Formed by Heat Treatment at Low pH. *Journal of Agricultural and Food Chemistry*, Vol. 58, No. 17, (August 2010), pp. 9800-9808, ISSN 0021-8561

Bauer, E.; Jakob, S. & Mosenthin, R. (2005). Principles of Physiology of Lipid Digestion. *Asian-Australasian Journal of Animal Sciences*, Vol. 18, No. 2, (January 2005), pp. 282-295, ISSN 1976-5517

Björck, I.; Granfeldt, Y.; Liljeberg, H.; Tovar, J. & Asp, N.-G. (1994). Food Properties Affecting the Digestion and Absorption of Carbohydrates. *American Journal of Clinical Nutrition*, Vol. 59, No. 3, (March 1994), pp. 699-705, ISSN 1938-3207

Centers for Disease Control and Prevention. (July 2011). U. S. Obesity Trends, In: *Overweight and Obesity*, 29.07.2011, Available from
http://www.cdc.gov/obesity/data/trends.html

Chen, L.; Remondetto, G.; Rouabhia, M. & Subirade, M. (2008). Kinetics of the Breakdown of Cross-Linked Soy Protein Films for Drug Delivery. *Biomaterial*, Vol. 29, No. 27, (September 2008), pp. 3750-3756, ISSN 0142-9612

Chicón, R.; Belloque, J.; Alonso, E. & López-Fadiño, R. (2008). Immunoreactivity and Digestibility of High-Pressure-Treated Whey Proteins. *International Dairy Journal*, Vol. 18, No. 4, (April 2008), pp. 367-376, ISSN 0958-6946

Cummings, J. & Englyst, H. (1995). Gastrointestinal Effects of Food Carbohydrate. *American Journal of Clinical Nutrition*, Vol. 61, No. 4, (April 1995), pp. 938-945, ISSN 1938-3207

Doherty, S.B.; Gee, V.L.; Ross, R.P.; Stanton, C.; Fitzgerald, G.F. & Brodkorb, A. (2011). Development of Whey Protein Micro-Beads as Potential Matrices for Probiotic Protection. *Food Hydrocolloids*, Vol. 25, No. 6, (August 2011), pp. 1604-1617, ISNN 0268-005X

Dona, A. C; Pages, G.; Gilbert, R.G & Kuchel, P. W. (2010). Digestion of Starch: In Vivo and In Vitro Kinetic Models Used to Characterize Oligosaccharide or Glucose Release. *Carbohydrate Polymers*, Vol. 80, No. 3, (May 2010), pp. 599-617, ISSN 0144-8617

Dupont, D.; Mandalari, G.; Mollé, D.; Jardin, J.; Rolet-Répécaud, O.; Duboz, G.; Léonil, J.; Mills, C.E.N. & Mackie, A.R. (2010). Food Processing Increases Casein Resistance to Simulated Infant Digestion. *Molecular Nutrition and Food Research*, Vol. 54, No. 11, (December 2010), pp. 1677-1689, ISSN 1613-4133

Englyst, K. & Englyst, H. (2005). Carbohydrate Bioavailability. *British Journal of Nutrition*, Vol. 94, No. 1, (August 2005), pp. 1–11, ISSN 1475-2662

Erickson, R. & Sim, Y. (1990). Digestion and Absorption of Dietary Protein. *Annual Review of Medicine*, Vol. 41, No. 1, (February 1990), pp. 133-139, ISSN 0066-4219

Fave, G.; Coste, T. & Armand, M. (2004). Physicochemical Properties of Lipids: New Strategies to Manage Fatty Acid Bioavailability. *Cellular and Molecular Biology*, Vol. 50, No. 7, (December 2004), pp. 815-831, ISSN 1165-158X

Fernandes, G.; Velangi, A. & Wolever, T. (2005). Glycemic Index of Potatoes Commonly Consumed in North America. *Journal of the American Dietetic Association*, Vol. 105, No. 4, (April 2005), pp. 557–562, ISSN 0002-8223

Friedman, M. (1996). Nutritional Value of Proteins from Different Food Sources. A Review. *Journal of Agricultural and Food Chemistry*, Vol. 44, No. 1, (January 1996), pp. 6-29, ISSN 1520-5118

German, J. & Dillard, C. (2006). Composition, Structure and Absorption of Milk Lipids: a Source of Energy, Fat-Soluble Nutrients and Bioactive Molecules. *Critical Reviews in Food Science and Nutrition*, Vol. 46, No. 1, (January 2006), pp. 57–92, ISSN 1549-7852

German, J.B. & Watzke, H.J. (2004). Personalizing Food for Health and Delight. *Comprehensive Reviews in Food Science and Food Safety*, Vol. 3, No. 4, (October 2004), pp. 145-151, ISSN 1541-4337

Giacco, R.; Brighenti, F.; Parillo, M.; Capuano, M.; Ciardullo, A.; Rivieccio, A.; Rivellese, A. & Riccardi, G. (2001). Characteristics of Some Wheat-Based Foods of the Italian Diet in Relation to their Influence on Postprandial Glucose Metabolism in Patients with Type 2 Diabetes. *British Journal of Nutrition*, Vol. 85, No. 1, (January 2001), pp. 33–40, ISSN 1475-2662

Goetze, O.; Steingoetter, A.; Menne, D.; van der Voort, I.R.; Kwiatek, M.A.; Boesiger, P.; Weishaupt, D.; Thumshirn, M; Fried, M. & Schwizer, W. (2007). The Effect of Macronutrients on Gastric Volume Responses and Gastric Emptying in Humans: a Magnetic Resonance Imaging Study. *American Journal of Physiology, Gastrointestinal and Liver Physiology*, Vol. 292, No. 1, (January 2007), pp. G11-G17, ISSN 0193-1857

Goodman, B.E. (2010). Insights into Digestion and Absorption of Major Nutrients in Humans. *Advances Physiology Education*, Vol. 34, No. 2, (June 2010), pp. 44-53, ISNN 1043-4046

Guzey, D. & McClements, D. (2006). Formation, Stability and Properties of Multilayer Emulsions for Application in the Food Industry. *Advances in Colloid and Interface Science*, Vol. 128-130, No. 1, (December 2006), pp. 227-248, ISSN 0001-8686

Hébrard, G.; Blanquet, S.; Beyssac, E.; Remondetto, G.; Subirade, M. & Alric, M. (2006). Use of Whey protein Beads as a New Carrier System for Recombinant Yeast in the Human Digestive Tract. *Journal of Biotechnology*, Vol. 127, No. 1, (December 2006), pp. 151-160, ISSN 0168-1656

Hernell, O.; Staggers, J. & Carey, M. (1990). Physical-Chemical Behavior of Dietary and Biliary Lipids During Intestinal Digestion and Absorption. 2. Phase Analysis and Aggregation States of Luminal Lipids During Duodenal Fat Digestion in Healthy Adult Human Beings. *Biochemistry*, Vol. 29, No. 8, (February 1990), pp. 2041-2056, ISSN 1520-4995

Hoebler, C.; Lecannu, G.; Belleville, C. ; Devaux, M.; Popineau, Y. & Barry, J. (2002). Development of an In Vitro System Simulating Bucco-Gastric Digestion to Assess the Physical and Chemical Changes of Food. *International Journal of Food Sciences and Nutrition*, Vol. 53, No. 5, (September 2002), pp. 389-402, ISSN 1465-3478

Jacobs, D.R. & Tapsell L.C. (2007). Food, not Nutrients, is the Fundamental Unit in Nutrition. *Nutrition Reviews*, Vol. 65, No. 10, (October 2007), pp. 439-450, ISSN 0029-6643

Jane, J-L. (2009). Structural Features of Starch Granules II. In: *Starch. Chemistry and Technology, Third Edition*, J. BeMiller & R. Whistler, (Ed.), 193-236, Academic Press, ISBN 978-012-7462-75-2, Burlington, MA, USA

Jurado, E.; Camacho, F.; Luzon, G.; Fernandez-Serrano, M. & Garcia-Roman, M. (2006). Kinetic Model for the Enzymatic Hydrolysis of Tributyrin in O/W Emulsions. *Chemical Engineering Science*, Vol. 61, No. 15, (August 2006), pp. 5010-5020, ISSN 0009-2509

Kim, E.H.J.; Petrie, J.R; Motoi, L.; Morgenstern, M.P.; Sutton, K.H; Mishra, S. & Simmons, L.D. (2008). Effect of Structural and Physicochemical Characteristics of the Protein Matrix in Pasta on In Vitro Starch Digestibility, *Food Biophysics*, Vol. 3, No. 2, (June 2008), pp. 229-234, ISSN 1557-1858

Li, Y. & McClements, D. (2010). New Mathematical Model for Interpreting pH-Stat Digestion Profiles: Impact of Lipid Droplet Characteristics on In Vitro Digestibility. *Journal of Agricultural and Food Chemistry*, Vol. 58, No. 13, (July 2010), pp. 8085-8092, ISSN 1520-5118

Li-Chan, E.C.Y. (2004). Properties of Proteins in Food Systems: an Introduction. In: *Proteins in Food Processing*, R.Y. Yada, (Ed.), 2-26, Woodhead Publisher Limited, ISBN 978-084-9325-36-6, Cambridge, England

Lucas, P.; Prinz, J.; Agrawal, K. & Bruce, I. (2002). Food Physics and Oral Physiology. *Food Quality and Preference*, Vol. 13, No. 4, (June 2002), pp. 203-213, ISSN 0950-3293

Lundin, L.; Golding, M. & Wooster T.J. (2008). Understanding Food Structure and Function in Developing Foods for Appetite Control. *Nutrition & Dietetics*, Vol. 65, No. 3, (June 2008), pp. S79-S85., ISSN 1747-0080

Mackie, A. & Macierzanga, A. (2010). Colloidal Aspect of Protein Digestion. *Current Opinion in Colloid and Interface Science*, Vol. 15, No. 1-2, (April 2010), pp. 102-108, ISSN 1359-0294

Malone, M.; Appelqvist, I. & Norton, I. (2003). Oral Behavior of Food Hydrocolloids and Emulsions. Part 1. Lubrication and Deposition Considerations. *Food Hydrocolloids*, Vol. 17, No. 6, (November 2003), pp. 763-773, ISSN 0268-005X

Maltais, A; Remondetto, G.E. & Subirade, M. (2009). Soy Protein Cold-Set Hydrogels as Controlled Delivery Devices for Nutraceutical Compounds. *Foods Hydrocolloids*, Vol. 23, No. 7, (October 2009), pp. 1647-1653, ISSN 0268-005X

Marciani, L; Gowland, P.A.; Spiller, R.C.; Manoj, P.; Moore, R.J.; Young, P. & Fillery-Travis, A.J. (2001a). Effect of Meal Viscosity and Nutrients on Satiety, Intragastric Dilution, and Emptying Assessed by MRI. *American Journal of Physiology, Gastrointestinal and Liver Physiology*, Vol. 280, No. 6, (June 2001), pp. G1227-G1233, ISSN 0193-1857

Marciani, L.; Gowland, P.; Fillery-Travis, A.; Manoj, P.; Wright, J.; Smith, A.; Young, P.; Moore, R. & Spiller, R. (2001b). Assessment of Antral Grinding of a Model Solid Meal with Echo-Planar Imaging. *American Journal of Physiology Gastrointestinal and Liver Physiology*, Vol. 280, No. 5, (May 2001), pp. 844-849, ISSN 1522-1547

Marciani, L.; Wickham, M.; Hills, B.; Wright, J.; Bush, D.; Faulks, R.; Fillery-Travis, A.; Spiller, R. & Gowland, P. (2004). Intragastric Oil-in-Water Emulsion Fat Fraction Measure During Inversion Recovery Echo-Planar Magnetic Resonance Imaging. *Journal of Food Science*, Vol. 69, No. 6, (August 2004), pp. 290-296, 1750-3841

Marciani, L.; Wickham, M.; Singh, G.; Bush, D.; Pick, B.; Cox, E.; Fillery-Traves, A.; Faulks, R.; Marsden, C.; Gowland, P.A. & Spiller, R.C. (2007). Enhancement of Intragastric Acid Stability of a Fat Emulsion Meals Delays Gastric Emptying and Increases Cholecystokinin Release and Gallbladder Contraction. *American Journal of Physiology, Gastrointestinal and Liver Physiology*, Vol. 292, No. 6, (June 2007), pp. G1607-G1613, ISSN 0193-1857

Mason, T.; Wilking, J.; Meleson, K.; Chang, C. & Graves, S. (2006). Nanoemulsions: Formation, Structure, and Physical Properties. *Journal of Physics: Condensed Matter*, Vol. 18, No. 41, (October 2006), pp. R635–R666, ISSN 1361-648X

Mason, W.R. (2009). Starch Use in Foods. In: *Starch. Chemistry and Technology, Third Edition*, J. BeMiller & R. Whistler, (Ed.), 745-795, Academic Press, ISBN 978-012-7462-75-2, Burlington, MA, USA

McClements, D. (2005). *Food Emulsions: Principles, Practice, and Techniques*, CRC Press, ISBN 978-0-8493-2023-1, Boca Raton, FL, United States

McClements, D.; Decker, E.; Park, Y. & Weiss, J. (2009). Structural Design Principles for Delivery of Bioactive Components in Nutraceuticals and Functional Foods, *Critical Reviews in Food Science and Nutrition*, Vol. 49, No. 6, (May 2009), pp. 577–606, ISSN 1549-7852

Mills, E. & Breiteneder, H. (2005). Food Allergy and Its Relevance to Industrial Food Proteins. *Biotechnology Advances*, Vol. 23, No. 6, (September 2005), pp. 409-414, ISSN 0734-9750

Monogioudi, E.; Faccio, G.; Lille, M.; Poutanen, K.; Buchert, J. & Mattinnen, M-L. (2011). Effect of Enzymatic Cross-Linking of β-Casein on Proteolysis by pepsin. *Food Hydrocolloids*, Vol. 25, No. 1, (January 2011), pp. 71-81, ISNN 0268-005X

Moreno, F. (2007). Gastrointestinal Digestion of Food Allergens: Effect on Their Allergenicity. *Biomedicine & Pharmacotherapy*, Vol. 61, No. 1, (January 2007), pp. 50-60, ISSN 0753-3322

Norton, I.; Fryer, P. & Moore, S. (2006). Product/Process Integration in Food Manufacture: Engineering Sustained Health. *American Institute of Chemical Engineers Journal*, Vol. 52, No. 5, (May 2006), pp. 1632-1640, ISSN 1547-5905

Norton, I.; Moore, S. & Fryer, P. (2007). Understanding Food Structure and Breakdown: Engineering Approaches to Obesity. *Obesity Reviews*, Vol. 8, No. 1, (March 2007), pp. 83-88, ISSN 1467-7881

Olbrich, C.; Kayser, O. & Muller, R. (2002). Lipase Degradation of Dynasan 114 and 116 Solid Lipid Nanoparticles (SLN) – Effect of Surfactants, Storage Time and

Crystallinity. *International Journal of Pharmaceutics*, Vol. 237, No. 1-2, (April 2002), pp. 119–128, ISSN 0378-5173

Osterholt, K.M.; Roe, L.S. & Rolls, B.J. (2007). Incorporation of Air into a Snack Food Reduces Energy Intake. *Appetite*, Vol. 48, No. 3, (May 2007), pp. 351-358, ISSN 0195-6663

Palzer, S. (2009). Food Structures for Nutrition, Health and Wellness. *Trends in Food Science and Technology*, Vol. 20, No. 5, (May 2009), pp. 194-200, ISSN 0924-2244

Parada, J. & Aguilera, J.M. (2007). Food Microstructure Affects the Bioavailability of Several Nutriens. *Journal of Food Science*, Vol. 72, No. 2, (March 2007), pp. R21-R32, ISSN 0022-1147

Parada, J. & Aguilera, J.M. (2011a). Review: Starch Matrices and the Glycemic Response. *Food Science and Technology International*, Vol. 17, No. 3, (June 2011), pp. 187-204, ISSN 1082-0132

Parada, J. & Aguilera, J. M. (2011b). Microstructure, Mechanical Properties, and Starch Digestibility of a Cooked Dough Made with Potato Starch and Wheat Gluten. *LWT – Food Science and Technology*, Vol. 44, No. 8, (October 2011), pp. 1739-1744, ISSN

Pérez, S.; Baldwin, P.M & Gallant, D.J. (2009). Structural Features of Starch Granules I. In: *Starch. Chemistry and Technology, Third Edition*, J. BeMiller & R. Whistler, (Ed.), 149-192, Academic Press, ISBN 978-012-7462-75-2, Burlington, MA, USA

Petitot, M.; Abecassi, J. & Micard, V. (2009). Structuring of Pasta Components During Processing: Impact of Starch and Protein on Digestibility and Allergenicity. *Trends in Food Science and Technology*, Vol. 20, No. 11-12, (December 2009), pp. 521-532, ISNN

Porter, J.; Trevaskis, N. & Charman, W. (2007). Lipids and Lipid-Based Formulations: Optimizing the Oral Delivery of Lipophilic Drugs. *Nature Reviews Drug Discovery*, Vol. 6, No. 3, (March 2007), pp. 231-248, ISSN 1474-1784

Remondetto. G.E.; Beyssac, E. & Subirade, M. (2004). Influence of the Microstructure of Biodegradable Whey Protein Hydrogels on Iron Release: an In Vitro Study. *Journal of Agricultural and Food Chemistry*, Vol. 52, No. 26, (November 2004), pp. 8137-8143, ISSN 0021-8561

Salminen, S.; Bouley, C.; Boutron-Ruault, M.; Cummings, J.; Franck, A.; Gibson, G.; Isolauri, E.; Moreau, M.; Roberfroid, M. & Rowland, I. (1998). Functional Food Science and Gastrointestinal Physiology and Function. *British Journal of Nutrition*, Vol. 80, No. 1, (August 1998), pp. 147–171, ISSN 1475-2662

Schmitt, C.; Bovay, C.; Rouvet, M.; Shojaei-Rami, S. & Kolodziejczyk, E. (2007). Whey Protein Soluble Aggregates from Heating with NaCl: Physicochemical, Interfacial, and Foaming Properties. *Langmuir*, Vol. 23, No. 8, (March 2007), pp. 4155-4166, ISSN 0743-7463

Schneeman, B. (2002). Gastrointestinal Physiology and Functions. *British Journal of Nutrition*, Vol. 88, No. 2, (August 2002), pp. 159-163, ISSN 1475-2662

Slaughter, S.; Ellis, P. & Butterworth, P. (2001). An Investigation of the Action of Porcine Pancreatic α-Amylase on Native and Gelatinized Starches. *Biochimica et Biophysica Acta*, Vol. 1525, No. 1-2, (February 2001), pp. 29-36, ISSN 0304-4165

Simopoulos, A. (1999). Essential Fatty Acids in Health and Chronic Disease. *American Journal of Clinical Nutrition*, Vol. 70, No. 3, (September 1999), pp. 560-569, ISSN 1938-3207

Singh, H. & Sarkar, A. (2011). Behavior of Protein-Stabilized Emulsions under various Physiological Conditions. *Advances in Colloid and Interface Science*, Vol. 165, No. 1, (June 2011), pp. 47-57, ISSN 0001-8686

Stănciuc, N.; van der Plancken, I.; Rotaru, G. & Hendrickx, M. (2008). Denaturation Impact in Susceptibility of Beta-Lactoglobulin to Enzymatic Hydrolysys: a Kinetic Study. *Revue Roumaine de Chimie*. Vol. 53, No. 10, (October 2008), pp. 921-929, ISSN 1843-7761

Suzuki, K.; Simpson, K.A.; Minnion, J.S.; Shillito, J.C. & Bloom, S.R. (2010). The Role of Gut Hormones and the Hypothalamus in Appetite Regulation. *Endocrine Journal*, Vol. 57, No. 5, (May 2010), pp. 359-372, ISSN 0918-8959

Tester, R.; Karkalas, J. & Qi, X. (2004). Starch Structure and Digestibility Enzyme-Substrate Relationship. *World's Poultry Science Journal*, Vol. 60, No. 2, (January 2004), pp. 186-195, ISSN 1743-4777

Troncoso, E. & Aguilera, J.M. (2009). Food Microstructure and Digestion. *Food Science and Technology Journal*, Vol. 23, No. 4, pp. 30-33

Ulrich, A. (2002). Biophysical Aspects of Using Liposomes as Delivery Vehicles. *Bioscience Reports*, Vol. 22, No. 2, (April 2002), pp. 129-150, ISSN 1573-4935

Untersmayr, E. & Jensen-Jarolim E. (2008). The Role of Protein Digestibility and Antacids on Food Allergy Outcomes. *Journal of Allergy and Clinical Immunology*, Vol. 121, No. 6, (June 2008), pp. 1301-1308, ISSN 0091-6749

Videira, M.; Botelho, M.; Santos, A.; Gouveia, L.; de Lima, J. & Almeida, A. (2002). Lymphatic Uptake of Pulmonary Delivered Radiolabelled Solid Lipid Nanoparticles. *Journal of Drug Targeting*, Vol. 10, No. 8, (September 2002), pp. 607-613, ISSN 1029-2330

Vingerhoeds, M.; Blijdenstein, T.; Zoet, F. & van Aken, G. (2005). Emulsion Flocculation Induced by Saliva and Mucin. *Food Hydrocolloids*, Vol. 19, No. 5, (September 2005), pp. 915-922, ISSN 0268-005X

Wansink, B. (2007). Helping Consumer Eat Less. *Food Technology*, Vol. 61, No.5, (May 2007), pp. 34-38, 0015-6639

Weisbrodt, N. (2001). Gastric Emptying, In: *Gastrointestinal Physiology*, L.R. Johnson, (Ed.), 37-45, Mosby, ISBN 978-0323012393, St. Louis, MI, USA

Whitcomb, D. & Lowe, M. (2007). Human Pancreatic Digestive Enzymes. *Digestive Diseases and Sciences*, Vol. 52, No. 1, (January 2007), pp. 1-17, ISSN 1573-2568

Wickham, M.; Garrood, M.; Leney, J.; Wilson, P. & Fillery-Travis, A. (1998). Modification of a Phospholipid Stabilized Emulsion Interface by Bile Salt: Effect on Pancreatic Lipase Activity. *Journal of Lipid Research*, Vol. 39, No. 3, (March 1998), pp. 623-632, ISSN 1539-7262

World Health Organization. (July 2011). Obesity, In: *Health Topics*, 29.07.2011, Available from http://www.who.int/mediacentre/factsheets/fs311/en/index.html

Zeece, M.; Huppertz, T. & Kelly, A. (2008). Effect of High-Pressure treatment on In Vitro Digestibility of β-Lactoglobulin. *Innovative Food Science and Emerging Technologies*, Vol. 9, No. 1, (January 2008), pp. 62-69, ISSN 1466-8564

6

Antihypertensive and Antioxidant Effects of Functional Foods Containing Chia (*Salvia hispanica*) Protein Hydrolysates

Ine M. Salazar-Vega, Maira R. Segura-Campos,
Luis A. Chel-Guerrero and David A. Betancur-Ancona
*Facultad de Ingeniería Química, Campus de Ciencias Exactas e Ingenierías,
Universidad Autónoma de Yucatán, Yucatán,
México*

1. Introduction

High blood pressure increases the risk of developing cardiovascular diseases such as arteriosclerosis, stroke and myocardial infarction. Angiotensin I-converting enzyme (ACE, dipeptidylcarboxypeptidase, EC 3.4.15.1) is a multifunctional, zinc-containing enzyme found in different tissues (Bougatef et al., 2010). Via the rennin-angiotensin system, ACE plays an important physiological role in regulating blood pressure by converting angiotensin I into the powerful vasoconstrictor angiotensin II and inactivating the vasodilator bradykinin. ACE inhibition mainly produces a hypotensive effect, but can also influence regulatory systems involved in immune defense and nervous system activity (Haque et al., 2009). Commercial ACE-inhibitors are widely used to control high blood pressure, but can have serious side-effects. Natural ACE-inhibitory peptides are a promising treatment alternative because they do not produce side-effects, although they are less potent (Cao et al., 2010).

Oxidation is a vital process in organisms and food stuffs. Oxidative metabolism is essential for cell survival but produces free radicals and other reactive oxygen species (ROS) which can cause oxidative changes. An excess of free radicals can overwhelm protective enzymes such as superoxide dismutase, catalase and peroxidase, causing destruction and lethal cellular effects (e.g., apoptosis) through oxidization of membrane lipids, cellular proteins, DNA, and enzymes which shut down cellular processes (Haque et al., 2009). Synthetic antioxidants such as butylatedhydroxyanisole (BHA) and butylatedhydroxytoluene (BHT) are used as food additives and preservatives. Antioxidant activity in these synthetic antioxidants is stronger than that found in natural compounds such as α-tocopherol and ascorbic acid, but they are strictly regulated due to their potential health hazards. Interest in the development and use of natural antioxidants as an alternative to synthetics has grown steadily; for instance, hydrolyzed proteins from many animal and plant sources have recently been found to exhibit antioxidant activity (Lee et al., 2010).

Native to southern Mexico, chia (*Salvia hispanica*) was a principal crop for ancient Mesoamerican cultures and has been under cultivation in the region for thousands of years. A recent evaluation of chia's properties and possible uses showed that defatted chia seeds have fiber (22 g/100 g) and protein (17 g/100 g) contents similar to those of other oilseeds currently used in the food industry (Vázquez-Ovando et al., 2009). Consumption of chia seeds provides numerous health benefits, but they are also a potential source of biologically-active (bioactive) peptides. Enzymatic hydrolysis is natural and safe, and effectively produces bioactive peptides from a variety of protein sources, including chia seeds. Chia protein hydrolysates with enhanced biological activity could prove an effective functional ingredient in a wide range of foods. The objective of present study was to evaluate ACE inhibitory and antioxidant activity in food products containing chia (*Salvia hispanica* L.) protein hydrolysates.

2. Material and methods

2.1 Materials

Chia (*S. hispanica*, L.) seeds were obtained in Yucatan state, Mexico. Reagents were analytical grade and purchased from J.T. Baker (Phillipsburg, NJ, USA), Sigma (Sigma Chemical Co., St. Louis, MO, USA), Merck (Darmstadt, Germany) and Bio-Rad (Bio-Rad Laboratories, Inc. Hercules, CA, USA). The Alcalase® 2.4L FG and Flavourzyme® 500MG enzymes were purchased from Novo Laboratories (Copenhagen, Denmark). Alcalase 2.4L is an endopeptidase from *Bacillus licheniformis*, with subtilisin Carlsberg as the major enzyme component and a specific activity of 2.4 Anson units (AU) per gram. One AU is the amount of enzyme which, under standard conditions, digests hemoglobin at an initial rate that produces an amount of thrichloroacetic acid-soluble product which produces the same color with Folin reagent as 1 meq of tyrosine released per minute. Optimal endopeptidase activity was obtained by application trials at pH 7.0. Flavourzyme 500 MG is an exopeptidase/endoprotease complex with an activity of 1.0 leucine aminopeptidase unit (LAPU) per gram. One LAPU is the amount of enzyme that hydrolyzes 1 mmol of leucine p-nitroanilide per minute. Optimal exopeptidase activity was obtained by application trials at pH 7.0.

2.2 Protein-rich fraction

Flour was produced from 6 Kg chia seed by first removing all impurities and damaged seeds, crushing the remaining sound seeds (Moulinex DPA 139) and then milling them (Krups 203 mill). Standard AOAC procedures were used to determine nitrogen (method 954.01), fat (method 920.39), ash (method 925.09), crude fiber (method 962.09), and moisture (method 925.09) contents in the milled seeds (AOAC, 1997). Nitrogen (N_2) content was quantified with a Kjeltec Digestion System (Tecator, Sweden) using cupric sulfate and potassium sulfate as catalysts. Protein content was calculated as nitrogen x 6.25. Fat content was obtained from a 1 h hexane extraction. Ash content was calculated from sample weight after burning at 550 °C for 2 h. Moisture content was measured based on sample weight loss after oven-drying at 110 °C for 2 h. Carbohydrate content was estimated as nitrogen-free extract (NFE). Oil extraction from the milled seeds was done with hexane in a Soxhlet system for 2 h. The remaining fraction was milled with 0.5 mm screen (Thomas-Wiley®, Model 4, Thomas Scientific, USA) and AOAC (1997) procedures used to determine proximate composition of the remaining flour. The defatted chia flour was dried in a Labline stove at 60 °C for 24 h. Defatted flour mill yield was calculated with the equation:

$$\text{Mill yield} = \frac{\text{Weight of 0.5 mm particle size flour}}{\text{Total weight of defatted flour}} \times 100 \tag{1}$$

Extraction of the protein-rich fraction was done by dry fractionation of the defatted flour according to Vázquez-Ovando et al. (2010). Briefly, 500 g flour was sifted for 20 min using a Tyler 100 mesh (140 μm screen) and a Ro-Tap® agitation system. Proximate composition was determined following AOAC (1997) procedures and yield calculated with the equation:

$$\text{Protein rich fraction yield} = \frac{\text{Protein rich fraction weight}}{0.5 \text{ particle size flour weight}} \times 100 \qquad (2)$$

2.3 Enzymatic hydrolysis of protein-rich fraction
The chia protein-rich fraction (44.62% crude protein) was sequentially hydrolyzed with Alcalase® for 60 min followed by Flavourzyme® for a total of up to 150 min. Degree of hydrolysis was recorded at 90, 120 and 150 min. Three hydrolysates were generated with these parameters: substrate concentration, 2%; enzyme/substrate ratio, 0.3 AU g^{-1} for Alcalase® and 50 LAPU g^{-1} for Flavourzyme®; pH, 7 for Alcalase® and 8 for Flavourzyme®; temperature, 50 °C. Hydrolysis was done in a reaction vessel equipped with a stirrer, thermometer and pH electrode. In all three treatments, the reaction was stopped by heating to 85 °C for 15 min, followed by centrifuging at 9880 xg for 20 min to remove the insoluble portion (Pedroche et al., 2002).

2.4 Degree of hydrolysis
Degree of hydrolysis (DH) was calculated by determining free amino groups with o-phthaldialdehyde following Nielsen et al. (2001): DH = $h/h_{tot} \times 100$; where h_{tot} is the total number of peptide bonds per protein equivalent, and h is the number of hydrolyzed bonds. The h_{tot} factor is dependent on raw material amino acid composition.

2.5 In Vitro biological activities
ACE inhibitory and antioxidant activities were evaluated in the chia (*S. hispanica*) protein hydrolysates. Hydrolysate protein content was previously determined using the bicinchoninic acid method (Sigma, 2006).

2.5.1 ACE inhibitory activity
Hydrolysate ACE inhibitory activity was analyzed with the method of Hayakari et al. (1978), which is based on the fact that ACE hydrolyzes hippuryl-L-histidyl-L-leucine (HHL) to yield hippuric acid and histidyl-leucine. This method relies on the colorimetric reaction of hippuric acid with 2,4,6-trichloro-s-triazine (TT) in a 0.5 mL incubation mixture containing 40 μmol potassium phosphate buffer (pH 8.3), 300 μmol sodium chloride, 40 μmol 3% HHL in potassium phosphate buffer (pH 8.3), and 100 mU/mL ACE. This mixture was incubated at 37 °C/45 min and the reaction terminated by addition of TT (3% v/v) in dioxane and 3 mL 0.2 M potassium phosphate buffer (pH 8.3). After centrifuging the reaction mixture at 10,000 $x\,g$ for 10 min, enzymatic activity was determined in the supernatant by measuring absorbance at 382 nm. All runs were done in triplicate. ACE inhibitory activity was quantified by a regression analysis of ACE inhibitory activity (%) versus peptide concentration, and IC_{50} values (i.e. the peptide concentration in μg protein/mL required to produce 50% ACE inhibition under the described conditions) defined and calculated as follows:

$$\text{ACE inhibitory activity (\%)} = \frac{A-B}{A-C} \times 100 \qquad (3)$$

Where: A represents absorbance in the presence of ACE and sample; B absorbance of the control and C absorbance of the reaction blank.

$$IC_{50} = \frac{50-b}{m} \qquad (4)$$

Where b is the intersection and m is the slope.

2.5.2 ABTS$^{\bullet+}$ (2,2'-azino-bis(3-ethylbenzothiazoline-6-sulfonic acid) decolorization assay

Antioxidant activity in the hydrolysates was analyzed following Pukalskas et al. (2002). ABTS$^{\bullet+}$ radical cation was produced by reacting ABTS with potassium persulfate. To prepare the stock solution, ABTS was dissolved at a 2 mM concentration in 50 mL phosphate-buffered saline (PBS) prepared from 4.0908 g NaCl, 0.1347 g KH$_2$PO$_4$, 0.7098 g Na$_2$HPO$_4$, and 0.0749 g KCl dissolved in 500 mL ultrapure water. If pH was lower than 7.4, it was adjusted with NaOH. A 70 mM K$_2$S$_4$O$_8$ solution in ultrapure water was prepared. ABTS radical cation was produced by reacting 10 mL of ABTS stock solution with 40 μL K$_2$S$_4$O$_8$ solution and allowing the mixture to stand in darkness at room temperature for 16-17 h before use. The radical was stable in this form for more than 2 days when stored in darkness at room temperature.

Antioxidant compound content in the hydrolysates was analyzed by diluting the ABTS$^{\bullet+}$ solution with PBS to an absorbance of 0.800 ± 0.030 AU at 734 nm. After adding 990 μL of diluted ABTS$^{\bullet+}$ solution (A 734 nm= 0.800 ± 0.030) to 10 μL antioxidant compound or Trolox standard (final concentration 0.5 -3.5 mM) in PBS, absorbance was read at ambient temperature exactly 6 min after initial mixing. All analyses were run in triplicate. The percentage decrease in absorbance at 734 nm was calculated and plotted as a function of the Trolox concentration for the standard reference data. The radical scavenging activity of the tested samples, expressed as inhibition percentage, was calculated with the equation:

$$\% \, Inhibition = \frac{A_B - A_A}{A_B} X \, 100 \qquad (5)$$

Where A_B is absorbance of the blank sample (t=0), and A_A is absorbance of the sample with antioxidant after 6 min.

The Trolox equivalent antioxidant coefficient (TEAC) was quantified by a regression analysis of % inhibition versus Trolox concentration using the following formula:

$$TEAC = \frac{\%I_M - b}{m} \qquad (6)$$

Where b is the intersection and m is the slope.

2.6 White bread and carrot cream containing chia protein hydrolysates

To test if the chia protein hydrolysates increased biological potential when added to food formulations, those were used as ingredients in preparing white bread and carrot cream, and the ACE inhibitory and antioxidant activity of these foods evaluated.

2.6.1 Biological potential and sensory evaluation of white bread containing chia protein hydrolysates

White bread was prepared following a standard formulation (Table 1) (Tosi et al., 2002), with inclusion levels of 0 mg (control), 1 mg and 3 mg chia protein hydrolysate/g flour.

Hydrolysates produced at 90, 120 and 150 min were used. Treatments (two replicates each) were formed based on inclusion level and hydrolysate preparation time (e.g. 1 mg/90 min, etc.), and distributed following a completely random design. Each treatment was prepared by first mixing the ingredients (Farinograph Brabender 811201) at 60 rpm for 10 min, simultaneously producing the corresponding farinograph. The "work input" value, or applied energy required (Bloksma, 1984), was calculated from the area under the curve (in which 1 cm² was equivalent to 454 J/kg). The resulting doughs were placed in a fermentation chamber at 25 °C and 75% relative humidity for 45 min. Before the second fermentation, the dough for each treatment was divided into two pieces (approximately 250 g) and each placed in a rectangular mold; each piece was treated as a replicate. The second fermentation was done for 75 min under the same temperature and humidity conditions. Finally, the fermented doughs were baked at 210 °C for 25 min.

Sensory evaluation of the baked white bread loaves was done by judges trained in evaluating baked goods. Evaluation factors and the corresponding maximum scores were: specific volume (15 points); cortex (15 points); texture (15 points); color (10 points); structure (10 points); scent (15 points); and flavor (20 points). Overall score intervals were: 40-50 "very bad"; 50-60 "bad"; 60-70 "regular"; 70-80 "good"; 80-90 "very good"; and 90-100 "excellent". For total nitrogen content, ACE inhibitory activity and antioxidant activity analyses, the bread was sliced, dried at 40 °C for 48 h and milled. Total nitrogen content was determined following the applicable AOAC (1997) method (954.01). To analyze ACE inhibitory activity, 10, 20, 30, 40 and 50 mg of milled bread were dissolved in 1 mL buffer mixture and centrifuged at 13,698 x g for 10 min. The supernatant (40 μl) was taken from each lot and processed according to Hayakari et al. (1978). After adding 990 μl diluted ABTS•⁺ solution to 50 mg of milled bread in PBS, antioxidant activity was determined according to Pukalskas et al. (2002).

Ingredients	Control (%)	Hydrolysate (%)	
		(1 mg/g)	(3 mg/g)
Flour	56.51	56.47	56.40
Water	33.33	33.31	33.28
Sugar	3.39	3.39	3.38
Yeast	2.82	2.82	2.82
Fat	1.69	1.69	1.69
Powdered milk	1.13	1.13	1.13
Salt	1.13	1.13	1.13
Hydrolysate	0.00	0.06	0.17

Table 1. Formulation of white bread made according to a standard formula (control) and with chia protein hydrolysate added at two levels (1 and 3 mg/g).

2.6.2 Biological potential and sensory evaluation of carrot cream containing chia protein hydrolysates

Carrot cream was prepared following a standard formulation (Table 2), with inclusion levels of 0 mg/g (control), 2.5 mg/g and 5 mg/g carrot. Hydrolysates produced at 90, 120 and 150 min were used. Treatments were formed based on inclusion level and hydrolysate preparation time (e.g. 2.5 mg/90 min, etc.), and distributed following a completely random design. Two replicates consisting of 330 g carrot cream were done per treatment. The carrots were washed, peeled and cooked in water at a 1:4 (p/v) ratio for 40 min. Broth and butter were dissolved in low fat milk and liquefied with the cooked carrots and the remaining ingredients. Finally, the mixture was boiled at 65 °C for 3 min.

Viscosity was determined for a commercial product (Campbell's®) and the hydrolysate-containing carrot creams using a Brookfield (DV-II) device with a No. 2 spindle, 0.5 to 20 rpm deformation velocity (γ) and a 24 °C temperature. A viscosity curve was generated from the γ log versus viscosity coefficient log (η), while the consistency index (k) and fluid behavior (n) were quantified by applying the potency law model: $\log \eta = \log k + (n-1) \log \gamma$. Brightness L* and chromaticity a*b* were determined with a Minolta colorimeter (CR200B). Differences in color (ΔE^*) between the control and hydrolysate-supplemented carrot creams was calculated with the equation (Alvarado & Aguilera, 2001): $\Delta E^* = [(\Delta L^*)^2 + (\Delta a^*)^2 + (\Delta b)^2]^{0.5}$. Biological potential was analyzed by first centrifuging the samples at 13,698 $x\ g$ for 30 min and then determining total nitrogen content (AOAC, 1997)(954.01 method), ACE inhibitory and antioxidant activity in the supernatant.

Using a completely random design, sensory evaluation was done of the control product and the hydrolysate-containing carrot creams with the highest biological activity. Acceptance level was evaluated by 80 untrained judges who indicated pleasure or displeasure levels along a 7-point hedonic scale including a medium point to indicate indifference (Torricella et al., 1989).

Ingredients	Control (%)	Hydrolysate (%)	
		2.5 mg/g	5 mg/g
Carrot	40.12	40.08	40.04
Low fat milk	38.58	38.54	38.50
Purified water	19.29	19.27	19.25
Butter	1.16	1.16	1.16
Broth	0.85	0.85	0.85
Hydrolysate	0	0.10	0.20

Table 2. Formulation of carrot cream made according to a standard formula (control) and with chia protein hydrolysate added at two levels (2.5 and 5 mg/g).

2.7 Statistical analysis

All results were analyzed using descriptive statistics with a central tendency and dispersion measures. One-way ANOVAs were run to evaluate protein extract hydrolysis data, *in vitro* ACE inhibitory, antioxidant and antimicrobial activities, and the sensory scores. A Duncan multiple range test was applied to identify differences between treatments. All analyses were done according to Montgomery (2004) and processed with the Statgraphics Plus ver. 5.1 software.

3. Results and discussion

3.1 Proximate composition

Proximate composition analysis showed that fiber was the principal component in the raw chia flour (Table 3), which coincides with the 40% fiber content reported elsewhere (Tosco, 2004). Its fat content was similar to the 33% reported by Ixtaina et al. (2010), and its protein and ash contents were near the 23% protein and 4.6% ash contents reported by Ayerza & Coates (2001). Nitrogen-free extract (NFE) in the raw chia flour was lower than the 7.42% reported by Salazar-Vega et al. (2009), probably due to the 25.2% fat content observed in that study. In the defatted chia flour, fiber decreased to 21.43% and fat to 13.44%, while protein content increased to 34.01%: as fat content decreased, crude protein content increased. Mill yield (0.5 mm particle size) from the defatted chia flour was 84.33%, which is lower than the 97.8% reported by Vázquez-Ovando et al. (2010). Dry fractionation yield of the defatted chia flour was 70.31% particles >140 μm and 29.68% particles <140 μm. Protein-rich fraction yield was higher than reported elsewhere (Vázquez-Ovando et al., 2009), probably due to lower initial moisture content in the processed flour, which increases the tendency to form particle masses and thus retain fine particles. The 44.62% protein content of the protein-rich fraction was higher than observed in the raw chia flour (23.99%) and defatted chia flour (34.01%).

Components	Chia flour	Defatted chia flour	Protein-rich fraction
Moisture	6.32[a]	6.17[a]	7.67[b]
Ash	4.32[a]	5.85[b]	8.84[c]
Crude fiber	35.85[b]	21.43[a]	11.48[c]
Fat	34.88[c]	13.44[b]	0.54[a]
Protein	23.99[a]	34.01[b]	44.62[c]
NFE	0.96[a]	25.27[b]	34.52[c]

Table 3. Proximate composition of chia (*Salvia hispanica* L.) flour, defatted flour and protein-rich fraction.[a-b] Different superscript letters in the same row indicate statistical difference (P < 0.05). Data are the mean of three replicates (% dry base).

3.2 Enzymatic hydrolysis of protein-rich fraction

The protein-rich fraction used to produce the protein hydrolysates was isolated by alkaline extraction and acid precipitation of proteins as described above. This fraction proved to be good starter material for hydrolysis. Production of extensive (i.e. >50% DH) hydrolysates requires use of more than one protease because a single enzyme cannot achieve such high DHs within a reasonable time period. For this reason, an Alcalase®-Flavourzyme® sequential system was used in the present study to produce an extensive hydrolysate. Protease and peptidase choice influences DH, peptide type and abundance, and consequently the amino acid profile of the resulting hydrolysate. The bacterial endoprotease Alcalase® is limited by its specificity, resulting in DHs no higher than 20 to 25%, depending on the substrate, but it can attain these DHs in a relatively short time under moderate conditions. In the present study, Alcalase® exhibited broad specificity and produced hydrolysates with 23% DH during 60 min reaction time. The fungal protease Flavourzyme® has broader specificity, which, when combined with its exopeptidase activity, can generate DH values as high as 50%. The highest DH in the present study (43.8%) was attained with Flavourzyme® at 150 min (Table 4), made possible in part by predigestion with Alcalase®, which increases the number of N-terminal sites, thus facilitating hydrolysis by Flavourzyme®. The 43.8% DH obtained here with the defatted chia hydrolysate was lower than the 65% reported by Pedroche et al. (2002) in chickpea hydrolysates produced sequentially with Alcalase® and Flavourzyme® at 150 min. Likewise, Clemente et al. (1999) reported that the combination of these enzymes in a two-step hydrolyzation process (3 h Alcalase® as endoprotease; 5 h Flavourzyme® as exoprotease) of chickpea produced DH >50%. In this study, the globular structure of globulins in the isolated protein limited the action of a single proteolytic enzyme, which is why sequential hydrolysis with an endoprotease and exoprotease apparently solves this problem. Cleavage of peptide bonds by the endopeptidase increases the number of peptide terminal sites open to exoprotease action. Imm & Lee (1999) reported that when using Flavourzyme® more efficient hydrolysis and higher DH can be achieved by allowing pH to drift. They suggested that a more effective approach would be initial hydrolysis with Alcalase® under optimum conditions followed by Flavourzyme® with pH being allowed to drift down to its pH 7.0 optimum. Using this technique for hydrolysis of rapeseed protein, Vioque et al. (1999) attained a 60% DH.

Hydrolysate (min)	DH (%)	IC_{50} mg/mL	TEAC (Mm/mg)
90	37.5[a]	44.01[a]	7.31[a]
120	40.5[b]	20.76[b]	4.66[b]
150	43.8[c]	8.86[c]	4.49[c]

Table 4. Degree of hydrolysis (DH), ACE inhibitory and antioxidant activities of chia (*Salvia hispanica*) protein hydrolysates produced at three hydrolysis times. [a-b]Different superscript letters in the same column indicate statistical difference (P < 0.05).

Controlled release of bioactive peptides from proteins via enzymatic hydrolysis is one of the most promising techniques for producing hydrolysates with potential applications in the pharmaceutical and food industries: hydrolysates with >10% DH have medical applications while those with <10% DH can be used to improve functional properties in flours or protein isolates (Pedroche et al. (2003). Several biological properties have been attributed to low-molecular-weight peptides, although producing them normally requires a combination of commercial enzyme preparations (Gilmartin & Jervis, 2002). When hydrolyzed sequentially with Alcalase® and Flavourzyme®, chia *S. hispanica* is an appropriate substrate for producing bioactive peptides with high DH (43.8%).

3.3 ACE inhibitory activity

ACE inhibitory activity of the chia protein hydrolysates produced with an Alcalase®-Flavourzyme® sequential system at 90, 120 and 150 min was measured and calculated as IC_{50} (Table 4). The fact that the alkaline proteases Alcalase® and Flavourzyme® have broad specificity and hydrolyze most peptide bonds, with a preference for those containing aromatic amino acid residues, has led to their use in producing protein hydrolysates with better functional and nutritional characteristics than the original proteins, and in generating bioactive peptides with ACE inhibitory activity (Segura-Campos et al., 2010). The chia protein hydrolysates produced with this sequential system exhibited ACE inhibitory activity, suggesting that the peptides released from the proteins are the agents behind inhibition. ACE inhibitory activity in the analyzed hydrolysates depended significantly on hydrolysis time, and therefore on DH. Bioactivity was highest in the hydrolysate produced at 150 min (IC_{50} = 8.86 µg protein/mL), followed by those produced at 120 min (IC_{50} = 20.76 µg/mL) and at 90 min (IC_{50} = 44.01 µg/mL). Kitts & Weiler (2003) found that peptides with antihypertensive activity consist of only two to nine amino acids and that most are di- or tripeptides, making them resistant to endopeptidase action in the digestive tract. The ACE inhibitory activity in the hydrolysates studied here was higher than reported by Segura et al. (2010) for *V. unguiculata* hydrolysates produced using a 60 min reaction time with Alcalase® (2564.7 µg/mL), Flavourzyme® (2634.3 µg/mL) or a pepsin-pancreatin sequential system

(1397.9 μg/mL). It was also higher than the 191 μg/mL reported by Pedroche et al. (2002) for chickpea protein isolates hydrolyzed sequentially with Alcalase® and Flavourzyme®. The chia protein hydrolysates' ACE inhibitory activity was many times higher than reported for *Phaseolus lunatus* and *Phaseolus vulgaris* hydrolysates produced with Alcalase® at 15 (437 and 591μg/mL), 30 (569 and 454 μg/mL), 45 (112 and 74μg/mL), 60 (254 and 61μg/mL), 75 (254 and 98 μg/mL) and 90 min (56 and 394 μg/mL), and with Flavourzyme® at 15 (287 and 401 μg/mL), 30 (239 and 151 μg/mL), 45 (265 and 127μg/mL), 60 (181 and 852 μg/mL) and 75 min (274 and 820 μg/mL). However, the *P. lunatus* hydrolysate produced with Flavourzyme® at 90 min had a lower IC_{50} value (6.9 μg/mL) and consequently higher ACE inhibitory activity than observed in the present study (Torruco-Uco et al., 2009).

The *in vitro* biological potential observed here in the enzymatically hydrolyzed chia proteins was higher than the 140 μg/mL reported by Li et al. (2007) for a rice protein hydrolysate produced with Alcalase®. After a single oral administration in spontaneously hypertensive rats (SHR), this rice hydrolysate exhibited an antihypertensive effect, suggesting its possible use as a physiologically functional food with potential benefits in the prevention and/or treatment of hypertension. Enzymatic hydrolysates from different protein sources, and IC_{50} values ranging from 200 to 246700 μg/mL, have also been shown to have *in vitro* ACE inhibitory activity as well as antihypertensive activity in SHR (Hong et al., 2005). Matsufuji et al. (1994) reported that peptides produced by enzymes such as Alcalase®, and which exhibit ACE inhibitory activity, may resist digestion by gastrointestinal proteases and therefore be absorbed in the small intestine, a quality also reported in a number of SHR studies. Based on the above, it is probable that the chia protein hydrolysates produced here with Alcalase®-Flavourzyme®, which exhibit ACE inhibitory activity, are capable of resisting gastrointestinal proteases and are therefore appropriate for application in food systems (e.g. functional foods) focused on people suffering arterial hypertension disorders. Further research will be needed, however, to determine if the peptide mixture exerts an *in vivo* antihypertensive effect because peptide ACE inhibitory potencies do not always correlate with their antihypertensive activities in SHR.

3.4 ABTS[•+] (2,2'-azino-bis(3-ethylbenzothiazoline-6-sulfonic acid) decolorization assay

Antioxidant activity of the chia protein hydrolysates, quantified and calculated as TEAC values (mM/mg), decreased as DH increased (Table 4). The highest TEAC value was for the hydrolysate produced at 90 min (7.31 mM/mg protein), followed by those produced at 120 min (4.66 mM/mg protein) and 150 min (4.49 mM/mg protein); the latter two did not differ (P<0.05). Increased antioxidant activity in hydrolyzed proteins has also been reported for dairy, soy, zein, potato, gelatin and egg yolk among other proteins. This increase has been linked to greater solvent exposure of amino acids (Elias et al., 2008), in other words, enzymatic hydrolysis probably increased exposure of antioxidant amino acids in the chia proteins, consequently providing them greater antioxidant activity. Extensive proteolysis of the chia protein hydrolysates at 120 and 150 min resulted in lower antioxidant activity because it may have generated free amino acids, which are not effective antioxidants. The increased antioxidant activity of peptides is related to unique properties provided by their chemical composition and physical properties. Peptides are potentially better food antioxidants than amino acids due to their higher free radical scavenging, metal chelation and aldehyde adduction activities. An increase in the ability of a protein hydrolysate to lower a free radical's reactivity is related to an increase in amino acid exposure. This leads to increased peptide-free radical reactions and an energy decrease in the scavenged free radical, both of which compromise a free radical's ability to oxidize lipids (Elias et al., 2008). The present results are lower than reported for *P. lunatus* hydrolysates produced with Alcalase® at 90 (9.89 mM/mg) or Flavourzyme® at 90 (11.55 mM/mg), and *P. vulgaris* hydrolysates produced with Alcalase® at 60 min (10.09 mM/mg) or Flavourzyme® at 45 min (8.42 mM/mg)(Torruco-Uco et al., 2009). They are also lower than *V. unguiculata* protein hydrolysates produced with Alcalase® (14.7 mM/mg), Flavourzyme® (14.5 mM/mg) or pepsin-pancreatin (14.3 mM/mg) for 90 min. However, the present results were higher than the 0.016 mM/mg reported by Raghavan et al. (2008) for tilapia protein hydrolysates. The above results show that chia protein hydrolysates undergo single-electron transfer reactions in the ABTS•+ reduction assay, which effectively measures total antioxidant activity of dietary antioxidants and foods. Under the analyzed conditions, the chia protein hydrolysates may have acted as electron donors and free radical sinks thus providing antioxidant protection. However, this purported antioxidant action needs to be confirmed for each peptide in different oxidant systems and under *in vitro* and *in vivo* conditions.

No relationship was observed between antioxidant activity and the hydrolysates with the highest ACE inhibitory activity. This suggests that peptide antioxidant activity may depend on the specific proteases used to produce them; the DH attained; the nature of the peptides released (e.g. molecular weight, composition and amino acid sequence); as well as the combined effects of their properties (e.g. capacity for free radical location, metallic ion chelation and/or electron donation) (Tang et al., 2009). Peptide size may also play a role since antihypertensive peptides are short, with only two to nine amino acids (are di- or tri-peptides), whereas antioxidant peptides contain from three to sixteen amino acid residues (Kitts & Weiler, 2003).

3.5 White bread and carrot cream containing chia protein hydrolysates
3.5.1 Biological potential and sensory evaluation of white bread containing chia protein hydrolysates

Addition of the chia protein hydrolysates (90, 120 and 150 min) to white bread resulted in products with higher ACE inhibitory activity than the control treatment. Bioactivity was higher (i.e. lower IC_{50} values) in the bread containing the hydrolysates produced at either 90 or 120 min, than in that containing the hydrolysate produced at 150 min. Hydrolysate inclusion level (i.e. 1 or 3 mg/g) had no effect (P>0.05) on product biological potential. Hydrolysate bioactivity (8.86-44.01µg protein/mL) declined notably after incorporation into the white bread (141.29-297.68 µg protein/mL), suggesting that fermentation and high temperatures during baking hydrolyzed the ACE inhibitory peptides and generated peptide fractions with lower antihypertensive potential. In contrast, antioxidant activity was unaffected by addition of the chia protein hydrolysates. As occurred with the IC_{50} values, hydrolysate TEAC values (7.31 mM/mg at 90 min; 4.66 mM/mg at 120 min; 4.49 mM/mg at 150 min) decreased after incorporation into the bread, with levels no higher than approximately 0.53 mM/mg (Table 5). Again, high temperature during baking probably lowered product biological potential by oxidating tryptophan and hystidine, or through methionine desulfuration.

Hydrolysis Time (min)	Inclusion level (mg/g)	IC_{50} (μg protein/ml)	TEAC (mM/mg protein)
Control	0	400.76[a]	0.53[a]
90	1	141.29[b]	0.53[a]
	3	155.88[b]	0.53[a]
120	1	163.14[b]	0.54[a]
	3	159.04[b]	0.53[a]
150	1	237.60[c]	0.53[a]
	3	297.68[c]	0.55[a]

Table 5. ACE inhibitory (IC_{50} values) and antioxidant (TEAC values) activity of white bread containing two levels (1 and 3 mg/g) of chia protein hydrolysates produced at three hydrolysis times (90, 120 and 150 min).[a-c] Different superscript letters in the same column indicate statistical difference ($P<0.05$). Data are the mean of three replicates.

Kneading of the bread dough containing chia protein hydrolysates required more ($P<0.05$) applied energy (26.1 to 28.7 kJ/kg) than for the control product (22.9 kJ/kg). Higher applied energy requirements were probably a result of the greater viscoelasticity in the hydrolysate-containing doughs due to the gum residuals, in which the protein-rich chia hydrolysate would have competed for water with the wheat flour protein and starch (Figure 1).

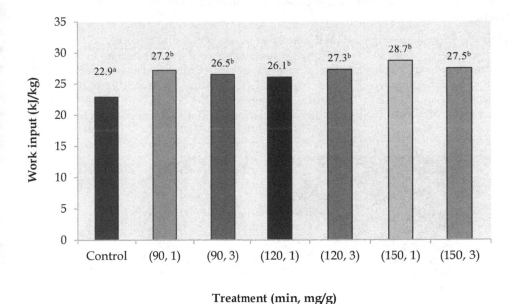

Fig. 1. Applied energy required during kneading of a control white bread and treatments containing different concentrations (1 and 3 mg/g) of chia protein hydrolysates produced at three hydrolysis times. [a-b]Different superscript letters indicate statistical difference ($P<0.05$)

Sensory evaluation of the hydrolysate-containing bread treatments resulted in scores of 80-90 ("very good") whereas the control was scored as 90-100 ("excellent") (Figure 2).

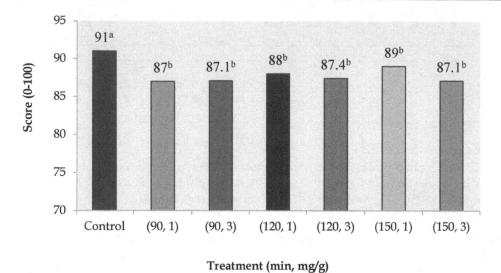

Treatment (min, mg/g)

Fig. 2. Scores generated by trained judges for sensory evaluation of a control white bread and treatments containing two concentrations (1 and 3 mg/g) of chia protein hydrolysates produced at three hydrolysis times. [a-b]Different superscript letters indicate statistical difference ($P<0.05$).

Differences in scores were attributed mainly to texture, color and structure factors (Figure 3). Crumbs were stickier in the hydrolysate-containing bread treatments than in the control, a difference which can be attributed to gum content. Crumb color was darker in the hydrolysate-containing bread treatments than in the control, probably due to hydrolysate inclusion level and Maillard reactions. Gum content in the chia protein hydrolysates also affected bread structure by producing a greater number of and larger-sized holes in the crumbs.

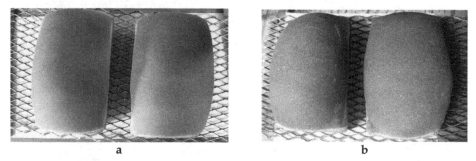

a   b

Fig. 3. White bread: a) Control b) White bread containing 3mg/g of chia hydrolysate

3.5.2 Biological potential and sensory evaluation of carrot cream containing chia protein hydrolysates

ACE inhibitory activity in the carrot cream improved markedly with addition of the chia protein hydrolysates (Table 6). An analogous improvement in ACE inhibitory activity was reported by Nakamura et al. (1995) in milk fermented with Calpis sour milk starter containing *Lactobacillus helveticus* and *Saccharomyces cerevisiae*, which they attributed to VPP and IPP peptides. Although biological potential did improve in the carrot creams, neither protein hydrolysate inclusion level (2.5 or 5 mg/g) nor hydrolysis time (90, 120 and 150 min) had a significant (P>0.05) effect. Addition of the chia protein hydrolysates (90 min, 120 min and 150 min) to carrot cream at both inclusion levels (2.5 and 5 mg/g) resulted in IC_{50} values as low as 0.24 µg/mL. These substantially lower values suggest that the peptides released from chia during hydrolysis with the Alcalase®-Flavourzyme® sequential system complemented the peptides (β-casokinins) released from the milk during carrot cream preparation, producing a higher ACE inhibitory activity than in the original hydrolysates or the carrot cream control treatment.

Antioxidant activity increased from 10.21 mM/mg in the carrot cream control treatment to 17.52-18.88 mM/mg in the treatments containing the chia protein hydrolysates. As occurred with ACE inhibitory activity, neither hydrolysate inclusion level (2.5 or 5 mg/g) nor hydrolysis time (90, 120 and 150 min) had a significant effect (P>0.05) on antioxidant activity. Again, this suggests that the higher antioxidant activity in the hydrolysate-containing carrot creams was due to the combined effect of the peptides released during hydrolysis of chia and the antioxidant potential of the carotenoids in the carrots included in the carrot cream.

Hydrolysis time (min)	Inclusion level (mg/g)	IC_{50} (µg protein/ml)	TEAC (mM/mg protein)
Control	0	27.67[a]	10.21[a]
90	2.5	1.23[b]	18.82[b]
	5	1.05[b]	18.54[b]
120	2.5	0.61[b]	18.88[b]
	5	0.24[b]	17.52[b]
150	2.5	1.29[b]	17.58[b]
	5	1.71[b]	18.60[b]

Table 6. ACE inhibitory (IC_{50}) and antioxidant (TEAC values) activities of carrot cream containing two levels (2.5 and 5 mg/g) of chia protein hydrolysates produced at three hydrolysis times (90, 120 and 150 min). [a-b]Different superscript letters in the same column indicate statistical difference (P<0.05). Data are the mean of three replicates.

Fluid behavior (n values) in the carrot creams indicated pseudoplastic properties, suggesting that apparent viscosity depended on deformation velocity rather than tension time (Table 7).

Their higher deformation velocity made these fluids thinner. The pseudoplastic behavior observed here was similar to that reported in other foods such as ice creams, yogurts, mustards, purees or sauces (Alvarado & Aguilera, 2001). No difference (P>0.05) in n and k values was observed between the carrot creams containing 2.5 mg/g hydrolysate (90, 120 or 150 min) and the control product. In contrast, the carrot creams containing 5 mg/g hydrolysate (90, 120 or 150 min) exhibited higher (P<0.05) k values and lower (P<0.05) n values than the control product, indicating that the hydrolysate-containing carrot creams had lower viscosity. This behavior was probably due to the amino acid composition of the chia protein hydrolysates, consisting mainly of hydrophobic residues, which may have limited their interaction with water.

Hydrolysis time (min)	Inclusion level (mg/g)	n	k (Pa s^n)
Control	0	0.46[a]	0.54[a]
90	2.5	0.46[a]	0.53[a]
	5	0.54b	0.43b
120	2.5	0.45[a]	0.54[a]
	5	0.56b	0.43b
150	2.5	0.44[a]	0.53[a]
	5	0.57b	0.44b
Commercial product	---	0.55b	0.41b

Table 7. Flow (n) and consistency (k) index values for carrot cream containing two levels (2.5 and 5 mg/g) of chia protein hydrolysates produced at three hydrolysis times (90, 120 and 150 min).[a-b] Different superscript letters indicate statistical difference (P<0.05). Data are the mean of three replicates.

Hydrolysis time had no effect (P>0.05) in the color (ΔE) values, but the carrot creams containing 2.5 mg/g hydrolysate exhibited lower (P<0.05) ΔE values than those containing 5 mg/g hydrolysate (Figure 4).

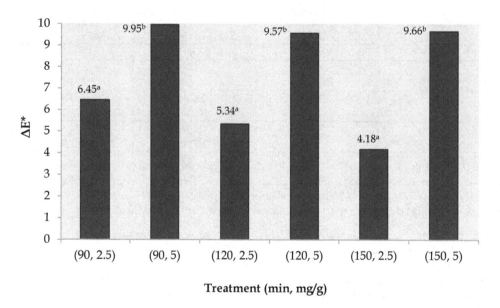

Fig. 4. Color (ΔE^*) values for carrot creams containing two concentrations (2.5 and 5 mg/g) of chia protein hydrolysates produced at three hydrolysis times. [a-b] Different superscript letters indicate statistical difference (P<0.05).

Because no statistical difference (P<0.05) was observed in the biological potential of the hydrolysate-containing carrot cream treatments (at both concentrations and all three hydrolysis times), sensory evaluation was done comparing the control product to the carrot creams containing chia protein hydrolysate produced at 90 min and incorporated at 2.5 and

5 mg/g (Figure 5). Control product scores were higher (P<0.05) than those for the carrot cream containing 2.5 mg/g hydrolysate, but not different (P>0.05) from those for the carrot cream containing 5 mg/g hydrolysate (Figure 6).

a)   b)   c)

Fig. 5. Carrot creams: a) Control, b) Carrot cream containing 2.5 mg/g of chia protein hydrolysate at 90 min, c) Carrot cream containing 5 mg/g of chia protein hydrolysate at 90 min.

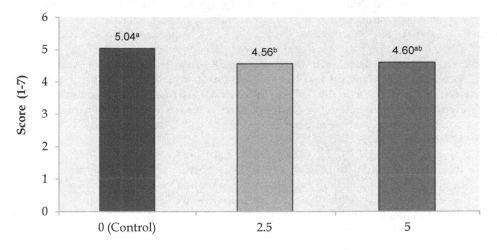

Hydrolysate concentration (mg/g)

Fig. 6. Scores generated by untrained judges for sensory evaluation of carrot cream containing three concentrations (0, 2.5 and 5 mg/g) of chia protein hydrolysates. [a-b] Different superscript letters indicate statistical difference (P<0.05).

4. Conclusions

Inclusion of the studied chia protein hydrolysates in white bread and carrot cream increased product biological potential without notably affecting product quality. Hydrolysis of a protein-rich fraction from *S. hispanica* with the Alcalase®-Flavourzyme® sequential system generated extensive hydrolysates with potential biological activity. This hydrolysis system produces low-molecular-weight hydrolysates, probably peptides, with ACE inhibitory and antioxidant activities and commercial potential as "health-enhancing ingredients" in the production of functional foods such as white bread and carrot cream.

5. Acknowledgments

The authors take this opportunity to thank the following persons for their special contributions: Ing. Hugo Sanchez, MC. Carlos Osella, Instituto de Tecnología de alimentos, Universidad Autónoma del litoral, Argentina.

6. References

Alvarado, J. & Aguilera, J. (2001). *Métodos para medir propiedades físicas en industrias de alimentos (2th Ed.)*. Acribia, S.A., ISBN 84-200-0956-3, Zaragoza, España.

AOAC., (1997). *Official methods of analysis of AOAC international*, (17th ed.). AOAC International, ISSN 1080-0344, Gaithersburg, M.D.

Ayerza, R., & Coates, W. (2001). *Chia seeds: new source of omega-3 fatty acids, natural antioxidants, and dietetic fiber*. Southwest Center for Natural Products Research & Commercialization, Office of Arid Lands Studies, Tucson, Arizona, USA.

Bloksma, A. (1984). *The Farinograph Handbook (3rd ed.)*. American Association of Cereal Chemists Inc., St. paul, Minnesota, USA.

Bougatef, A., Balti, R., Nedjar-Arroume, N., Ravallec, R., Adje, E.Y., Souissi, N., Lassoued, I., Guillochon, D., Nasri, M. (2010). Evaluation of angiotensin I-converting enzyme (ACE) inhibitory activities of smooth hound (*Mustelus mustelus*) muscle protein hydrolysates generated by gastrointestinal proteases: identification of the most potent active peptide. *European Food Research and Technology*, Vol. 231, No.1, pp. 127-135, ISSN 1438-2377.

Cao, W., Zhang, C., Hong, P., Ji, H., Hao, J. (2010). Purification and identification of an ACE inhibitory peptide from the peptic hydrolysates of *Acetes chinensis* and its antihypertensive effects in spontaneously hypertensive rats. *International Journal of Food Science and Technology*, Vol.45, No.1, pp. 959-965, ISSN 0950-5423.

Clemente, A., Vioque, J., Sánchez-Vioque, R., Pedroche, J., Millán, F. (1999). Production of extensive chickpea (*Cicer arietinum* L.) protein hydrolysates with reduced antigenic activity. *Journal of Agricultural and Food Chemistry*, Vol.47, No.9, pp. 3776-3781, ISSN 0021-8561.

Elias, R.J., Kellerby, S.S, Decker, E.A. (2008). Antioxidant activity of proteins and peptides. *Critical Reviews in Food Science and Nutrition*, Vol. 48, No.5, pp.430-411, ISSN 1040-8398.

Gilmartin, L. & Jervis, L. (2002). Production of cod (*Gadus morhua*) muscle hydrolysates: influence of combinations of commercial enzyme preparation on hydrolysate peptide size range. *Journal of Agricultural and Food Chemistry*, Vol.50, No.19, pp. 5417-5423, ISSN 0021-8561.

Haque, E., Chand, R., Kapila, S. (2009). Biofunctional properties of bioactive peptides of milk origin. *Food Reviews International*, Vol.25, No. 1, pp. 28-43, ISSN 8755-9129.

Hayakari, M., Kondo, Y., Izumi, H. (1978). A rapid and simple espectrophotometric assay of angiotensin-converting enzyme. *Analytical Biochemistry*, Vol.84, No.1, pp.361-369, ISSN 0003-2697.

Hong, L.G., Wei, L., Liu, H., Hui, S.Y. (2005). Mung-bean protein hydrolysates obtained with Alcalase® exhibit angiotensin I-converting enzyme inhibitory activity. *Food Science and Technology International*, Vol.11, No.4, pp. 281-287, ISSN 1082-0132.

Imm, J.Y., & Lee, C. M. (1999).Production of seafood flavor from Red Hake (*Urophycis chuss*) by enzymatic hydrolysis. *Journal of Agricultural and Food Chemistry*, Vol. 47, No.6, pp. 2360-2366, ISSN 0021-8561.

Ixtaina, V.Y., Vega, A., Nolasco, S.M., Tomás, M.C., Gimeno, M., Bárzana, E., Tecante, A. (2010). Supercritical carbón dioxide extraction of oil from Mexican chia seed (Salvia hispanica L.); chracterization and process optimization. *Journal of Supercritical Fluids*, Vol.55, No.1, pp.192-199, ISSN 0896-8446.

Kitts, D., & Weiler, K. (2003). Bioactive proteins and peptides form food sources. Applications of bioprocesses used in isolation and recovery. *Current Pharmaceutical Design*, Vol.9, No.16, pp. 1309-1325, ISSN 1381-6128.

Lee, J.K., Yun, J.H., Jeon, J.K., Kim, S.K., Byun, H.G. (2010). Effect of antioxidant peptide isolated from *Brachionus calciflorus*. *Journal of the Korean Society for Applied Biological Chemistry*, Vol.53, No.2, pp.192-197, ISSN 1738-2203.

Li, G.H., Qu, M.R., Wan, J.Z., You, J.M. (2007). Antihypertensive effect of rice protein hydrolysate with in vitro angiotensin I-converting enzyme inhibitory activity in spontaneously hypertensive rats. *Asia Pacific Journal of Clinical Nutrition*, Vol.16, No.1, pp.275-280, ISSN 0964-7058 .

Matsufuji, H., Matsui, T., Seki, E., Osajima, K., Nakashima, M., Osajima, Y. (1994). Angiotensin I-converting enzyme inhibitory peptides in an alkaline protease hydrolysate derived from sardine muscle. *Bioscience, Biotechnology and Biochemistry*, Vol.58, No.1, pp.2244–2245, ISSN 0916-8451.

Montgomery, D. (2004). *Diseño y análisis de experimentos (2th Ed.)*. Limusa-Wiley, ISBN 968-18-6156-6, México, D.F.

Nakamura, Y., Yamamoto, N., Sakai, K., Okubo, A., Yamazaki, S., Takano, T. (1995). Purification and characterization of angiotensin I-converting enzyme inhibitors from sour milk. *Journal of Dairy Science*,Vol.78, No.1, pp. 777-783, ISSN 0022-0302.

Nielsen, P., Petersen, D., Dammann, C. (2001). Improved method for determining food protein degree of hydrolysis. *Journal of Food Science*, Vol.66, No.5, pp. 642-646, ISSN 0022-1147.

Pedroche, J., Yust, M.M, Girón-Calle, J., Alaiz, M., Millán, F., Vioque, J. Utilisation of chickpea protein isolates for production of peptides with angiotension I-converting enzyme (ACE) inhibitory activity. *Journal of the Science of Food and Agriculture*, Vol.82, No.1, pp. 960-965, ISSN 0022-5142.

Pedroche, J., Yust, M., Girón-Calle, J., Vioque, J., Alaiz, M., Millán, F. (2003). Plant protein hydrolysates and tailor-made foods. *Electronic Journal of Environmental, Agricultural and Food Chemistry*, Vol.2, No.1, pp. 233-235, ISSN 1579-4377.

Pukalskas, A., Van Beek, T., Venskutonis, R., Linssen, J., Van Veldhuizen, A., Groot, A. (2002). Identification of radical scavengers in sweet grass (*Hierochloe odorata*). *Journal of Agricultural and Food Chemistry*, Vol.50, No.10, pp. 2914-2919, ISSN 0021-8561.

Raghavan, S., Kristinsson, G.H., Leewengurgh, C. (2008). Radical scavenging and reducing ability of tilapia (*Oreichromis niloticus*) protein hydrolysates. *Journal of Agricultural and Food Chemistry*, Vol.56, No. 21, pp.10359-10367, ISSN 0021-8561.

Salazar-Vega, I.M., Rosado-Rubio, G., Chel-Guerrero, L.A., Betancur-Ancona, D.A., & Castellanos-Ruelas, A.F. (2009). Composición en ácido graso alfa linolénico (ω3) en el huevo y carne de aves empleando chia (*Salvia hispanica*) en el alimento. *Interciencia*, Vol.34, No.3, pp. 209-213, ISSN 0378-1844.

Segura-Campos, M.R., Chel-Guerrero, L.A. Betancur-Ancona, D.A. (2010). Angiotensin-I converting enzyme inhibitory and antioxidant activities of peptide fractions extracted by ultrafiltration of cowpea *Vigna unguiculata* hydrolysates. *Journal of the Science of Food and Agriculture,*Vol.90, No.1, pp. 2512-2518, ISSN 0022-5142.

Sigma (2006).*Technical bulletin OF Sigma BCA assay*. Product Code BCA, B9643, Available from http:// www. Sigmaaldrich.com/sigma/bulletin/b9643bul.pdf.

Tang, C.H., Peng, J., Zhen, D. W., Chen, Z. (2009). Physicochemical and antioxidant properties of buckwheat (*Fagopyrum esculentum Moench*) protein hydrolysates. *Food Chemistry*, Vol.115, No.2, pp. 672-687, ISSN 0308-8146.

Torricella, M., Zamora U., Pulido A. (1989). *Evaluación Sensorial aplicada al desarrollo de la calidad en la industria sanitaria (1th ed.)*. Editorial universitaria, Instituto de Investigaciones para la Industria Alimentaria, ISBN 978-959-16-0577-1, Habana, Cuba.

Torruco-Uco, J., Chel-Guerrero, L., Martínez-Ayala, A., Dávila-Ortíz, G., Betancur-Ancona, D. (2009). Angiotensin-I converting enzyme inhibitory and antioxidant activities of protein hydrolysates from *Phaseolus lunatus* and *Phaseolus vulgaris*. LWT-Food Science and Technology, Vol.42, No. 1, pp. 1597-1604, ISSN 0023-6438.

Tosco, G. (2004). Los beneficios de la chía en humanos y animales. Nutrimentos de la semilla de chía y su relación con los requerimientos humanos diarios. *Actualidades Ornitológicas*, No.119, pp.1-70.

Tosi, E., Ré, E., Masciarelli, R., Sánchez, H., Osella, C., De la Torre, M. (2002). Whole and defatted hyperproteic amaranth flours tested as wheat flour supplementation in mold breads. *LWT-Food Science and Technology*, Vol.35, No.5, pp. 472-475, ISSN 0023-6438.

Vioque, J., Sánchez-Vioque, R., Clemente, A., Pedroche, J., Bautista, J., Millan, F. (1999). Production and characterization of an extensive rapeseed protein hydrolysate. *Journal of the American Oil Chemists Society*, Vol.76, No.7, pp. 819-823, ISSN 0003-021X.

Vázquez-Ovando, A., Rosado-Rubio, G., Chel-Guerrero, L., Betancur-Ancona, D.A. (2009). Physicochemical properties of a fibrous fraction from chia (*Salvia hispánica* L.). *LWT-Food Science and Technology*, Vol.42, No.1, pp. 168-173, ISSN 0023-6438.

Vázquez-Ovando, J.A., Rosado-Rubio, J.G., Chel-Guerrero, L.A., Betancur-Ancona, D.A. (2010). Dry processing of chia (*Salvia hispanica* L.) flour: chemical and characterization of fiber and protein. *CYTA-Journal of Food*, Vol.8, No.2, pp. 117-127, ISSN 1947-6345.

7

Allium Species, Ancient Health Food for the Future?

Najjaa Hanen[1], Sami Fattouch[2], Emna Ammar[2] and Mohamed Neffati[1]
[1]Laboratoire d'Ecologie Pastorale, Institut des Régions Arides
[2]National Institute of Applied Sciences and Technology (INSAT);
University of Carthage, Tunis
[3]UR Study & Management of Urban and Coastal Environments,
National Engineering School in Sfax
Tunisia

1. Introduction

Allium is the largest and the most important representative genus of the *Alliaceae* family that comprises 700 species, widely distributed in the northern hemisphere, North America, North Africa, Europe and Asia (Tsiaganis et al., 2006). Besides the well known garlic and onion, several other species are widely grown for culinary use and for folk medicine including leek (*Allium porrum* L.), scallion (*Allium fistulosum* L.), shallot (*Allium ascalonicum* Hort.), wild garlic (*Allium ursinum* L.), garlic (*Allium sativum*) and onion (*Allium cepa*) (Lanzotti, 2006; Tsiaganis et al., 2006). Its consumption is attributed to several factors, mainly heavy promotion that links flavour and health. The powerful and unusual flavors of many of these plants and their possible nutritional impact and medical applications have attracted the attention of plant physiologists, chemists, nutritionists, and medical researchers (Graham and Graham, 1987).

Allium roseum is a very polymorphous, widespread species that is represented by 12 different taxa: 4 varieties, 4 subvarieties and 4 forms in North Africa (Cuénod, 1954; Le Floc'h, 1983). In Tunisia, the same authors mentioned the presence of only three varieties: var. *grandiflorum*, var. *perrotii* and var. *odoratissimum*. The *odoratissimum* variety is an endemic taxon in North Africa and a perennial spontaneous weed (Cuénod, 1954). Its flowering stem is about 30-60 cm, leaves are fleshy and very small, flowers are wide, rosy or white coloured and its odour is eyelet (Jendoubi et al., 2001).

In Southern Tunisia, local people on the extension area where *A. roseum* or rosy garlic occurs have extensively developed uses for this species both as a cooking ingredient and a sauce (Najjaa et al., 2011a). *A. roseum* leaves are the main edible part, with a distinctive pungent odour and strong flavour. Besides its culinary use, rosy garlic is also used in folk medicine. Le Floc'h (1983) reported its use for the treatment of headaches and rheumatism. It is also used for the treatment of bronchitis, colds as an inhalation, fever diminution and as an appetizer.

While several studies have provided information about *A. roseum*, detailed studies documenting compositional, nutritional and functional properties are very limited, if not lacking. The objective of the present study was to characterize chemical composition, nutritional properties, bioactive components, and antioxidant and antimicrobial activities of *A. roseum* grown in Tunisia and to infer their role in human nutrition.

All samples were collected on the same day. To preserve freshness, *A. roseum* samples were transported to the laboratory on ice. Upon arrival, leaves of *A. roseum* samples were cleaned to remove all foreign matter and washed with distilled water. For the analysis of moisture content, pH, carotenoids, vitamin C, and anthocyanidins, samples were frozen and stored at –80 °C until analysis. A second portion of samples was kept fresh, and the leaves were extracted with methanol, then evaporated at 40 °C using a rotary evaporator under high vacuum. The resulting crude extract was used for total phenolic content and flavonoids composition analysis. Fresh leaves were also used to prepare an aqueous extract for allicin determination. The third portion of *A. roseum* samples was air dried in the shade for 24 h at 45 °C. From these air-dried samples, plants were chopped into pieces ≤ to 5 cm and ground in a Sorvall Omnimixer into a fine powdery consistency and used for other chemical analyses. The powder sample was packed in a hermetic glass vessel and stored at +4 °C for subsequent analyses. In this study, all methodologies used and data presented are in accordance with FAO standards (Greenfield and Southgate, 2003).

The *A. roseum* methanolic extracts were tested against a panel of spoilage and pathogenic bacteria strains, including the Gram positive *Staphylococcus aureus* ATCC 25923, and *Enterococcus fecalis* ATCC 29212, and the Gram negative *Escherichia coli* ATCC 25922, as well as the yeast *Candida albicans* ATCC 90028, *Candida glabrata* ATCC 90030, *Candida kreusei* ATCC 6258, *Candida parapsilosis* ATCC 22019. The cultures were incubated at 37°C (27°C for the yeast) for 18 h and then diluted in Nutrient Broth to obtain 10^6 CFU/mL.

The recommended methods of the Association of Official Analytical Chemists (AOAC, 1995) were adopted to determine the level of crude protein, water content, ash, carbohydrates, fibres and lipids. Nitrogen content was determined using the Kjeldahl method (AOAC, 1995) and multiplied by a factor of 6.25 to determine the total protein content. Water content was estimated by drying the sample to a constant weight at 70±2 °C (AOAC, 2002). Ash was determined by the incineration of 1.0 g sample in a muffle furnace, at 550 °C for 6 h (Alfawz, 2006). Fibres content was quantified using 2 g sample previously boiled with diluted H_2SO_4 (0.3) using Wende method (AOAC, 1995). Soluble carbohydrates were determined by the phenol-sulphuric acid colorimetric method (AOAC, 1995). Total carbohydrates were calculated by difference as follows: Carbohydrates (%) = 100 - [Proteins (%) + Lipids (%) + Ash (%) + Fibres (%)], according to Alfawz (2006). The ash obtained underwent an acidic hydrolysis and the minerals (Ca, Na, K, Mg, Fe, Zn, Cu, Ni, Mn, Pb, Cd and Cr) were determined separatively, using an atomic absorption spectrophotometer (Hitachi Z6100). Phosphorus content was determined using a spectrophotometer method, based on phosphoric molybdovanadate absorption at 730 nm according to Falade, Otemuyiwa, Oladipo, Oyedapo, Akinpelu and Adewusi (2005).

Fatty acids composition was analysed by gas chromatography. The fatty acids were previously methylated to esters using a born trifluoride methanol complex (14% w/v). The mixture was held one hour at 100 °C.

For the bioactive compound, the total phenolic content of *A. roseum* methanolic extract was determined using Folin-Ciocalteu reagent (Fattouch et al., 2007). Total flavonoids were measured by a colorimetric assay according to Galvez, Martin-Cordero, Houghton and Ayuso (2005). The vitamin C content was analyzed with 2, 6-dichloroindophenol titrimetric method. Total anthocyanidins were determined using the Reay, Fletcher and Thomas (1998) method. Colorimetric quantification of total carotenoids was determined, as described by Mackinney (1941). Allicin determination was based on Miron et al. (2002) method.

Various concentrations (1 µg/ml up to 10 mg/ml) of *A. roseum* extracts were used to determine the antimicrobial and antifungal activity. Minimum inhibitory concentration (MIC) values were determined by a micro-titre plate dilution method.

The assessment of radical scavenging activity was determined using ABTS (2, 2'-Azino-(bis-3-ethylbenzthiazoline-6-sulfonic acid) di-ammonium salt) radical scavenging activity of the methanolic extracts was determined according to Re, Pellegrini, Proreggente, Pannala, Yang and Rice-Evans (1999). The results were expressed in terms of Trolox equivalent antioxidant capacity (TEAC). The antioxidant activity of the extracts was also evaluated using the DPPH (2, 2-diphenyl-2-picrylhydrazyl) free radical specrophotometrically according to Fattouch et al. (2007).

All analysis mentioned were effected with quality control, than a proper sampling plan was followed with representative samples from the geographic area studied and sufficient replications of the sample were used to ensure statistically reliable and valid data. The analyses of the nutrient contents samples were made in our laboratory where ISO/CEI 17025 (2005) was respected to assure the quality of results.

2. Results and discussion

2.1 Nutritional composition
The proximate chemical and nutritional composition of *A. roseum* edible part collected from Tunisia is listed in Table 1.

Components	Mean Value*
Soluble carbohydrates (g/100 g DW[a])	32.80 ± 0.21
Protein (g/100 g DW[a])	22.70 ± 1.51
Fibre (g/100 g DW[a])	12.30 ± 0.05
Ash (g/100 g DW[a])	7.20 ± 1.31
Fat (g/100 g DW[a])	3.60 ± 0.29

*Values are means ± SD, $n = 3$.
[a]: Dry weight

Table 1. Content of soluble carbohydrates, protein, fibre, ash and fat in *Allium roseum* L. expressed as g/100 g of dry weight basis. (Moisture content 81.2 ± 2.6 g/100 g fresh weight).

2.1.1 Water content and pH
Water content is important because it affects the plant's properties. Compared to related vegetables, the *A. roseum* water content is lower than that of *A. porrum* varing from 83 to 89% (Tirilly and Bourgeois, 1999) and the *A. cepa* (89%) (Dini et al., 2008). The *A. roseum* is rather

neutral (pH = 6.80 ± 0.05) compared to that of garlic (pH = 6.05) (Haciseferoğullari et al., 2005).

2.1.2 Sugars, proteins, fibres and lipids

Soluble carbohydrates represent the most abundant A. *roseum* leaves nutrients class (> 30%); as has also been observed in onion bulbs (Moreau et al., 1996) and garlic (Haciseferoğullari et al., 2005). The total carbohydrates content in this species calculated by difference is 54.2 g/100 g DW. Compared to the carbohydrates content of A. *porrum* (5 to 11%) (Tirilly and Bourgeois, 1999) and to aerial parts of other *Alliums* (5 to 12%) (Brewster, 1994), the leaves of A. *roseum* are rich sources of soluble carbohydrates. Dietary fibres are considered as unavailable carbohydrates, but nonetheless they still play a very important role in maintaining good health. Interestedly, A. *roseum* aerial part fibres content was higher than that reported for A. *cepa* bulb (1.7%), the edible part of the vegetable. Rosy garlic leaves proteins rate is relatively high compared to A. *sativum* bulbs (9.3%) and A. *cepa* (1.7%) (Haciseferoğullari et al., 2005; Dini et al., 2008). Fats accounted for 0.68% of the fresh weight of A. *roseum*, making them the least abundant class of nutrients. Yet this was higher than typical values of <0.5% fresh weight basis for most plant tissues, and also compared to most *Allium* plants. Where onion, leek and garlic contain 0.15%, 0.25% and 0.42%, respectively, as reported by Haciseferoğullari et al., (2005) and Tsiaganis et al., (2006).

2.1.3 Minerals

A. *roseum* is characterized by high ash content (Table 1) including macro and micro elements (Table 2). *Allium* ash content ranges from 0.6 and 1.0% and higher values are associated to high dry matter content (Brewster, 1994). The mineral element composition of A. *roseum* exhibited a higher concentration of potassium than calcium and magnesium (Table 2). Minerals are important as constituents of bones, teeth, soft tissues, haemoglobin, muscle, blood and nerve cells and are vital to overall mental and physical well being (Jouanny, 1988). The high content of potassium in A. *roseum* is nutritionally significant in since it contributes to the control of hypertension which results in excessive excretion of potassium (Dini et al., 2008). Calcium is found at relatively high concentration in A. *roseum* (Table 2). Onion leaf calcium concentration (2540 mg/100 g fresh weight) (Boukari et al., 2001) is much higher than that of A. *roseum* leaves but bulb calcium concentration (45 mg/100 g fresh weight) (Adrian et al., 1995) is much lower. Therefore, calcium concentration in A. *roseum* is between that of onion leaves and bulbs. A. *roseum* can be considered as a source of calcium for human nutrition. This is important since calcium mineral deficiency is a world-wide problem; particularly in developing countries where the daily average intake is very low, ranging between 300 and 500 mg for adults (Boukari et al., 2001). The low sodium content of A. *roseum* and consequently low Na/K ratio (0.03) is another indication that A. *roseum* consumption would reduce the incidence of hypertension (Iqbal et al., 2006). A. *roseum* leaves also contains several oligo-elements including iron, zinc, copper and manganese. These values are similar to, but higher than, those of Haciseferoğullari et al., (2005) in A. *sativum*. The iron content of 'rosy garlic' was somewhat higher than that of A. *cepa* (8.1 mg/100 g) (Moreau et al., 1996). The magnesium, iron and phosphorous levels are adequate. Cadmium, lead and chromium were below the detection limit, as observed by Moreau et al., (1996) for onion.

Component	Concentration*
Major elements	
Potassium	1530.500 ± 0.036
Calcium	712.500 ± 0.048
Magnesium	101.900 ± 0.007
Sodium	46.500 ± 0.003
Na:K ratio	0.030
Anions	
Chlorides	724.00 ± 0.01
Sulfates	437.00 ± 0.03
Phosphates	219.00 ± 6.40
Nitrates	< 0.03
Heavy metals	
Iron	10.110 ± 0.002
Manganese	2.000 ± 0.001
Zinc	1.800 ± 0.001
Copper	1.100 ± 0.006
Nickel	< 0.013
Lead	< 0.035
Cadmium	< 0.007
Chromium	< 0.006

*Values are means ± SD, $n = 3$.

Table 2. The mineral content of *Allium roseum* expressed in mg/100 g of dry weight basis

2.1.4 Fatty acid composition

The fatty acid composition of *A. roseum* leaves is given in Table 3. Chromatographic analysis revealed twelve compounds. Unsaturated fatty acids accounted for most of the fatty acids (85 %) and were represented mainly by linolenic, linoleic, oleic and gadoleic. Five saturated acids (palmitic, stearic, myristic, arachidic and margaric), accounted for ~15% of the total fatty acids. Myristoleic, palmitoleic and heptadecanoic acids were found as minor compounds. The overall fatty acid profile of *A. roseum* reveals a good source of the nutritionally essential linolenic and oleic acids (Zia-Ul-Haq et al., 2007). Linoleic and linolenic acids are the most important essential fatty acids required for growth, physiological functions and maintenance (Pugalenthi et al., 2004). While the major fatty acid in *A. roseum* was linolenic acid, linoleic acid was most abundant in onion, garlic and leek where it represents about 50% of the total (Tsiaganis et al., 2006). The same authors demonstrated that garlic oils contain relatively high levels of linoleic acid, and that myristoleic acid ($C_{14:1}$) was absent in onion. As a consequence, the *A. roseum* fatty acid composition quality is comparable to that of *A. sativum* (Tsiaganis et al., 2006). It could be concluded that fatty acid composition varies within the species. We may note that the most abundant fatty acids are similar to those found in the oil of other *Allium* species. The less abundant fatty acids are present in *A. roseum* but at a lower concentration than reported by Tirilly and Bourgeois (1999) in *A. porrum*, Moreau et al. (1996) in *A. cepa* and Tsiaganis et al. (2006) in *A. sativum*. Overall, the *A. roseum* fatty acid composition was not qualitatively different from that of the other species.

Fatty acid	Percentage*
Myristic ($C_{14:0}$)	0.78 ± 0.11
Myristoleic ($C_{14:1}$)	0.12 ± 0.08
Palmitic ($C_{16:0}$)	12.82 ± 0.33
Palmitoleic ($C_{16:1}$)	0.50 ± 0.18
Margaric ($C_{17:0}$)	0.16 ± 0.13
Heptadecanoic ($C_{17:1}$)	0.15 ± 0.12
Stearic ($C_{18:0}$)	0.98 ± 0.23
Oleic ($C_{18:1}$)	2.87 ± 0.70
Linoleic ($C_{18:2}$)	25.68 ± 0.44
Linolenic ($C_{18:3}$)	52.68 ± 0.41
Arachidic ($C_{20:0}$)	0.19 ± 0.17
Gadoleic ($C_{20:1}$)	2.55 ± 0.32
Total	
Saturated	14.93 ± 0.11
Monounsaturated	6.19 ± 0.11
Polyunsaturated	78.37 ± 0.11

*Values are means ± SD, n = 3.

Table 3. Fatty acid composition of *Allium roseum*

2.2 Bioactive compounds
The content of potential health-promoting substances, flavonoid, total phenolic content, vitamin C, tannin, anthocyanidin, carotenoids and allicin in the wild *A. roseum* growing in the arid region of Tunisia is listed in Table 4.

2.2.1 Phenolic compounds, flavonoids, anthocyanidins, carotenoids and vitamin C contents
Total phenolic content of *A. roseum* expressed in equivalent catechol was higher than that reported for garlic (61.8 mg/100 g FW) and onion (31.0 mg/100 g FW) (Kaur and Kapoor, 2002). Although, shallots had the highest total phenolic content (114.7 mg/100 g) among the bulb onion varieties tested by Lanzotti (2006). However, this content is lower than that of rosy garlic leaves. Significant correlations were observed between the total phenolic content of *A. roseum*, and antioxidant activity, suggesting that phenolic compounds would be the major contributors to the antioxidant capacity of *A. roseum* (Najjaa et al., 2011a). Moreover, tannins and flavonoids were detected by several other authors and are usually less abundant in several other species of *Allium* (*A. cepa, A. ascalonicum, A. sativum*) (Bozin et al., 2008; Zielinskaa et al., 2008). Flavonoids are important secondary plant metabolites. The flavonoids content of rosy garlic is seven time that of garlic (0.5 mg/100 g) (Miron et al., 2002). *Allium* species are among the richest sources of dietary flavonoids and contribute significantly to the overall intake of flavonoids (Slimestad et al., 2007). *In vitro* and *in vivo* pharmacological tests have shown that flavonoids exhibit the following variety of actions: (i) antioxidative (Boyle et al., 2000); (ii) reduction of cardiovascular disease (Janssen et al., 1998) and (iii) reduction of carcinogenic activity (Steiner, 1997).

The high *A. roseum* vitamin C content (1523.35 mg/100 g DW) may be an important reason that it has been reputedly used as a traditional Tunisian medicine for treating rheumatism and cold. Furthermore, its high vitamin C content confers considerable nutritional value. *A. roseum* leaves had high anthocyanidin content (1239.62 µg/100 g DW). Much is known about the anthocyanins of *A. cepa* bulbs, and leaves of *A. victorialis* and *A. schoenoprasum* (Terahara et al., 1994; Fossena et al., 2000; Slimestad et al., 2007). Moreover, *A. roseum* had a typical carotenoids content (Table 4) of leafy vegetables, which is higher than those of legumes and fruits (Combris et al., 2007).

Substances	Mean value*
Phenolic compounds (mg CA/100g DW [ab])	736.65 ± 1.51
Flavonoids (mg CE/g DW [ac])	3.37 ± 0.32
Anthocyanidin (µg CE/100 g DW [ac])	1239.62 ± 6.79
Vitamin C (mg/100 g DW [a])	1523.35 ± 74.72
Total Carotenoids (µg/100 g DW[a])	242.25 ± 48.84
Allicin (mg / 100g DW [a])	657.00 ± 0.49

*Values are means ± standard deviations of triplicate determination (Mean ± SD (n = 3)).
[a] DW = dry weight
[b] Total phenolic contents expressed as as mg catechol (CA) equivalents per gram of dry weight
[c] Total flavonoid and anthocyanidin content were expressed as mg catechin (CE) /100 g dry weight

Table 4. *Allium roseum* L. var. *odoratissimum* bioactive substances content.

2.2.2 Allicin content
Garlic antibacterial bioactive principal was identified as diallylthiosulphinate and was given allicin as trivial name since 1944. This bioactive substance is also detected in *A. roseum* with a concentration equivalent to 0.0328 µg/mL. This result is similar to that mentioned by Miron et al. (2002) in garlic (0.0308 µg/mL). Allicin (diallylthiosulfinate) is the most abundant organosulfurous compound, representing about 70% of the overall thiosulfinates formed upon garlic cloves crushing (Miron et al., 2002).

2.3 Antioxidant activity
The antioxidant activities of leaf extracts were assessed and confirmed using two functional analytical methods based on the radicals (ABTS and DPPH) scavenging potential, as recommended by Sànchez-Alonso et al., (2007). A good correlation was found between DPPH and ABTS methods (R^2=0.827), indicating that these two methods gave consistent results. The extracts obtained were all able to inhibit the DPPH, as well as ABTS radicals (Table 5). The antioxidant potential was 378.89 mg Trolox/100g DW with the DPPH method, and 399.99 mg Trolox/100g DW with the ABTS. In comparison to previous data based on the ABTS scavenging capacity, *A. roseum* leaf extracts were comparable or higher than other investigated species known to be rich in antioxidants including strawberry (25.9), raspberry (18.5), red cabbage (13.8), broccoli (6.5), and spinach (7.6) (Proteggente et al., 2002). Significant correlations were observed between the TPC of *A. roseum*, and antioxidant activity (R^2=0.828 for TPC vs. DPPH and R^2=0.925 for TPC vs. ABTS), suggesting that polyphenolic compounds are the major contributors to the antioxidant capacity of *A. roseum*.

Regarding the favourable redox potentials and relative stability of their phenoxyl radical, these biomolecules are considered to be human health promoting antioxidants (Acuna et al., 2002).

Extracts	DPPH (mg Trolox /100g DW)	ABTS (mg/100g DW)
Methanol (75%)	378.80±5.55	399.90± 4.59

Table 5. Free radical scavenging activity of A. roseum

2.4 Antibacterial activity

The *in vitro* antibacterial effects of the A. roseum extracts obtained with the methanolic extract values are presented in Table 6. The results showed that A. roseum extracts have great potential as antimicrobial agent against the tested bacteria. C. albicans and C. glabrata, were the most sensitive tested organisms to the extract with the MIC values were 0.63 and 2.5 mg/ml, respectively.

The strong antifungal activity was observed against C. albicans and C. glabrata may be related to the high level of polyphenols content. Cai et al. (2000) showed that several classes of polyphenols such as phenolic acids, flavonoids and tannins serve as plant defence mechanism against pathogenic microorganisms. In fact, the site and the number of hydroxyl groups on the phenol components increased the toxicity against the microorganisms.

Strains	MIC (mg/ml)
Escherichia Coli ATCC 25922	10±1.20
Enterococcus Faecalis ATCC29212	10±0.57
Staphylococcus aureus ATCC 25923	10±0.60
Candida albicans ATCC 90028	0,63±1.85
Candida glabrata ATCC 90030	2,5±1.20
Candida kreusei ATCC 6258	10±2.13
Candida parapsilosis ATCC 22019	10±1.41

MIC, Minimum Inhibitory Concentrations as (mg ml^{-1}).

Table 6. Minimal inhibitory concentrations of extracts of A. roseum on bacterial growth

3. Conclusion

This study revealed that A. *roseum* var. *odoratissimum* growing in Tunisia had a high soluble carbohydrates, crude protein and dietary fibre contents, compared to other *Alliums*. Its mineral content was high in potassium, and calcium. The mineral composition of 'rosy garlic' is sufficient in Ca, P, K, Cu, Fe, Zn and Mg so that it can meet many macronutrient and micronutrient requirements of the human diets. As a consequence, a diet based on A. *roseum* would help in preventing deficiencies in potassium, calcium, iron and magnesium. Furthermore, edible part oil included 15% saturated and 85% unsaturated fatty acids. Linolenic acid and palmitic acid were the most abundant unsaturated and saturated fatty acids, respectively. This fatty composition confers to the A. *roseum* oil considerable nutritional value, acting on physiological functions and reducing cardiovascular, cancer and arthroscleroses diseases occurrence risk. The most abundant phytonutrients found in A.

roseum (polyphenolic compounds, flavonoids, anthyacinidins, vitamin C and allicin) exhibit a positive effect on human health as antioxidants and antibacterial compounds. Since the chemical composition of *A. roseum* has not been reported before, this report provides a starting point for comparison to the other *Allium* genus vegetables and it confirms the potentially important positive nutritional value that *A. roseum* can have on human health. Since *A. roseum* is a rich source of many important nutrients and bioactive compounds responsible for many promising health beneficial physiological effects, it may be considered a nutraceutical that serves as a natural source of necessary components to help fulfil our daily nutritional needs and as a functional food as well as in ethnomedecine .

4. References

Acuna, U.M., Atha, D.E., Nee, M.H., & Kennelly, E. J. (2002). Antioxidant capacities often edible North American plants. *Phytotherapy Research, 16,* 63-65.

Adrian, J., Potus, J., & Frangne, R. (1995). La science alimentaire de a à z. Paris: Lavoisier, 2ème Edition.

Alfawz, M.A. (2006). Chemical composition of hummayd (*Rumex visicarius*) grown in Saudi Arabia. *Journal of Food Composition and Analysis, 19,* 552-555.

AOAC. (1995). Official Methods of Analysis, 16th Edition. AOAC International. (Chapter 12, Tecn. 960.52).Washington: Cuniff, P. (Ed.), p. 7

AOAC. (2002). Official methods of analysis of AOAC International. 17th edition. Gaithersburg , MD, USA, Association of Analytical Communities.

Besbes, S., Blecker, C., Deroanne, C., Drira, N.E., & Attia, H. (2004). Date seeds: chemical composition and characteristic profiles of the lipid fraction. *Food Chemistry, 84,* 577-584.

Boukari, I., Shier, N.W., Xinia, E.F.R., Frisch, J., Bruce, A.W., Pawloski, L., & Fly, A.D. (2001). Calcium analysis of selected Western African foods. *Journal of Food Composition and Analysis, 14,* 37-42.

Boyle, S.P., Dobson, V.L., Duthie, S.J., Kyle, J.A.M., & Collins, A.R. (2000). Absorption and DNA protective effects of flavonoid glycosides from onion meal. *European Journal of Nutrition, 39,* 213-223.

Bozin, B., Mimica-Dukic, N., Samojlik, I., Goran, A., & Igic, R. (2008). Phenolics as antioxidants in garlic (*Allium sativum* L. *Alliaceae*). *Food Chemistry, 111,* 925-929.

Brewster, J.L. (1994). Onions and other vegetable *Alliums*. Wallingford: CAB International, UK.

Cai, Y., Luo, Q., Sun, M.,& Carke, H. (2002). Antioxidant activity and phenolic compounds of 112 food and plants. *Journal of chromatography A, 975,* 71-93.

Combris, P., Amiot-Carlin, M., & Cavaillet, F. (2007). Les fruits et légumes dans l'alimentation. Enjeux et déterminants de la consommation. Rennes, INRR : Expertise scientifique collective Inra.

Cuénod, A. (1954). Flore analytique et synoptique de la Tunisie. Cryptogames vasculaires, Gymnospermes et Monocotylédones, Tunis: S.E.F.A.N.

Dini, I. Carlo Tenore, G., & Dini, A. (2008). Chemical composition, nutritional value and antioxidant properties of *Allium caepa* L. Var. tropeana (red onion) seeds. *Food Chemistry, 107,* 613-621.

Doner, G., & Ege, A. (2004). Evaluation of digestion procedures for the determination of iron and zinc in biscuits by flame atomic absorption spectrometry. *Analytica Chimica Acta, 520*, 217-222.

Falade, O.S., Otemuyiwa, I.O., Oladipo, A., Oyedapo, OO.., Akinpelu, B.A., & Adewusi, S.R.A. (2005). The chemical composition and membrane stability activity of some herbs used in local therapy for anemia. *Journal of Ethnopharmacology, 102*, 15-22.

Fattouch, S., Caboni, P., Coroneo, V., Tuberoso, C.I.G., Angioni, A., Dessi, S, Marzouki, N., & Cabras, P. (2007). Antimicrobial activity of Tunisian Quince. (Cydonia oblonga miller) Pulp and peel polyphenolic extracts. *Journal of Agriculture and Food Chemistry, 55*, 963-969.

Fossen, T., Slimestad, R., Overstedal D.O., & Andersena, M. (2000). Covalent anthocyanin-favonol complexes from flowers of chive, *Allium schoenoprasum. Phytochemistry, 54*, 317-323.

Gálvez, C.J., Martin-Cordero, C., Houghton, A.J., &Ayuso, M.J. (2005). Antioxidant Activity of methanol extracts obtained from *Plantago* species. *Journal of Agricultural and Food Chemistry, 53*, 1927–1933.

Graham, H.D., &Graham, E.J.F. (1987). Inhibiton of *Aspergillus parasiticus* growth and toxin production by garlic. *Journal of Food Safety, 8*, 101-108.

Greenfield, H., & Southgate, D.A.T, (2003). Food composition data: production, management and use. Rome: Part 1. 2nd Edition. FAO.

Haciseferoğullari, H., Özcan, M., Demir, F., & Çalişir, S. (2005). Some nutritional and technological properties of garlic (*Allium sativum* L.). *Journal of Food Engineering, 68*, 463-469.

International Standardization Organization (ISO). (2005). Exigences générales concernant la compétence des laboratoires d'étalonnages et d'essais (ISO/CEI 17025), ISO, Genève, 29 p.

Iqbal, A., Khalil, I.A., Ateeq, N., & Khan, M.S. (2006). Nutritional quality of important food legumes. *Food Chemistry, 97*, 331-335.

Janssen, K., Mensink, R.P., Cox, F.J., Harryvan, J.L., Hovenier, R., Hollman, P.C., & Katan, M.B. (1998). Effects of the flavonoids quercetin and apigenin on hemostasis in healthy volunteers: results from an *in vitro* and a dietary supplemented study. *American Journal of Clinical Nutrition, 67*, 255-262.

Jendoubi, R., Neffati, M., Henchi, B., &Yobi, A. (2001). Système de reproduction et variabilité morphologique chez *Allium roseum* L. *Plant Genetic Research Newsletter, 127*, 29-34.

Jouanny, J.(1988). Notions essentielles de matières médicales homéopathiques. France : Bouen.

Kaur, C., & Kapoor, H.C. (2002). Antioxidant activity and total phenolic content of some Asian vegetables. *International Journal of Food Science and Technology, 37*, 153-161.

Lanzotti, V. (2006). The analysis of onion and garlic. *Journal of Chromatography A, 1112*, 3-22.

Le Floc'h, E. (1983). Contribution à une étude ethnobotanique de la flore tunisienne. Programme flore et végétation tunisienne. Tunis : Ministère de l'Enseignement Supérieur et de la Recherche Scientifique.

Mackinney, G. (1941). Absorption of light by chlorophyll solution. *Journal of Biological Chemistry, 140*, 315–322.

Messiaen, C.M., Cohat, J., Pichon, M., Leroux, J.P., &Beyries, A., (1993). Les *Allium* alimentaires reproduits par voie végétative, INRA, Paris, pp. 150-225.

Miron, T., Shin, I., Feigenblat, G., Weiner, L., Mirelman, D., Wilchek, M., & Rbinkov, A. (2002). A spectrophotometric assay far allicin, alliin, and alliinase (alliin lyase) with a chromogenic thiol: reaction of 4–mercaptopyridine with thiosulfinate. *Analytical Biochemistry, 307,* 76-83.

Moreau, B., Le Bohec, J., & Guerber–Cahuzac, B. (1996). L'oignon de garde. Paris: Lavoisier,.

Najjaa, H., Neffati, M., Zouari, S., & Ammar, E. (2007). Essential oil composition and antibacterial activity of different extracts of a North African endemic species. *Comptes Rendus de Chimie, 10,* 820-826.

Najjaa, H., Zerria, K., Fattouch, S., Ammar, E., & Neffati, M. (2009). Antioxidant and antimicrobial activities of *Allium roseum* "Lazoul", a wild edible endemic species in North Africa. *International Journal of Food Properties* (LJFP-2009-0049.R2) (In press).

Najjaa, H., Zouari, S., Ammar, E., & Neffati, M. (2009b). Phytochemical screening and antibacterial properties of *Allium roseum* L. a wild edible species in North Africa. *Journal of Food Biochemistry,* 35: 699–714.

Proteggente, A.R., Pannala, A.S, Pagana, G., Van Buren, L., Wagner, E., & Wiseman, S. (2002). The antioxidant activity of regularly consumed fruit and vegetables reflects their phenolic and vitamin C composition. *Free Radical research, 36,* 217-233.

Pugalenthi, M., Vadivel, V., Gurumoorthi, P., & Janardhanan, K. (2004). Comparative nutritional evaluation of little known legumes, *Tamarindus indica, Erythrina indica* and *Sesbania bispinosa. Tropical and Subtropical Agroecosystems, 4,* 107–123.

Re, R. Pellegrini, N., Proreggente, A. , Pannala, A. , Yang, A. , & Rice-Evans, C. (1999). Antioxidant activity applying an improved ABTS radical cation decolorizing assay. *Free radicals in biology and medicine, 26,* 1231-1237.

Reay, P.F., Fletcher, R.H., &Thomas, V.J.G. (1998). Chlorophylls, carotenoids and anthocyanin concentrations in the skin of "Gala" apples during maturation and the influence of foliar applications of nitrogen and magnesium. *Journal of the Science of Food and Agricultural, 76,* 63–71.

Sànchez-Alonso, I., Jimenez-Escrig, A., Saura-Calixto, F., & Borderias, A. J. (2007). Effect of grape antioxidant dietary fibre on the prevention of lipid oxidation in minced fish: evaluation by different methodologies. *Food Chemistry, 101,* 372-378.

Slimestad, R., Fossen, T., &Vagen, I.M. (2007). Onions: a source of unique dietary flavonoids. *Journal of Agricultural and Food Chemistry, 55,* 10067–10080.

Steiner, M. (1997). The role of flavonoids and garlic in cancer prevention. In H. Ohigashi, *Food Factors for Cancer Prevention,* (pp. 222-225). New York: Springer.

Terahara, N., Yamaguchi, M., & Honda, T. (1994). Malonylated anthocyanins from bulbs of red onions, *Allium cepa* L. *Bioscience Biotechnology Biochemistry, 58,* 1324–1325.

Tirilly, Y., Bourgeois, C.M. (1999). Technologie des légumes. Paris : Lavoisier.

Trémolières, A. (1998). Les lipides végétaux : voies de biosynthèse des glycérolipides. Paris : Université De Boeck.

Tsiaganis, M.C., Laskari, K., & Melissari, E. (2006). Fatty acid composition of *Allium* species lipids. *Journal of Food Composition and Analysis, 19,* 620-627.

Zia-Ul-Haq, M., Iqbal, S., Ahmad, S., Imran, M., Niaz, A., & Bhanger, M.I. (2007). Nutritional and compositional study of Desi chickpea (*Cicer arietinum* L.) cultivars grown in Punjab. *Pakistan. Food Chemistry, 105,* 1357–1363.

Zielinskaa, D., Nagels, L., & Piskuła, M.K. (2008). Determination of quercetin and its glucosides in onion by electrochemical methods. *Analytica Chimica Acta, 617,* 22–31.

Zou, ZM., Yu, D.Q., & Cong, P.Z. (1999). Research progress in the chemical constituents and pharmacological actions of *Allium* species. *Acta Pharmacologica Sinica, 34,* 395-400.

8

Wine as Food and Medicine

Heidi Riedel, Nay Min Min Thaw Saw, Divine N. Akumo,
Onur Kütük and Iryna Smetanska
Technical University Berlin, Department of Food Technology and Food Chemistry,
Methods of Food Biotechnology
Germany

"Wine is the most civilized thing in the world." - Ernest Hemingway

1. Introduction

The name "Wine" is derived from the Latin word vinum, "wine" or "(grape) vine". Wine is the earliest domesticated fruit crops and is defined as an alcoholic beverage which is produced by the fermentation of grape juice. Grapes are small berries with a semi-translucent flesh, whitish bloom and a smooth skin, whereby some berries contain edible seeds and others are seedless. Grape berries have a natural chemical balance which allows a completely fermentation without the addition of sugar, acid, enzymes or other nutrients. It is a rich source of vitamins, many essential amino acids, minerals, fatty acid and others. Grapevine, botanically called *Vitis vinifera*, has a wide range of different species whereby wine is a mixture of one or more varieties (Bouquet et al. 2006). Pinot Noir, Chardonnay, or Merlot for example are predominated by grapes with a minimum of 75 or 85% grape by law and the result is a varietal as opposed to a blended wine. Nevertheless, blended wines are not of minor value compared to varietal wines. Wines from the Bordeaux, Rioja or Tuscany regions are one of the most valuable and expensive wines which are a mixture of many different grape varieties of the same vintage. Wines of high quality are not permitted to be labeled as varietal names because of the 'cépage' (grape mix) which is restricted by law. Red Bordeaux wines are a composition of four different grapes including, but not exclusively, Cabernet Sauvignon and Merlot. Red and white Burgundy are made from a single grape variety and they use their regional label because of marketing strategies and historical reasons. To name some few native North American grapes like *Vitis labrusca, Vitis aestivalis, Vitis rupestris, Vitis rotundifolia* and *Vitis riparia* which are usually used for eating as fruit or made into grape juice, jam, or jelly sometimes into wine for example Concord wine (*Vitis labrusca species*). The most common vineyards worldwide are planted with the European vinifera vines that have been grafted with native species of North America. This is because grape species from North America are resistant against phylloxera (Granett et al. 2001). The theory of "terroir" is defined by the variety of the grape, orientation and topography of the vineyard, elevation, type and chemistry of soil, the climate and seasonal conditions under which grapes are grown. Among wine products, there is a high variety which is due to the fermentation and aging processes. Many winemakers with small production volume prefer growing and using production methods that preserve the unique sensory properties like aroma and the taste of their terroir.

The main challenge of the producers is to reduce differences in sources of grapes by using wine making technology such as micro-oxygenation, tannin filtration, cross-flow filtration, thin film evaporation, and spinning cone. Grapevine is one of the major fruits in the world which is grown in temperate regions on the northern and southern hemisphere mostly between the 30th and 50th parallel. Grapevine is cultivated in large fields because of their economic value (Bouquet et al. 2006). The most popular wine regions of the world are France, Italy, Northern California, Germany, Australia, South Africa, Chile and Portugal respectively. The genus grape (*Vitis vinifera*) contains around 60 species which are common in the temperate zones with some species growing in the tropical region. In 2009 the production of grape was about 69 million tons (especially for wine, juice and raisins) compared to data from 1995 with only 55 million tons (Data from the Organisation Internationale de la Vigne et du Vin (OIV)). Grapes need a minimum of 1500 hours of sunshine to ripen fully, red wine needs more radiation than white.

1.1 Some historical facts of wine

Wine and grape are one of the oldest fruits and the traditional winemaking processes are an ancient art, which began as early as 1,000 B.C. Archaeological investigations and discoveries attest that the wine production by fermenting processes, took place from early as 6000 BC. Other studies from China show that grapes were used together with rice to produce fermented juices as early as 7000 BC. Some research studies document the origin home countries of wine to the Balkan Range along the coast of the Black Sea. Wine is mentioned in historical literature documents as Iliad and Odyssey by Homer. In Greco-Roman mythology, Dionysius is adored as the god of wine. He is also known as Bacchus whereby Dionysius is regarded as the patron of vine events. One of the most important grape wine producers in Europe is Turkey as well as other neighboring countries around the Mediterranean Sea. Especially Anatolia was described as the origin place of viticulture and wine making. One of the first traces of the cultivation of grape wine was in the Early Bronze Age around the Mediterranean basin (Gorny 1996). Many archaeological investigations prove the early domestication of wine in the East (This et al. 2006). During the Bronze Age in the Mediterranean were olive, fig and grape the most common fruits. Scientists discovered many evidences like grape pips in a shrine, wine shop with jars and cups from the Bronze Age (Refai 2002). In Europe around the Mediterranean area between Black Sea and Caspic Sea grape was cultivated and used for winemaking 4000 BC (Monti 1999). The wild grapevine specie Vitis vinifera ssp. Sylvestris Gmelin was grown from Portugal to Turkmenistan and the north of Tunesia. About 8400 years seeds of the oldest wild grape in Turkey were discovered in a valley near Urfa (This et al. 2006). Specific investigations of the chlorotype showed a higher diversity of the wild grape population in the central and eastern parts than in the western areas of the Mediterranean (Arroyo-Garcia et al. 2006). All domesticated grapevine species originated from the wild type Vitis vinifera ssp. sylvestris whereby Vitis vinifera L. is the only native species of Eurasia and appeared 65 million years ago. To enhance the yield of grapevine hermaphrodite genotypes were selected for domestication procedures with intensive pigmentation and techniques for propagation (Terral et al.).Nowadays the skills of the winemaking process are considered for intellectual persons. Special famous events only about wine are exhibits, expos, and auctions worldwide. The Boston Wine Expo is one major annual convention where top wine producer exhibit, sell their goods and show new technologies. Such expos serve as venue for the world's top producers to exhibit and sell their good. Persons with high interests as well as wine collectors attend such exhibitions to exchange ideas and share their passions for wine. Wine is associated with education, lifestyle and class and that is the reason why wine is always included in special occasions. What did early wine producers start out with, and how did they change grapevines in the course of domesticating them? How does the evolutionary history of grapevines affect grape growers today?

2. Classification of wine

The naming of wines has a long tradition and is based on their grape variety or by their place of production. European wines are labeled after their place of production like Bordeaux, Rioja and Chianti as well as the type of grapes used such as Pinot, Chardonnay and Merlot. All other wines from all over the world are generally named for the grape variety. Non- European wine labels become more and more famous and the market recognition get more stability. Some examples include Napa Valley, Barossa Valley, Willamette Valley, Cafayate, Marlborough, and Walla Walla just to name a few. Sensory properties like the taste of a wine depends on the grape species and the blend, and furthermore on the ground and climatic conditions (terroir).

2.1 Red wine
The color of wine is caused by the presence or absence of the grape skin during the fermentation process. Grapes with colored juice like Alicante Bouchet became popular as colorants so called "teinturier". The basic natural products of red wine are red or black grapes, but the intensive red color originates from maceration, which is a process whereby the skin is left in contact with the juice during fermentation. In the following table 1 are listed some red wine varieties, their country origin and characteristics.

Wine variety	Country of origin	Characteristics
Aleatico	Italy	Dark skinned grape, fragrant, very rare
Alicante Bouschet	France	Red skinned grape
Cabernet Sauvignon	France (Bordeaux)	principal component of Bordeaux reds
Concord	America	Most important variety in US, belongs to Vitis labrusca
Dolcetto	Italy	Wine is soft and fruity
Pinotage	South Africa	Wild grown hybrid variety

Table 1. Varieties of Red Wine

Dependent on the grape specie, climatic conditions during the ripening process and many other external factors can influence the sugar and alcohol content of the wine. In the following table 2 is shown the nutritional value of red table wine.

Energy 80 Kcal, 360 KJ	
Carbohydrates	26 g
Sugar	0.6 g
Fat	0.0 g
Protein	0.1 g
Alcohol	10.6 g

Table 2. Nutritional value of red table wine per 100 g

2.2 White wine

White wine can be produced from any color of grape as the skin is separated from the juice during fermentation. The following table 3 shows some examples white wine, their country origin and characteristics.

Wine variety	Country of origin	Characteristics
Chardonnay	France	Widely grown throughout the world.
Frontignac	Greece	highly fragrant
Gewürztraminer	France (Alsace)	highly fragrant and spicy
Muscadelle	France (Bordeaux)	An aromatic variety.
Picolit	Italy (Friuli region)	Used to make sweet white wine.
Riesling	Germany	noble variety producing some of the world's greatest wines.
Sauvignon Blanc	France (Bordeaux)	A highly aromatic variety.

Table 3. Varieties of White Wine

2.3 Rosé wine

Rosé wine is produced from different very dark red grape-varieties whereby it is not a blending of red and white wine. In recent times many wine dressers mix a special amount of white wine with red wine.

2.4 Sparkling wines

Sparkling wines contains carbon dioxide which is naturally made due to the fermentation process; champagne for example. To achieve this sparkling effect, the wine has to ferment two times. The first time in an uncovered container that carbon dioxide can escape into the environment. In a second step the wine is in a sealed fermentation container so that the gas remains in the wine. In the following table 4 are listed some famous sparkling wines from different countries.

Wine variety	Country of origin	Characteristics
Chardonnay	France	Basic component of champagne
Macabeo	Middle East	Basic component of the Spanish sparkling wine Cava.
Muscat Blanc À Petits Grains	Greece	Used for Italian sparkling wines known as Asti.
Prosecco	Italy (Veneto)	Used for Prosecco, an Italian sparkling wine.

Table 4. Varieties of Sparkling Wines

2.5 Table wine

The characteristic of table wines is that the alcohol content is not higher than 14% in the U.S. whereas in Europe, the alcohol range of light wine must be between 8.5% and 14% by volume. Depending on the color of the wine, table wines are classified as "white", "red" or "rosé".

2.6 Dessert wine

The sugar range in dessert wines can be from slightly sweet (less than 50 g/L sugar) to very sweet wines (more than 400 g/L sugar). For example wines such as Spätlese are produced from grapes harvested after they reached the maximum ripeness. Dried grape wines like Recioto and Vin Santo are made from partially raisined grapes after harvesting. Botrytized wines are produced from grapes infected with Botrytis cinerea; some examples include Sauternes from Bordeaux, Bonnezeaux and Quarts de Chaume, Tokaji Aszú from Hungary, and Beerenauslese from Germany and Austria.

2.7 Fortified wine

Fortified wines are sweet with high alcoholic content because their fermentation process stopped by the addition of spirit like brandy. To popular fortified wines belong Port, Madeira, Tokay and Banyuls.

3. Social and cultural aspects of wine

Wine has a long tradition as cultural beverage and is a popular social gathering since ancient times. Wine was a favorite drink among Roman emperors, Greek scholars, monks living in monasteries and other civilizations. Monks and royalty preferred to drink wine, while beer was only used from the workers. Egyptians investigated the wine regardingly in that quality and developed the first arbors and pruning methods. One path of wine history could follow the developments and science of grape growing and wine production; another might trace the spread of wine commerce through civilization, but there would be many crossovers and detours between them. However the timeline is followed, clearly wine and history have greatly influenced one another. Fossil vines, 60-million-years-old, are the earliest scientific evidence of grapes. The earliest written account of viniculture is in the Old Testament of the Bible which tells us that Noah planted a vineyard and made wine. As cultivated fermentable crops, honey and grain are older than grapes, although neither mead nor beer has had anywhere near the social impact of wine over recorded time. This unique alcoholic drink is enjoyed by people from all walks of life up until contemporary times. The social background of wine includes gatherings, parties, religious rites, special occasions, and even casual events. Wine experts believe that wine is more than a product, it is a culture. It is not just a commodity; it is a collector's item. The main reason why wine is strict regarded to social tools is because of historical distingue purpose. Wine has special characteristics and qualities that make it a favorite among works of art, poetries, and other literary pieces. Winemaking and oenophilists investigate technological novelties and processes are constantly being invented to reach the perfection in wine production.

4. Grapevine in food industry

There are many different ways in which grape fruits can be used and they include; fresh, preserved, dried into raisins or crushed for juice or wine (Wellness Encyclopedia of Food and Nutrition, 1992). Grape berries are sensitive fruits and should be carefully handled during the winemaking process because once in a bottle, it will develop with time. The long period of wine process is affected by different external factors which are listed below. The optimal temperature during the ripening process of the wine should be between 12 and 15 °C as well as the humidity should be between 70 and 80%. The circulation of fresh air in the wine cellar should avert any odour from moisture, chemicals; wood fruits etc. to avoid negative side effects in the wine. Also light, vibration and noise will ruin potentially good wine because it disturbs the development process. Wine matures with time and every wine needs different periods for their developing process. The maximum storage time of wine because after reaching the peak the wine will degrade (table 5).

Wine type	Maximum storage time (years)
Dry White	1-8
Sweet White	2-8
Rosé	1-3
Young Red	0-4
Mature Red	1-20
Champagne	3-10
Sparkling Wine	1-6
Sherry & Port	1-20

Table 5. Different types of wine according to their storage time

5. Factors influencing on the phenolic synthesis and wine quality

"The sun, with all those planets revolving around it and dependent on it, can still ripen a bunch of grapes as if it had nothing else in the universe to do."
Galilée (Galileo Galilei), 1564-1642

The cultivation and winemaking process needs a long term experience and optimal conditions for the production of a high quality wine. The quality of wine depends on the maturity of berries or the so called "sugar ripeness", that is the content of soluble substances in the fruit as well as sugar/ acid balance (Kliewer 1964, Coombe 1960). Thanks to new investigations with the use of analytical chemistry like gas chromatography, volatile compounds most especially aromas which are typical for Sauvignon blanc wine can be analyzed. As earlier mentioned, grape is a rich source for phenolic substances and these compounds can be measured since the 80s by HPLC. The initial fruit chemistry measurements can be differentiated from the sensory perception of wine because of the biochemical and chemical transformation of compounds during fermentation. Because many factors influence the entire process of winemaking, thus for a good quality a long term experience is needed to go through all the steps of winery and vineyard respectively. Since thousands of years, humans have tried to manipulate the natural growth habits of grapevines to make the plant productive and economically effective for agriculture. The main focus was in breeding new varieties of grapes to maximize production, fruit quality and economic efficient (Coombe 1960, Bogs et al. 2006).

Different kind of factors such as genotype, environment and cultural practices have an influence on the phenolic biosynthesis and accumulation through the ripening process of grape berries. The quantity of phenolic compounds and also the composition has an influence on the wine quality. One characteristic of grape are the high concentrations of anthocyanin which can be used as chemical marker for the classification of red-grape varieties and wines. Furthermore the intravarietal heterogeneity can be used as characteristic which induces a very different behavior among the different clones. The research interest focuses on high productive clones with a loss of color and phenolic compounds. Rootstock genotypes are associated with water, gas exchange status and canopy growth as well as yield (Koundouras et al. 2006). That is why rootstocks have an influence on the composition of phenolic compounds and also on the time of harvesting (Koundouras et al. 2006).

Grape and the quality of wine is also influences by many different environmental factors such as topography, agro-pedology, climate), which are described as "terroir". The amount and composition of phenolics depends also on the sunlight exposure and the temperature which acts on the grape berries during the ripening process (Cohen et al. 2008). Low temperatures at night have a positive effect on the anthocyanin accumulation whereas high temperatures cause a decrease of their concentration. Otherwise the accumulation of colored pigments seems to be increased linearly with increasing sunlight exposure. Berries which are exposed ultraviolet light and temperature are related with metabolic reactions and alterations in anthocyanin composition (Joscelyne et al. 2007). Water availability is one of the most important factors which are responsible for the wine quality. Vine water status cause accumulation of phenolic compounds in grape berries with positive reactions of water deficit on berry phenolic composition (Qian et al. 2009). Environmental factors like rainfall, soil water storage capacity, as well as soil characteristics such as soil depth, structure, texture and fertility affect phenolic composition (Mateus et al. 2003). Different agricultural practices during the berry ripening process are also responsible for the synthesis of phenolic substances. Another interesting aspect are cultural aspects such as training system, row vine spacing, pruning, bunch thinning, bud and leaf removal and also the management of fertilization and water irrigation (Poni et al. 2009) with their special influence on phenolic biosynthesis and accumulation. Another interesting effect on the phenolic compounds during the ripening of grape berries are different agronomic techniques and growth conditions such as conventional, organic or biodynamic systems (Vian et al. 2006). Finally, other the vine age and pathogenesis (Amati et al. 1996).

6. Management of fruit quality

Considering the fact that grapes grow in a wide range of temperate climates on the northern and southern hemisphere, special modifications in breeding resistant and tolerant to environmental factors have been achieved to extend the margins of production. However the standard approach to grapes production is intensive agricultural management to balance between climate and site (soil conditions, topography) on one hand and vine biology and preferred fruit quality on the other hand. For high quality end products, the choice of cultivar (clone, rootstock) as well as the genetic potential of the grape plants are from high importance and can influence the quality of fruits. The yield and growth of wine depends on many environmental factors like climate, soil, water, nutrients, just to name a few. In some areas, vineyards have to be watered by irrigation which is by inter- annual changes in soil water. Nevertheless, the yield and fruit quality is inconsistent, most especially red varieties are growing under the principle "regulated deficit irrigation" which suppress the vegetative growth and influence directly the fruit quality. That means water deficit cause smaller berries, early sugar maturity and modifications in phenolic contents. Temperature, solar radiation, intensity of sunlight reaching the fruit and air movement have an influence on the metabolism in grape berries and furthermore the fruit quality. The orientation of the vineyards should be north-south because of the daily solar cycle. Recent investigations showed that the amount for vines in a loose vertical canopy and under water stress can be about 42 to 45 % of the daily total solar radiation. On the other hand fruits of east-west oriented vineyards are either shaded north-facing fruits or south-facing fruits with high sunlight intensities during the day.

7. Polyphenolics in wine

Grapevine and their wide range of phenolic compounds represent a large group of biomolecules with an important role in enology. Sensory properties of wine such as taste, astringency, bitterness as well as color are caused by many different phenolic substances accumulated in grape (Kammerer et al. 2004). High antioxidant potential of grapevine is also related with other health promoting compounds and that is why the consumption amounts of red wine is increasing (Pitsavos et al. 2005). Wine "experts" explain this phenomenon as the "French paradox" which describes the high life expectancy of French people in regard to their diets, exceptionally high in fats. Many nutritionists believe that this phenomenon is caused by the high consumption of red wine which contains high amounts of antioxidants and flavonoids (Renaud and de Lorgeril 1992). In a recent study the blood drawn of 20 probands before and after drinking wine were analyzed. After drinking it, higher levels of nitric oxide was found (nitric oxide reduce clots), as well as a reduction in platelet aggregation. Furthermore an increase of alpha-tocopherol, which is connected with the antioxidant vitamin E, and the total amount of antioxidants in blood were found to be 50% higher. The consumption of wine causes other effects like the protection of LDL cholesterol from oxidation. The negative properties of oxidized LDL are responsible for arteries damaging their walls and an increased risk of atherosclerosis (Iriti and Faoro 2006). Actual investigations found that some phenolic compounds which are accumulated in grape skins inhibit protein tyrosine kinases. These enzymes regulate cells, inhibiting production of endothelin-1 which seems to be a key component in several heart conditions (da Luz et al. 1999). Furthermore scientists got a special interest to investigate the composition of phenolic compounds in grape because of their anticarcenogenic properties (Block 1992) as well as neuroprotective effect (Ma et al.). (Monagas et al. 2005)) explains the acidy character of the phenolic function and due to the nucleophilic characteristics of the aromatic group is responsible for their reactivity.

Phenolic compounds are a large group of secondary metabolites which can be classified in various ways whereby the most common separation is in flavonoids and non-flavonoids (Table 6). There are more specific and detailed families in each group whereby the chemicals structure of the compound is responsible for properties such as color, aroma and taste (Fournand et al. 2006).

Phenolics are formed from the essential amino acid, phenylalanine within the phenylpropanoid biosynthetic pathway. More than 4,000 phenolic compounds have been identified whereby their role in plants is linked to several functions: protection from UV light, pigmentation, defense against pathogens (antifungal properties), nodule production, attraction of pollinators as well as dispersion of seed (Gould et al. 2000). Grapes accumulate a high variety of polyphenolics as describes below. Phenolic acids are classified into hydroxybenzoic and hydroxycinnamic acids whereby hydroxybenzoic acids derive from benzoic acid. Grapes pulp accumulates high amounts of gallic acid (Lu and Serrero 1999) as well as flavan-3-ols in grape (Singleton and Esau 1969). In the following table 7 are listed more phenolic acids as and their derivates (Monagas et al. 2005, Rentzsch et al. 2007).

Phenolic compounds	
Flavonoids	Non- Flavonoids
	Phenolic acids
Flavonols	• Benzoic acids
	• Cinnamic acids
Flavononols and flavones	Stilbenes
Flavanols	
• Catechins	
• Condensed Tannins	
• Procyanidins	
• Prodelphinidins	
Anthocyanins	

Table 6. Classification and composition of phenolic compounds in grape (adapted from (Kontoudakis et al.)

Benzoic acid	Hydroxycinnamic acid	Hydroxycinnamic ester
p-Hydroxybenzoic	Sinapic	Trans-feruloyltartaric acid (fertaric acid)
Protocatechuic	p-Cumaric	Trans-p-coumaroyltartaric acid (coutaric acid)
Vanillic	Caffeic	Trans-caffeoyltartaric acid (caftaric acid)
Gentisic	Ferulic	
Syringic		
Gallic		
Salicilic		

Table 7. Common phenolic acids in grape and derivates

Hydroxicinnamic acids are located and accumulated in the vacuoles of the skin and pulp as tartaric esters (Ribereau Gayon 1965). The highest amounts of principal hydroxicinnamic acids found in grapes are the caftaric, cutaric and fertaric acids in *trans* form as well as lower contents of the *cis* form (Singleton and Esau 1969). Further bioactive important compound in grape are stilbenes whereby *trans*-resveratrol is the most abundant. Stilbenes in grape can occur in oligomeric and polymeric form (Rentzsch et al. 2007). Plants synthesize stilbenes as defense reaction against infections by fungis or UV irradiation especially in leaves, roots and skin. *Vitis rotundifolia* seeds accumulated high contents of stilbene (Langcake and Pryce 1977, Adrian et al. 2000). Stilbenes play no important role in the organoleptic characteristics of wine but their importance in human health due to their antioxidative, anticarcinogenic potential and neuroprotective effects is from high interest (Nassiri-Asl and Hosseinzadeh 2009). Flavonoids are another large group of compounds synthesizes in grape and are divided in four subclasses. The class name depends on the base of the oxidation state of the pyran ring: the flavonols, the flavanonols and flavones, the flavanols and the anthocyanins (Souquet et al. 2000). Flavanols and anthocyanins have the highest concentrations in wine and play an important role of the quality of red wine. Flavonols are yellow colored pigments which are responsible for the protection against UV light and are mainly accumulated in grape skin (Mane et al. 2007) but also detected in grape pulp (Pereira et al. 2006). In table 8 are listed the most abundant flavonols in grape.

The most abundant phenolics of the group flavanonols and flavones are astilbin and engeletin. They were found in high amounts in the skin and wine of white grapes, grape pomace and in stems (Souquet et al. 2000) but also in red wine (Vitrac et al. 2000). The chemical structure of flavones is similar with flavonols. Some examples of flavones in grape are apigenin, baicalein and luteolin (Wang and Huang 2004). Grape contains also high amounts of flavanols or flavan-3-ols in seed, skin and stem (Gomez-Miguez et al. 2006, Souquet et al. 2000). Flavanols were found in the monomeric, oligomeric and polymeric form and are responsible for organoleptic characteristics in grape wine. The common name of flavan-3-ols monomers is "catechin"(Escribano-Bailon et al. 2001). Anthocyanins (greek: purple flower) are a very well investigated large group of phenolics, are mainly accumulated in skin of red grapes and are responsible for red wine color (Castillo-Munoz et al. 2009). Cyanidin, Delphinidin, Peonidin, Petunidin and Malvidin are the most important anthocyanins in grape (Monagas et al. 2005).

Flavonol	
Kaempferol	Myricetin
Kaempferol-3-O -glucoside	Myricetin-3-O -glucoside
Kaempferol-3-O -galactoside	Myricetin-3-O -glucuronide
Kaempferol-3-O -glucuronide	Isorhamnetin-3-O -glucoside
Quercetin	Isorhamnetin
Quercetin-3-O -glucoside	Quercetin-3-O -glucoronide

Table 8. Classification of flavonols

8. Accumulation of metabolic compounds during ripening

Basically, the maturity of grape berries depends on solar radiation which is indirectly by driving temperature of berries. That means that the sugar accumulation of berries is slowly under low temperatures and the ripening process will be slow. Organic acids will be metabolized more slowly and that cause high concentrations of acids. High temperatures cause various biochemical thresholds which limit metabolic reactions and is commonly called "the vine shuts down". Grape berries are a rich source of phenolic compounds which are important plant secondary metabolites and are produced through photosynthesis (primary metabolites). The composition of phenolic substances is responsible for sensory properties in plant derived food including grapes and wine (Sandhu and Gu). Every wine contains a typical aroma, color, taste and mouth feel. The three major phenolics in grape are anthocyanins and flavonols (Ferrer et al. 2008). Flavonols are useful for UV radiation absorption and anti- microbial properties as response to wounds. The synthesis of these substances in berries is promoted in the vineyard due to solar radiation that is why they are accumulated in the exocarp (surface layer) (Versari et al. 2001). Anthocyanins cause the color modifications of fruits during ripening with variations from red, purple to black what is typical for the color of red wine. Anthocyanins form complexes with tannins and flavonols that contribute in the stability of the pigments. They are also synthesized in darkness but berries exposed for a long time to solar radiation during the ripening period will synthesize and accumulate higher contents of plant pigments. Astringent sensory properties of wine are the consequence of condensed tannins which are typical compounds of seed and skin of grapes. The decision on the optimal harvesting time depends on the climate, fruit developmental stage because high contents of condensed tannins in young berries have a higher biological benefit because of their bitterness and astringency.

9. Health promoting compounds in grape

The interests of scientists in the potential for specific phytochemicals increased in the last years because of their medicinal importance. Actual investigations show that the consumption of fresh fruits and vegetables are important to prevent chronic diseases and provides essential nourishment to humans. Plant geneticists around the world are able to breed special cultivars with higher nutraceuticals value to increase the concentration of health-promoting compounds in grape (God et al. 2007). Grape has many different varieties in health promoting compounds and is rich in bioactive metabolites like phenolic compounds that act as antioxidants or resveratrol which serve as chemo-preventative. Plant phenolics are attractive for researchers and the industry due to their techno-functional and putative beneficial bio-functional properties (Kammerer et al. 2004). The health benefit in wine comes from their high content in flavonoids and phytonutrients which are responsible for the color in grapes. Quercetin and Resveratrol are two important stilbenes in grapes which are responsible for heart- protection effects (Frankel et al. 1993). These bioactive phenolic compounds can reduce blood clotting due to their antiaggregant effect and protect low-density lipoproteins (LDL) cholesterol from oxidative reactions which can cause arterial damage (Albini et al.). The resveratrol (3-4'-5-trihydroxystilbene), is a bioactive secondary metabolite with high importance in medicine and pharmaceutics. Many studies showed that *trans*-resveratrol has a high antioxidative potential and reduces the risks of coronary heart disease and prevent any formation of cancer cells (Cui et al. 2002). The results of different investigations demonstrated that the concentrations of anthocyanins, phenolics and resveratrol differ significantly among cultivars and breeding lines. Grape (*Vitis vinifera* L.) has large amounts of phenolic compounds whereby the highest accumulation is in the skins and seeds (Rodriguez *et al.*, 2006; Poudel *et al.*, 2008). The most abundant phenolic compounds in grape skins are flavonols. The seeds contain high amounts of monomeric phenolic compounds like (+)-catechins, (-)-epicatechin and (-)-epicatechin-3-Ogallate as well as dimeric, trimeric and tetrameric procyanidins which do have antimutagenic and antiviral effects (Kammerer et al. 2004).

In *in vitro* tests, it was verified that phenolic compounds inhibit the oxidation of low-density lipoproteins (LDL) (Fauconneau et al. 1997). Phenolic substances are from high importance in the quality of grapes and wines. They can be classified in two groups: non-flavonoid (hydroxybenzoic, hydroxycinnamic acids, stilbenes) and flavonoid compounds (anthocyanins, flavan-3-ols, flavonols). Anthocyanins are a large group within the family of phenolics which are responsible for the pigmentation- coloration in grapes. Anthocyanins produce in a reaction with flavanols are more stable pigments (Butkhup and Samappito 2008). Flavan- 3-ols (monomeric catechins and proanthocyanidins) create a further large class of phenolic compounds which are responsible for astringent and bitter properties. They are responsible for the browning process in grape (Macheix et al. 1991). Special phenolic compounds are part in the phenomenon of co-pigmentation. Another large group of flavonoids are flavonols like quercetin, myricetin, kaempferol, isorhamnetin as well as their glycosides which have a high potential such as antioxidants. Grape became very attractant because of their high contents of phenolics especially antioxidant properties and their beneficial effects on human health (Vitseva et al. 2005). Scientific investigations verified the health benefits of catechins and procyanidins and the use of grape extract as an antioxidant supplement in the dietary food (Guendez et al. 2005). Special antioxidants can be used as food preservation due to their protective effects against microorganisms (Vattem et al. 2005). Phenolic compounds which have microbiological effects are found in seeds, skins and stem extracts of *Vitis vinifera* (Chidambara Murthy et al. 2002). Further phytonutrient in grape with health benefits are saponins which reduce blood cholesterol by binding its molecules and preventing their absorption, as well as reducing inflammation and cancer-protecting effect. Scientists from the University of California found that alcohol improve the bioavailability of saponins that means making them more easily absorbed in wine. The content of saponins in red wines is 3 to 10 times higher than the amount of saponins contained in white wines. Besides there is a positive association between the alcohol concentration and saponin concentration; that means stronger wines having more saponins. The highest content of saponins was found in red Zinfandel (16% alcohol) followed by Syrah, Pinot noir and Cabernet Sauvignon. White varieties like Sauvignon blanc and Chardonnay contain much less saponins. Furthermore Grape berries contain Pterostilbene with antioxidant activity which has also positive effects against cancer and cholesterol.

10. Resveratrol

Resveratrol is in the human diet and daily life of high importance because of their protection against benzopyrene which is the main substance in cigarette smoke and provokes lung cancer. Resveratrol inhibits the cell receptor aryl hydrocarbon receptor (AhR) on which benzopyrene and carcinogenic polycyclic aromatic hydrocarbons are not able to bind to cell membranes; because by binding they cause an expression of several cancer-promoting genes (Halls and Yu 2008). Researchers found out that Resveratrol inhibits the production of endothelin-1 and influence hearth cells by inhibiting antiotensin II which is a very powerful vasocontricting hormone. They are equally known to prevention the differentiation of fibroblasts into myofibroblasts because of the collagen production. Resveratrol is one of the main components in red wine but they contain several other phytochemicals like catechins and epicatechins in high concentrations (Anekonda 2006). These two phenolic compounds have a high potency to reduce the activity of COX-1 and COX-2. The anti- microbial effect of Resveratrol was carried out in Turkey where an extract from grape seeds, skin and stems showed anti-microbial effects against 14 bacteria like Escherichia coli and Staphylococcus aureus with the only resistant bacterium being Yersinia enterocolica at a concentration of 2.5%.

11. Byproducts of winery

Many byproducts and wastes contain polyphenols with a high potential for the application as food antioxidants and preventive agents against several diseases as well as due to a cost-efficient recovery. Valorization from grape by-products contains different bioactive substances like pigments, flavonoids, stilbenes and phenolic acids which could be used as natural antioxidants or colorants. Polyphenolic substances are extracted into wine, but the highest concentration remains in the vinification waste (pomace, stems, and seeds), which form over 13% of the processed grape weight (Torres et al. 2002). The majority of grape byproducts remains in the waste during the vinification process, it is also an animal food as source of some products such as ethanol, alcoholic beverages, tartaric and citric acid, grape seed oil and dietary fiber (Torres et al. 2002). Residues of the winemaking process so called pomace a very attractive residual sources of valuable bioactive compounds, even though it is still underutilized (Kammerer et al. 2004). For example fractions of Parellada grape (Vitis vinifera) containing oligomers with galloylation ca. 30% which are the most potential free radical scavengers and antioxidants (Torres et al. 2002). Several substances from grape pomace like ethanol, tartaric, malic and citric acids as well as of seed oil can be recovered. Winery byproducts are from high importance in the food industry and also for agricultural purposes because pomace is used as soil conditioner or for compost production. Red and white grape varieties were investigated in their phenolic potential of the press residues. The results showed over 35 different compounds especially anthocyanins, phenolic acids, non-anthocyanin, flavonoids and stilbenes. Anthocyanins were found in high concentrations in red grape pomace up to 132 g/kg dry matter. Some compounds found in pomace showed significant differences in their content in dependence of the cultivar-specific differences, grape ripening stage, microclimatic and phytosanitary conditions (Kammerer et al. 2004). Seed extracts of *Vitis vinifera* contain high contents of flavan-3-ols and their derivatives. The main compounds of pomace and stem extracts are significant amounts of flavonoids, stilbenes, and phenolic acids. Stems contain high concentrations of *trans*-resveratrol and ε-viniferin (Anastasiadi et al. 2009). Pomace was also used for the recovery of phenolics on laboratory scale. Until now for the extraction of anthocyanin was acted using acidified and sulfited solvents. Sulfite cannot be quantitatively removed and allergenic reaction after the consumption of sulfited food was visible. A new pectinolytic and cellulolytic method of pomace enhance the release of phenolic compounds (Maier et al. 2006). By the optimization of this extraction process the yields of extracted phenolic acids, non-anthocyanin flavonoids and anthocyanins reaching 91.9, 92.4 and 63.6 %. Pomace is also a rich source for the edible oil from the seed and rich in polyphenols with high antioxidant activity. To obtain the phenolic compounds from the press residues in high contents is very easy, whereby they can be applied as supplements of functional or enriched foods (Maier et al. 2006). A new technology with resin adsorption allowes a high level of purification and concentration of anthocyanin extracts from a Cabernet Mitos´ grape pomace. These processes are common in industrial processes, for example to debitter citrus juices or to stabilize and standardize juice concentrates (Kammerer et al. 2004). Furthermore, resin adsorption can be also an effective method to concentrate plant phenolics, to fractionate extracts and to enrich some compounds. This novel technique can also contribute to the production of valuable plant extracts with health-beneficial properties. High-speed countercurrent chromatography (HSCCC) is also a new technique which is useful for the analytical and preparative part in winemaking process as well as for the isolation of bioactive compounds from crude plant extracts. By using the HSCCC method phenolic extracts from a ´Lemberger´ grape pomace were analyzed and some compounds like caftaric, coutaric and fertaric acids were isolated. Purified extracts of caftaric, coutaric, and fertaric acids were up to 97.0 %, 97.2 % and 90.4 % and the end product from 10 g of grape pomace were 62, 48 and 23 %. These byproducts especially from grape waste can be used for the recovery of bioactive beneficial compounds to increase the profit of conventional processing techniques and to maximize sustainable agricultural production (Pinelo et al. 2005). Grape has a wide variety of polyphenols whereby flavonoids are the best investigated compounds and known to have antibacterial activities. These metabolites are produced and accumulated due to their interaction with extracellular soluble proteins and/or bacterial cell walls (Cowan 1999). Furthermore catechins inhibit in vitro the action of many bacteria like *Vibrio cholerae*, *Streptococcus mutans*, and *Shigella*, whereas (-)-epigallocatechin gallate is a potent Gram-positive bactericidal that acts by damaging the respective bacterial membranes (Ikigai et al. 1993).

12. Grape phenolics and their nutritional pharmacological effects

Recent investigations show that the addition of biomolecules in food supports human health. Especially glycosylated, esterified, thiolyated, or hydroxylated forms of bioactive compounds displays their health benefit in metabolic activities combined with several diseases. All these bioactive plant food components are mainly found in whole grains, fruits and vegetables (Klotzbach-Shimomura 2001). There are thousands of bioactive food compounds derived from plants like the polyphenols, phytosterols, carotenoids, tocopherols, tocotrienols, isothiocyanates and diallyl- (di, tri)sulfide compounds, fiber, and fruto-ogliosaccharide. Grape is a well known and investigated plant because of their social importance and useful metabolites especially flavonoids in humans health. Polyphenols belong to the major substances in grape which are a widely distributed group of biomolecules. Bioflavonoids belongs to health promoting compounds in grape and their recent scientific interest confirm the importance in our daily healthy diet. They play an important role in longlivety, cancer prevention and heart disease. Many studies investigated the content of phenolics and their effect on human cancer cells. (Yi et al. 2005) reported that muscadine grapes are rich in phenolic compounds and show a high potential on human liver cancer cells HepG2. The phenolic content of four cultivars of muscadine grapes ('Carlos', 'Ison', 'Noble', and 'Supreme') were investigated and separated into phenolic acids, tannins, flavonols, and anthocyanins. Extracted phenolic acids of muscadine grapes inhibited the HepG2 cell population growth in about 50% at concentration of 1-2 mg/mL. Anthocyanins showed the greatest positive effects regardingly apoptosis as well as cell viability at concentrations of 70-150 and 100-300 µg/mL .

13. Bioavailability of phenolic compounds

Polyphenols are very important micronutrients in our diet, and play a key role in the prevention of cancer and cardiovascular diseases. The health benefit of polyphenol depends on the consumed amount as well as their bioavailability (Manach et al. 2003). The main interests are the antioxidant properties of polyphenols as well as their appearence in the human diet and their role in the prevention of cancer, cardiovascular and neurodegenerative diseases (Scalbert et al. 2005). Polyphenols are responsible for the activity of a wide range of enzymes and cell receptors (Middleton et al. 2000). Polyphenols are not absorbed with equal efficacy because they are extensively metabolized by intestinal and hepatic enzymes as well as the intestinal micro flora.

14. Wine in biotechnological approach

Plant cell cultures are a potential alternative to traditional agriculture for the industrial production of valuable bioactive secondary metabolites. Especially pharmaceutical compounds and food additives like flavors, fragrances and colorants), perfumes and dyes are produced and accumulated in plant cell cultures (DiCosmo and Misawa 1995). Thereby the anthocyanin production can explain the basic mechanisms of biosynthesis of secondary metabolites, their transport as well as their accumulation in specific plant tissue. Plant pigments like anthocyanins are the large group of water-soluble pigments which are responsible for many colors. These pigments are also used in acidic solutions for the stabilization of the red color in soft drinks, sugar confectionary, jams and bakery products. As mentioned before grape pomaces is the major source of anthocyanins for commercial purposes and wastes from juice and wine industries (Curtin et al. 2003). Crude preparations of anthocyanins are relatively inexpensive and are used extensively in the food industry. The costs of pure anthocyanins are around US$ 1,250–2,000/kg. Cell suspension cultures of *Vitis vinifera* produce high contents of anthocyanins after cessation of cell division (Kakegawa et al. 2005). Furthermore the external influence and stimulation of the anthocyanin biosynthesis is regulated by the amino acid phenylalanine which is accumulated in the plant cells (Shimada et al. 2005). End products of the flavonoid biosynthesis pathway include anthocyanin pigments. Plant pigment extracts contain mixtures of various anthocyanin molecules, which differ in their levels of hydroxylation, methylation, and acylation. The major anthocyanins which are produced and accumulated in *V. vinifera* cell culture are cyanidin 3-glucoside (Cy3G), peonidin 3-glucoside (Pn3G), malvidin 3-glucoside (Mv3G) as well as the acylated versions of these compounds, cyaniding 3-p-coumaroylglucoside (Cy3CG), peonidin 3-p-coumaroylglucoside (Pn3CG) and malvidin 3-coumaroylglucoside (Mv3CG) (Conn et al. 2003). The production of anthocyanins by plant in- vitro cultures has been estimated in different plant species. Many investigation studies are using a cell line as model system for the production of secondary metabolites especially anthocyanin, because of the color that enables production. (Sano et al. 2005), of Nippon Paint Co. in Japan, investigated the production of anthocyanins. High osmotic potential in *Vitis vinifera* L. (grape) cell suspension cultures caused an increase in the anthocyanin production. By addition of sucrose or mannitol in the medium the osmotic pressure and the anthocyanin concentration accumulated was increased (Zhao et al.).

15. References

Adrian, M., Jeandet, P., Douillet-Breuil, A. C., Tesson, L. & Bessis, R. (2000). Stilbene content of mature Vitis vinifera berries in response to UV-C elicitation. J Agric Food Chem, 48, 6103-5.

Albini, A., Pennesi, G., Donatelli, F., Cammarota, R., De Flora, S. & Noonan, D. M. Cardiotoxicity of anticancer drugs: the need for cardio-oncology and cardio-oncological prevention. J Natl Cancer Inst, 102, 14-25.

Amati, A., Piva, A., Castellari, M. & Arfelli, G. (1996). Preliminary studies on the effect of Oidium tuckeri on the phenolic composition of grapes and wine. . Vitis, 35, 149-150.

Anastasiadi, M., Chorianopoulos, N. G., Nychas, G. J. & Haroutounian, S. A. (2009). Antilisterial activities of polyphenol-rich extracts of grapes and vinification byproducts. J Agric Food Chem, 57, 457-63.

Anekonda, T. S. (2006). Resveratrol--a boon for treating Alzheimer's disease? Brain Res Rev, 52, 316-26.

Arroyo-Garcia, R., Ruiz-Garcia, L., Bolling, L., Ocete, R., Lopez, M. A., Arnold, C., Ergul, A., Soylemezoglu, G., Uzun, H. I., Cabello, F., Ibanez, J., Aradhya, M. K., Atanassov, A., Atanassov, I., Balint, S., Cenis, J. L., Costantini, L., Goris-Lavets, S., Grando, M. S., Klein, B. Y., Mcgovern, P. E., Merdinoglu, D., Pejic, I., Pelsy, F., Primikirios, N., Risovannaya, V., Roubelakis-Angelakis, K. A., Snoussi, H., Sotiri, P., Tamhankar, S., This, P., Troshin, L., Malpica, J. M., Lefort, F. & Martinez-Zapater, J. M. (2006).

Multiple origins of cultivated grapevine (Vitis vinifera L. ssp. sativa) based on chloroplast DNA polymorphisms. Mol Ecol, 15, 3707-14.

Block, G. (1992). The data support a role for antioxidants in reducing cancer risk. Nutr Rev, 50, 207-13.

Bogs, J., Ebadi, A., Mcdavid, D. & Robinson, S. P. (2006). Identification of the flavonoid hydroxylases from grapevine and their regulation during fruit development. Plant Physiol, 140, 279-91.

Bouquet, A., Torregrosa, L., Iocco, P. & Thomas, M. R. (2006). Grapevine (Vitis vinifera L.). Methods Mol Biol, 344, 273-85.

Butkhup, L. & Samappito, S. (2008). An analysis on flavonoids contents in Mao Luang fruits of fifteen cultivars (Antidesma bunius), grown in northeast Thailand. Pak J Biol Sci, 11, 996-1002.

Castillo-Munoz, N., Fernandez-Gonzalez, M., Gomez-Alonso, S., Garcia-Romero, E. & Hermosin-Gutierrez, I. (2009). Red-color related phenolic composition of Garnacha Tintorera (Vitis vinifera L.) grapes and red wines. J Agric Food Chem, 57, 7883-91.

Chidambara Murthy, K. N., Singh, R. P. & Jayaprakasha, G. K. (2002). Antioxidant activities of grape (Vitis vinifera) pomace extracts. J Agric Food Chem, 50, 5909-14.

Cohen, S. D., Tarara, J. M. & Kennedy, J. A. (2008). Assessing the impact of temperature on grape phenolic metabolism. Anal Chim Acta, 621, 57-67.

Conn, S., Zhang, W. & Franco, C. (2003). Anthocyanic vacuolar inclusions (AVIs) selectively bind acylated anthocyanins in Vitis vinifera L. (grapevine) suspension culture. Biotechnol Lett, 25, 835-9.

Coombe, B. G. (1960). Relationship of Growth and Development to Changes in Sugars, Auxins, and Gibberellins in Fruit of Seeded and Seedless Varieties of Vitis Vinifera. Plant Physiol, 35, 241-50.

Cowan, M. M. (1999). Plant products as antimicrobial agents. Clin Microbiol Rev, 12, 564-82.

Cui, J., Tosaki, A., Bertelli, A. A., Bertelli, A., Maulik, N. & Das, D. K. (2002). Cardioprotection with white wine. Drugs Exp Clin Res, 28, 1-10.

Curtin, C., Zhang, W. & Franco, C. (2003). Manipulating anthocyanin composition in Vitis vinifera suspension cultures by elicitation with jasmonic acid and light irradiation. Biotechnol Lett, 25, 1131-5.

Da Luz, P. L., Serrano Junior, C. V., Chacra, A. P., Monteiro, H. P., Yoshida, V. M., Furtado, M., Ferreira, S., Gutierrez, P. & Pileggi, F. (1999). The effect of red wine on experimental atherosclerosis: lipid-independent protection. Exp Mol Pathol, 65, 150-9.

Dicosmo, F. & Misawa, M. (1995). Plant cell and tissue culture: alternatives for metabolite production. Biotechnol Adv, 13, 425-53.

Escribano-Bailon, T., Alvarez-Garcia, M., Rivas-Gonzalo, J. C., Heredia, F. J. & Santos-Buelga, C. (2001). Color and stability of pigments derived from the acetaldehyde-mediated condensation between malvidin 3-O-glucoside and (+)-catechin. J Agric Food Chem, 49, 1213-7.

Fauconneau, B., Waffo-Teguo, P., Huguet, F., Barrier, L., Decendit, A. & Merillon, J. M. (1997). Comparative study of radical scavenger and antioxidant properties of phenolic compounds from Vitis vinifera cell cultures using in vitro tests. Life Sci, 61, 2103-10.

Ferrer, J. L., Austin, M. B., Stewart, C., Jr. & Noel, J. P. (2008). Structure and function of enzymes involved in the biosynthesis of phenylpropanoids. Plant Physiol Biochem, 46, 356-70.

Fournand, D., Vicens, A., Sidhoum, L., Souquet, J. M., Moutounet, M. & Cheynier, V. (2006). Accumulation and extractability of grape skin tannins and anthocyanins at different advanced physiological stages. J Agric Food Chem, 54, 7331-8.

Frankel, E. N., Waterhouse, A. L. & Kinsella, J. E. (1993). Inhibition of human LDL oxidation by resveratrol. Lancet, 341, 1103-4.

God, J. M., Tate, P. & Larcom, L. L. (2007). Anticancer effects of four varieties of muscadine grape. J Med Food, 10, 54-9.

Gomez-Miguez, M., Gonzalez-Manzano, S., Escribano-Bailon, M. T., Heredia, F. J. & Santos-Buelga, C. (2006). Influence of different phenolic copigments on the color of malvidin 3-glucoside. J Agric Food Chem, 54, 5422-9.

Gorny, R. T. (1996). Viniculture and Ancient Anatolia. In: The Origins and Ancient History of Wine, Mc Govern, P.E., S.J. Fleming and S.H. Katz (Eds.). Gordon and Breach Publisher, Amsterdam, 133.

Gould, K. S., Markham, K. R., Smith, R. H. & Goris, J. J. (2000). Functional role of anthocyanins in the leaves of Quintinia serrata A. Cunn. J Exp Bot, 51, 1107-15.

Granett, J., Walker, M. A., Kocsis, L. & Omer, A. D. (2001). Biology and management of grape phylloxera. Annu Rev Entomol, 46, 387-412.

Guendez, R., Kallithraka, S., Makris, D. P. & Kefalas, P. (2005). An analytical survey of the polyphenols of seeds of varieties of grape (Vitis vinifera) cultivated in Greece: implications for exploitation as a source of value-added phytochemicals. Phytochem Anal, 16, 17-23.

Halls, C. & Yu, O. (2008). Potential for metabolic engineering of resveratrol biosynthesis. Trends Biotechnol, 26, 77-81.

Ikigai, H., Nakae, T., Hara, Y. & Shimamura, T. (1993). Bactericidal catechins damage the lipid bilayer. Biochim Biophys Acta, 1147, 132-6.

Iriti, M. & Faoro, F. (2006). Grape phytochemicals: a bouquet of old and new nutraceuticals for human health. Med Hypotheses, 67, 833-8.

Joscelyne, V. L., Downey, M. O., Mazza, M. & Bastian, S. E. (2007). Partial shading of Cabernet Sauvignon and Shiraz vines altered wine color and mouthfeel attributes, but increased exposure had little impact. J Agric Food Chem, 55, 10888-96.

Kakegawa, K., Ishii, T. & Matsunaga, T. (2005). Effects of boron deficiency in cell suspension cultures of Populus alba L. Plant Cell Rep, 23, 573-8.

Kammerer, D., Claus, A., Carle, R. & Schieber, A. (2004). Polyphenol screening of pomace from red and white grape varieties (Vitis vinifera L.) by HPLC-DAD-MS/MS. J Agric Food Chem, 52, 4360-7.

Kliewer, W. M. (1964). Influence of Environment on Metabolism of Organic Acids and Carbohydrates in Vitis Vinifera. I. Temperature. Plant Physiol, 39, 869-80.

Klotzbach-Shimomura, K. (2001). Herbs and health: safety and effectiveness. J Nutr Educ, 33, 354-5.

Kontoudakis, N., Esteruelas, M., Fort, F., Canals, J. M. & Zamora, F. Comparison of methods for estimating phenolic maturity in grapes: correlation between predicted and obtained parameters. Anal Chim Acta, 660, 127-33.

Koundouras, S., Marinos, V., Gkoulioti, A., Kotseridis, Y. & Van Leeuwen, C. (2006). Influence of vineyard location and vine water status on fruit maturation of nonirrigated cv. Agiorgitiko (Vitis vinifera L.). Effects on wine phenolic and aroma components. J Agric Food Chem, 54, 5077-86.

Langcake, P. & Pryce, R. J. (1977). A new class of phytoalexins from grapevines. Experientia, 33, 151-2.

Lu, R. & Serrero, G. (1999). Resveratrol, a natural product derived from grape, exhibits antiestrogenic activity and inhibits the growth of human breast cancer cells. J Cell Physiol, 179, 297-304.

Ma, L., Cao, T. T., Kandpal, G., Warren, L., Fred Hess, J., Seabrook, G. R. & Ray, W. J. Genome-wide microarray analysis of the differential neuroprotective effects of antioxidants in neuroblastoma cells overexpressing the familial Parkinson's disease alpha-synuclein A53T mutation. Neurochem Res, 35, 130-42.

Macheix, J. J., Sapis, J. C. & Fleuriet, A. (1991). Phenolic compounds and polyphenoloxidase in relation to browning in grapes and wines. Crit Rev Food Sci Nutr, 30, 441-86.

Maier, T., Sanzenbacher, S., Kammerer, D. R., Berardini, N., Conrad, J., Beifuss, U., Carle, R. & Schieber, A. (2006). Isolation of hydroxycinnamoyltartaric acids from grape pomace by high-speed counter-current chromatography. J Chromatogr A, 1128, 61-7.

Manach, C., Morand, C., Gil-Izquierdo, A., Bouteloup-Demange, C. & Remesy, C. (2003). Bioavailability in humans of the flavanones hesperidin and narirutin after the ingestion of two doses of orange juice. Eur J Clin Nutr, 57, 235-42.

Mane, C., Souquet, J. M., Olle, D., Verries, C., Veran, F., Mazerolles, G., Cheynier, V. & Fulcrand, H. (2007). Optimization of simultaneous flavanol, phenolic acid, and anthocyanin extraction from grapes using an experimental design: application to the characterization of champagne grape varieties. J Agric Food Chem, 55, 7224-33.

Mateus, N., Silva, A. M., Rivas-Gonzalo, J. C., Santos-Buelga, C. & De Freitas, V. (2003). A new class of blue anthocyanin-derived pigments isolated from red wines. J Agric Food Chem, 51, 1919-23.

Middleton, E., Jr., Kandaswami, C. & Theoharides, T. C. (2000). The effects of plant flavonoids on mammalian cells: implications for inflammation, heart disease, and cancer. Pharmacol Rev, 52, 673-751.

Monagas, M., Bartolome, B. & Gomez-Cordoves, C. (2005). Updated knowledge about the presence of phenolic compounds in wine. Crit Rev Food Sci Nutr, 45, 85-118.

Monti, A. (1999). [Vine and wine in history and in law from the 11th to the 19th centuries]. Arch Stor Ital, 157, 357-65.

Nassiri-Asl, M. & Hosseinzadeh, H. (2009). Review of the pharmacological effects of Vitis vinifera (Grape) and its bioactive compounds. Phytother Res, 23, 1197-204.

Pereira, G. E., Gaudillere, J. P., Pieri, P., Hilbert, G., Maucourt, M., Deborde, C., Moing, A. & Rolin, D. (2006). Microclimate influence on mineral and metabolic profiles of grape berries. J Agric Food Chem, 54, 6765-75.

Pinelo, M., Rubilar, M., Jerez, M., Sineiro, J. & Nunez, M. J. (2005). Effect of solvent, temperature, and solvent-to-solid ratio on the total phenolic content and antiradical activity of extracts from different components of grape pomace. J Agric Food Chem, 53, 2111-7.

Pitsavos, C., Makrilakis, K., Panagiotakos, D. B., Chrysohoou, C., Ioannidis, I., Dimosthenopoulos, C., Stefanadis, C. & Katsilambros, N. (2005). The J-shape effect of alcohol intake on the risk of developing acute coronary syndromes in diabetic subjects: the CARDIO2000 II Study. Diabet Med, 22, 243-8.

Poni, S., Bernizzoni, F., Civardi, S. & Libelli, N. (2009). Effects of pre-bloom leaf removal on growth of berry tissues and must composition in two red vitis vinifera L. cultivars. Australian Journal of Grape and Wine Research 15 (2), 185-193.

Qian, M. C., Fang, Y. & Shellie, K. (2009). Volatile composition of Merlot wine from different vine water status. J Agric Food Chem, 57, 7459-63.

Refai, M. (2002). Incidence and control of brucellosis in the Near East region. Vet Microbiol, 90, 81-110.

Renaud, S. & De Lorgeril, M. (1992). Wine, alcohol, platelets, and the French paradox for coronary heart disease. Lancet, 339, 1523-6.

Rentzsch, M., Schwarz, M., Winterhalter, P. & Hermosin-Gutierrez, I. (2007). Formation of hydroxyphenyl-pyranoanthocyanins in Grenache wines: precursor levels and evolution during aging. J Agric Food Chem, 55, 4883-8.

Ribereau Gayon, P. (1965). [Identification of Esters of Cinnamic Acid and Tartaric Acid in the Limbs and Berries of V. Vinifera]. C R Hebd Seances Acad Sci, 260, 341-3.

Sandhu, A. K. & Gu, L. Antioxidant capacity, phenolic content, and profiling of phenolic compounds in the seeds, skin, and pulp of Vitis rotundifolia (Muscadine Grapes) As determined by HPLC-DAD-ESI-MS(n). J Agric Food Chem, 58, 4681-92.

Sano, T., Oda, E., Yamashita, T., Naemura, A., Ijiri, Y., Yamakoshi, J. & Yamamoto, J. (2005). Anti-thrombotic effect of proanthocyanidin, a purified ingredient of grape seed. Thromb Res, 115, 115-21.

Scalbert, A., Manach, C., Morand, C., Remesy, C. & Jimenez, L. (2005). Dietary polyphenols and the prevention of diseases. Crit Rev Food Sci Nutr, 45, 287-306.

Shimada, S., Inoue, Y. T. & Sakuta, M. (2005). Anthocyanidin synthase in non-anthocyanin-producing Caryophyllales species. Plant J, 44, 950-9.

Singleton, V. L. & Esau, P. (1969). Phenolic substances in grapes and wine, and their significance. Adv Food Res Suppl, 1, 1-261.

Souquet, J. M., Labarbe, B., Le Guerneve, C., Cheynier, V. & Moutounet, M. (2000). Phenolic composition of grape stems. J Agric Food Chem, 48, 1076-80.

Terral, J. F., Tabard, E., Bouby, L., Ivorra, S., Pastor, T., Figueiral, I., Picq, S., Chevance, J. B., Jung, C., Fabre, L., Tardy, C., Compan, M., Bacilieri, R., Lacombe, T. & This, P. Evolution and history of grapevine (Vitis vinifera) under domestication: new morphometric perspectives to understand seed domestication syndrome and reveal origins of ancient European cultivars. Ann Bot, 105, 443-55.

This, P., Lacombe, T. & Thomas, M. R. (2006). Historical origins and genetic diversity of wine grapes. Trends Genet, 22, 511-9.

Torres, J. L., Varela, B., Garcia, M. T., Carilla, J., Matito, C., Centelles, J. J., Cascante, M., Sort, X. & Bobet, R. (2002). Valorization of grape (Vitis vinifera) byproducts. Antioxidant and biological properties of polyphenolic fractions differing in procyanidin composition and flavonol content. J Agric Food Chem, 50, 7548-55.

Vattem, D. A., Ghaedian, R. & Shetty, K. (2005). Enhancing health benefits of berries through phenolic antioxidant enrichment: focus on cranberry. Asia Pac J Clin Nutr, 14, 120-30.

Versari, A., Parpinello, G. P., Tornielli, G. B., Ferrarini, R. & Giulivo, C. (2001). Stilbene compounds and stilbene synthase expression during ripening, wilting, and UV treatment in grape cv. Corvina. J Agric Food Chem, 49, 5531-6.

Vian, M. A., Tomao, V., Coulomb, P. O., Lacombe, J. M. & Dangles, O. (2006). Comparison of the anthocyanin composition during ripening of Syrah grapes grown using organic or conventional agricultural practices. J Agric Food Chem, 54, 5230-5.

Vitrac, X., Larronde, F., Krisa, S., Decendit, A., Deffieux, G. & Merillon, J. M. (2000). Sugar sensing and Ca2+-calmodulin requirement in Vitis vinifera cells producing anthocyanins. Phytochemistry, 53, 659-65.

Vitseva, O., Varghese, S., Chakrabarti, S., Folts, J. D. & Freedman, J. E. (2005). Grape seed and skin extracts inhibit platelet function and release of reactive oxygen intermediates. J Cardiovasc Pharmacol, 46, 445-51.

Wang, S. P. & Huang, K. J. (2004). Determination of flavonoids by high-performance liquid chromatography and capillary electrophoresis. J Chromatogr A, 1032, 273-9.

Yi, W., Fischer, J. & Akoh, C. C. (2005). Study of anticancer activities of muscadine grape phenolics in vitro. J Agric Food Chem, 53, 8804-12.

Zhao, Q., Duan, C. Q. & Wang, J. Anthocyanins Profile of Grape Berries of Vitis amurensis, Its Hybrids and Their Wines. Int J Mol Sci, 11, 2212-28.

Part 2

Quality Control

9

Food Quality Control:
History, Present and Future

Ihegwuagu Nnemeka Edith[1] and Emeje Martins Ochubiojo[2]
[1]Agricultural Research Council of Nigeria
[2]National Institute for Pharmaceutical Research and Development
Nigeria

1. Introduction

Food is any substance which when consumed provides nutritional support for the body. It may be of plant or animal origin, containing the known five essential nutrients namely, carbohydrates, fats, proteins, vitamins and minerals. Usually after consumption, food undergoes different metabolic processes that eventually lead to the production of energy, maintenance of life, and/or stimulation of growth (Aguilera 1999). The history of early man shows that, people obtained food substances through hunting, gathering, and agriculture. The assurance and protection of food quality has always been important to man. This is evident from the fact that, one of the earliest laws known to man was that of Food. Right from the Garden of Eden, there was a law guiding the consumption of food. In our time too, governments over many centuries have endeavoured to provide for the safety and wholesomeness of man's food by legal provisions, (Alsberg 1970; Jango-Cohen 2005). In spite of these provisions, adulteration of foods has increased and the detection of these adulterants has proved more difficult, essentially because of the sophisticated methods being used in the adulteration. The birth of modern chemistry in the early nineteenth century made possible the production of materials possessing properties similar to normal foods which, when fraudulently used, did not readily attract the attention of the unsuspecting consumer. However, modern analytical techniques are now available to detect adulterants in foods. In this modern age, most of the food consumed by man and animals alike is supplied by the food industry. The food industry is operated at different levels by either local, national or multinational corporations, that use either subsistence or intensive farming and industrial agriculture to maximize output. Whichever applies the importance of ensuring the quality of food and food products from the food industry cannot be over emphasized. In this chapter, we look at past and present attempts by scientists, individuals and/or governments to control food quality and prospect what the future might be.

2. Food quality control

Quality control is the maintenance of quality at levels and tolerance limits acceptable to the buyer while minimizing the cost for the vendor. Scientifically, quality control of food refers to the utilization of technological, physical, chemical, microbiological, nutritional and sensory parameters to achieve the wholesome food. These quality factors depend on specific attributes such as sensory properties, based on flavor, color, aroma, taste, texture and quantitative properties namely; percentage of sugar, protein, fibre etc. as well as hidden attributes likes peroxides, free fatty acids, enzyme (Adu-Amankwa 1999; Radomir 2004,Raju and Onishi 2002;).

Although, quality attributes are many, not all need to be considered at every point in time for every particular product. It is important to always determine how far relatively a factor is in relation to the total quality of the product. The quality attribute of a particular product is based on the composition of the product, expected deteriorative reactions, packaging used, shelf life required and the type of consumers. The most important element and ultimate goal in food quality control is protecting the consumer. To ensure standardization of these procedures, food laws and regulations cover the related acts affecting the marketing, production, labeling, food additive used, dietary supplements, enforcement of General Manufacturing Practice (GMP), Hazard Analysis and Critical Control Point (HACCP), federal laws and regulations, factory inspections and import/export inspections (Adamson 2004; Gravani 1986, Lasztity et al, 2004, Roe 1956).

3. Importance of food quality control (QC)

The most important quality factor of processed food is safety and reliability followed by "deliciousness" and "appropriate price" (Raju and Onishi 2002). The colossal loss a food industry will record if defective products were rejected or recalled, as well as the damaging effect on the company's image and public trust justifies the need for food quality control. For this reason, quality assurance should be a corporate goal, and should stem from the uppermost management level to the least staff of the industry. The Plan, Do, Check, Action (PDCA) cycle should be used when quality control is implemented.

4. History of food quality control

Many years ago, about 2500 years BC Mosaic and Egyptian laws had provisions to prevent the contamination of meat. Also, more than 2000 years ago, India already had regulations prohibiting the adulteration of grains and edible fats. In the actual sense, the laws of Moses contained decrees on food that are quite similar to certain aspects of modern food laws. Books of the Old Testament prohibited the consumption of meat from animals that died other than those intentionally slaughtered, perhaps consciously or otherwise, this was to ensure that contaminated meats were not consumed. There were also regulated weights and measures in foods and other commodities. Other ancient food regulations are referred to in Chinese, Hindu, Greek, and Roman literatures. Early records according to Lasztity et al (2004), classical writers also referred to the control of beer and inspection of wines in Athens, "to ensure purity and soundness of these products." The Roman government also provided state control over food supplies to protect consumers against bad quality and fraud. Even when most traders prefer to deal honestly and fairly, history has shown the need to ensure quality as evident by laws to protect purchasers and honest traders against those who refuse to adhere to accepted codes of good practice. It has been found (Adamson 2004) that, when food was scarce, resulting in expected increase in demand, fraudulent practices were prevalent.

4.1 The middle ages

This was the era of formation of trade guilds, especially the European communities with their powerful influence on the regulation of commerce (Petró-Turza and Földesi 2004).

These were groups of tradesmen of particular specialties whose purpose was to provide control and general supervision over the honesty and integrity of their members and the quality of their products. For example, in 1419, a proclamation was issued prohibiting the adulteration or mixing of wine from different geographical areas (Adamson 2004). Many countries had their own peculiar way of controlling food quality. For example, in 1649 a Commonwealth statute was enacted to regulate the quality of butter (Petró-Turza and Földesi 2004). In France the most interesting and complete economic document of the Middle Ages available on the subject is the *Livre des Métiers*, which, in the thirteenth century, outlined a code of the comparative practices of the trade guilds of Paris.

By the seventeenth and eighteenth centuries, chemistry was being used as an analytical tool in the fight against food adulteration. Robert Boyle, using the principles of specific gravity, established the foundation for the scientific detection of the adulteration of food (Adamson 2004). Not much has changed, just the degree and level of sophistication of fraud, and the analytical techniques used to detect it.

4.2 Industrial revolution in the nineteenth century

Records (Roe1956; Alsberg 1970) show that, there were some sporadic standardization activities in ancient cultures and during the middle ages, a more organized standardization however, actually developed in the second half of the nineteenth century. This was when the industry reached a state of development and it was obvious that, unification was indispensable for large projects such as the railway. Efforts were made to find uniform measures for length and weight, but the meter and kilogram were not accepted as those standards until 1870s (Roe1956; Alsberg 1970). The period of industrial revolution was a time of tremendous expansion in many fields, which had a particular bearing on food production, regulations, and control too. Rapid changes from a rural to an urbanized society and from a domestic to a factory production system placed a strain on food production and distribution. Expectedly, the period created many public health problems, particularly in the industrialized centres, which were ill prepared to accommodate the masses that flocked to them. There was much poverty, and the uncontrolled development of industrial towns led to appalling conditions similar to those which can still be seen in urban areas in some parts of the world today. There were attendant calls for reform and improvements in matters relating to public health including essential food supplies within the crowded and unhygienic industrial centres (Petró-Turza and Földesi 2004). In England in 1820, Friederich Accum's *A Treatise on Adulteration of Foods and Culinary Poisons* highlighted the fraudulent practices that endangered public health. Unfortunately, at that time, the knowledge and understanding of hygiene and the dangers of food adulteration was so low and limited that this work was not considered. Many earlier papers that tried to draw attention to the adulteration of some foodstuffs were also been disregarded (Industrial Revolution 6th edn. 2004).

Worthy of note however, was the setting up of a municipal service for the control of foodstuffs and beverages in Amsterdam in 1858. This was closely followed in England 1860, by the enactment of the first comprehensive modern food law in the world; "An Act to Prevent the Adulteration of Food and Drink." In addition to this being the first such act, it provided for a scientific approach to food problems by the appointment of an analyst whose sole duty, was "to examine the purity of articles of food and drink." (FAO 1999).

Seven years later in Budapest, Hungary, a municipal service was set up for control of drinking water, this culminated in the establishment of an institute for control of food. During this period, similar types of laws appeared in Belgium, Italy, Austria, Hungary, and the Scandinavian countries. This was also a period of the foundation of institutions serving food inspection and food quality control. To give an idea of the volume of activity of such institutions, it may be mentioned that, according to a report from the Food and Drug Inspection of the State Board of Health of Massachusetts, since the passage of the law in 1882 until 1907, more than 176 000 food samples were controlled, and more than 11 000 were found to be adulterated. Although the main food control activity at this time was in the industrialized nations of Western Europe, many other countries such as Australia, Canada, and the US also enacted food laws. Although Australia did not enact a national food law, each state had the power to enact food laws and has remained ever since (Alsberg 1970; Hutt& Merril 1991).

4.3 The twentieth century

The most prominent and substantive development at this period took place in India. The country amended its food adulteration control to ensure the purity of articles of food sold throughout the country between 1919 and 1941. In 1954, India enacted the Prevention of Food Adulteration Act and this Act, with its later amendments, is still in force (Roe 1956; Lasztity 2004). In the Far East, food control was slow to appear, and it was not until the 1940s or as late as the 1960s that food control measures were introduced. During this period, many Latin American countries also enacted food laws, even though the legal systems of most Latin American nations are based upon those of Spain and Portugal, significant differences have developed in their food laws. Efforts are now being made to harmonize these differences.

In Africa, food laws were of little significance until the second half of the twentieth century Lasztity et al (2004). Independent states that started to emerge in the late 1950s were influenced in many matters, including food control services, by the European countries with which they had been closely associated (Petró-Turza and Földesi 2004). For example, French territories had developed French food enactments, while British territories followed British procedures. Food legislation was often inherited in total by the newly independent states. Although it is obvious that, there is the need for major adjustments today as situations now are quite different from those for which the enactments had originally been designed, several factors ranging from corruption to lack of visionary leadership has contributed largely to the inability of many African countries to make reasonable headway. This is beyond the scope of this book. Worthy of mention however, is the great lack of skilled or trained personnel in developing countries to draw up food laws suited to the nation's particular circumstances, or the scientific or technical staff necessary for food analysis, sampling, and efficient inspection (Lasztity et al, 2004). There is also a shortage of materials and equipment, and many other problems connected with the inauguration and operation of an effective food control system.

5. The food supply chain

The food supply chain is a net chain that food moves along the peasant household, the processing industry, the distribution centre, the wholesaler, the retailer as well as consumer

(Thirupathi 2006). In general, the food supply chain is composed of five links: product material supplying link, the production processing link, the packing storage and transporting link, the sale link and the consumer expending link with each link involving related sub-links and the different organization carriers.

The supply chain is a networked structure, composed of flow of the physical, the information, the finance, the technical, the standardized, the security and the value-added connections (Grunert 2005; Thirupathi 2006).

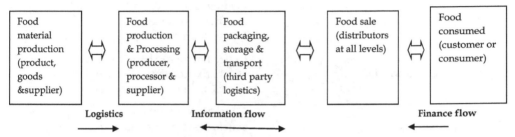

Fig. 1. Food supply chain basic structure (Thirupathi 2006)

5.1 Quality procedures
Along the supply chain, food is unavoidably exposed to numerous hazards. Therefore, knowing the risk factors at each phase of the supply chain assists in ensuring that an effective and comprehensive quality system is put in place. To guarantee food quality, it is important to ensure that, all the steps of the supply chain are carried out strictly, with care, and according to the standard operating procedures. At each phase of the supply chain, the potential risks, responsibilities, and how they can best be addressed must be fully explored.

6. Food contamination

Contamination of food can occur at any of the phases of the food supply chain and these will be expounded under the following broad categories:
1. Physical
2. Chemical
3. Microbiological
4. Other contaminants

6.1 Physical contamination
One of the major physical contaminations is adulteration. It is the mixing of inferior quality material with the superior product, thereby reducing the nature, quality and originality in taste, color, odor and nutritional value and ultimately causing ill effects on the health of the consumer (Thirupathi 2006). The main motive of adulteration and of course the "adultratee" is to gain undue advantage and most often profits. Almost all the food stuffs being sold in the market are prone to adulteration, but main food products that are often heavily adulterated are spices, milk products, edible oil, beverages drinks, sweets, pulses, sugar, processed foods, rice and cereal products like flour. The table below shows some common foodstuff and their adulterants (Lucey 2006).

6.2 Chemical contamination

Chemicals, which elicit harmful effects when consumed by animals or humans, are said to be toxic. The use of chemicals in the production and processing of food and food products not only affects the quality, but also disguises the deterioration and constitutes deliberate adulteration which is potentially very harmful to the health (Wilm 2003). It is advised that food additives like colouring matter, preservatives, artificial sweetening agents, antioxidants, emulsifiers/stabilizer, flavors/flavouring enhancers etc., if used should be of approved quality and processed under good manufacturing practices.

Types of Foodstuff	Main adulterants (Chemical formula and type)
Fish ,Meat & Dairy Products, Eggs and Poultry.	PCBs (polychlorinated biphenyls)
Citrus and other fruits	Juice Production and flavourings (multiple organic acids)
Oil and fat solvents	Dichloropropane ($CH_3CHClCH_3Cl$).
Alcohol Beverages	Ethanol (CH_2CH_2OH).
In Asian-style sauces such as soy, oyster, mushroom sauces, etc.	3-MCPD (3-monochloropropane-1,2-diol)
Potato tubers.	Glycoalkaloids (GA)
Unclean soft wheat for use. in non-staple foods and baby foods	Deoxynivalenol (Vomitoxin)
Fruit brandies and liqueurs, wines, distilled spirits.	Ethyl carbamate

Table 2. List of chemical adulterants in food

6.3 Microbiological contamination

Microbiological contamination of food is perhaps the most prevalent health problem in the contemporary world (Wilm 2003; Thirupathi 2006). To ensure good quality and safe food therefore, microbiological criteria should be established and freedom from pathogenic microorganisms must be ensured, including the raw materials, ingredients and finished products at any stage of production/processing. Accordingly the microbiological examination of the foods products has to be adopted widely. The microbiological criteria must be applied to define the distinction between acceptable and unacceptable foods (Thirupathi 2006). Food poisoning often results from the consumption of old, used, residual, fermented or spoiled food, as these may be contaminated with bacteria or other microorganisms, hence toxic. Infants and children are more susceptible to food poisoning and care should always be taken when giving them food. Gastroenteritis is caused by food contaminated with the *enterococcus, streptococcus faecalis*, which is frequently found in the human intestinal tract (Thirupathi 2006). Food poisoning may also be caused by inadequately refrigerated food contaminated with microorganisms such as *Clostridium perfringens* which grows well in the alimentary canal producing the poisoning within 8-12 hours after the ingestion of contaminated food. (Wilm 2003). *Bacillus cereus*, a gram positive, aerobic, spore-forming organism has been reported to be the etiologic agent in numerous food poisoning outbreaks. Some of the fatal effects of microbial contamination of food include; liver cancer is which is high in some countries due to aflatoxin. Flavism (hemolytic anemia) is caused by eating broad beans or by inhaling the pollen of its flower. In severe cases death may occur within 24-48 hours of the onset of the attack. Lathyrism is a disease, which paralyses the lower limbs (Thirupathi 2006).

6.4 Other contaminants

Metals are one of the many unintentional contaminants of food. When present beyond trace amounts, they are toxic. They find their way into food through air, water, soil, industrial pollution and other routes including food utensils. Common examples include, enamelware of poor quality which contributes antimony and galvanized utensils leaching zinc (Thirupathi 2006). A major source of tin contamination is tin plate, which is used for making containers for all types of processed foods. It has been shown that, a small quantity of metal is added when food is cooked in aluminum utensils (Wilm 2003). Although, copper is an essential trace element required by the human body, but copper-contaminated food is toxic.

Fumigants used to sterilize food under conditions in which steam heating is impractical contaminate food. This is because, they may react with food constituents to produce or destroy essential nutrients. For example, ethylene oxide, a commonly used fumigant, reacts with inorganic chloride to form ethylene chloro hydride, which is toxic (Wilm 2003; Thirupathi 2006). Some solvents like trichloro ethylene used for the extraction of oil from oil seeds react with the foodstuff being processed resulting in the formation of toxic products.

During processing of food, lipids may undergo numerous changes on prolonged heating, oxidative and polymerization reactions could take place, thereby decreasing the value of the processed products.

Smoking of meat and fish for preservation and flavouring is an old practice. But this processing contaminates the food with polycyclic aromatic hydrocarbons such as benzopyrene, many of which are carcinogenic.

Lubricants, packing materials etc. also contaminate foods.

A number of chemicals are intentionally added to foods to improve their nutritional value, maintain freshness, impact desirable properties or aid in processing. They also contaminate food if excessive in quantity (Thirupathi et al, 2006).

7. Methods of food quality control

In addition to ensuring safe and health food for the consumer, product manufacturers and service industries have realized that competition in a global market require a continual and committed effort towards the improvement of product and service quality. Therefore, they follow the process improvement cycle comprising **PLAN** (plan improvement), **DO** (implement plan for improvement), **CHECK** (analyze collected data) and **ACT** (take action). Quality control process consists of raw materials, in-process, product and service. The major factors in process that cause variability in quality of finished product are people, equipment and methods or technologies employed in the process. Use of proper statistical process control methods is also vital for assurance of the product quality. Usually, the value of quality characteristics is used to provide feedback on how processes may be improved. Statistical quality control comprises the following procedure:

a. Finished product is measured
b. Sampling occurs for days or weeks
c. Lot is either accepted or rejected based on information from sample

Contrary to statistical quality control, statistical process control methods focus on identifying factors in process that cause variability in finished product, eliminating the effect of these factors before worse product is manufactured, and control charts give on-line feedback of information about process. (Raju 2002). Food quality control measures have continuously improved since the 20th century, owing largely to the implementation of good practices, quality systems and increased traceability in food production. Ever since microorganisms were discovered in our environment and linked to typhoid fever and other diseases that have plagued humanity, public health authorities have been concerned with the accumulation of filth and foul odours in urban areas (Raju 2002). The first early inspection systems based on sensory evaluations were legally enforced at the beginning of the 20th century. Initial bacteriological techniques to detect pathogenic bacteria in foods, such as shellfish, appeared soon after (Raju 2002). From that point on, the food and beverage industry has applied stricter product inspection procedures and more and more effective production methods to conserve the freshness of natural raw materials. Today, the establishment of good manufacturing practices (GMP) and good hygienic practices (GHP) in many countries has significantly reduced the risk of spoilage and pathogenic microorganisms in modern food products. In addition to complying with national and international food regulations, food manufacturers are required to follow international quality standards, such as ISO as well as the Hazard Analysis Critical Control Point (HACCP) system.

In recent years, there has been an increasing focus on traceability in food production (Raju 2002). This has followed public concerns arising from cases of food contaminations and the development of foods containing ingredients derived from genetically modified (GM) crops. In the light of the increasing need for food more rapid food testing, it became clear that, the traditional microbiological detection and identification methods for food borne pathogens was no longer effective, because, it was time consuming and laborious to perform, and are increasingly unable to meet the demands for rapid quality control. A rapid method is generally characterized as a test giving quicker results than the standard accepted method of isolation and biochemical and/or serological identification (Raju 2002). Some of the newer and more rapid methods of food quality control are:

7.1 Ion mobility spectrometry (IMS) or differential mobility spectrometry (DMS)

These methods are used in the identification and quantification of analytes with high sensitivity. The selectivity can even be increased - as necessary for the analyses of complex mixtures - using pre-separation techniques such as gas chromatography or Multi-Capillary Columns (MCC). The method is suitable for application in the field of food quality and safety -including storage, process and quality control as well as the characterization of food stuffs (Vautz 2006).

7.2 Electronic-nose

This is an instrument which comprises an array of electronic chemical sensors with partial specificity or broad-band chemical selectivity and an appropriate pattern recognition system, capable of recognizing simple or complex odours (Gardner and Bartlett's 1994). This method is used to detect the bacterial growth on foods such as meat and fresh vegetables. It can also be used to test the freshness of fish. It is used in the process control of cheese, sausage, beer, and bread manufacture as well as for detection of off-flavors in milk and dairy products. This technique has other applications too. The advantages of the electronic nose can be attributed to its rapidity, objectivity, versatility, non requirement for the sample to be pretreated and ease of use Natale et al (1997).

7.3 Immunochemical methods
This method is based on antigen – antibody interaction. The antibodies are highly specific for the antigen (analyte), and secondly, the antigen, the antibody, or an antiglobulin may be conjugated to an enzyme that produces an intensely colored or fluorescent product in the presence of the enzyme substrate to enhance the detectability of the analyte in an amplification step. Toxins produced by *E. coli, Clostridium, Salmonella and Shigella* have also been similarly detected (Mason et al, 1961).

7.4 Enzyme Immunoassay (EIA)
Microorganisms are often characterized and identified by the presence of unique protein carbohydrate markers also called antigens, located within the body or the flagella of the cell (Greiner and Konietzny, 2008). Detection of these unique antigens has been a cornerstone of diagnostic microbiology for many years. In recent years, EIA using monoclonal antibodies have made available rapid and consistent microbiological detection systems. The most widely used systems employ a sandwich technique using antibody attached to a polystyrene matrix to which the sample is added. Post incubation, a second antibody, which is specific for the organism and has been tagged with an enzyme, is added. The addition of enzyme substrate to the mixture completes the EIA. The presence of the specific organism results in a colorimetric change in the enzyme substrate, which may be observed visually or with a spectrophotometer. Most EIA are very specific but lack sensitivity. Normal sensitivity has been reported to be in the range of 106 org/ml (Greiner and Konietzny, 2008).

7.5 Biosensors
Biosensor is usually a device or instrument comprising a biological sensing element coupled to a transducer for signal processing (Songa et al, 2009). Biological sensing elements include enzymes, organelles, antibodies, whole cells, DNA, and tissues. There are different types, conductance bioluminescence enzyme sensors utilizing potentiometric, amperometric, electrochemical, optoelectric, calorimetric, or piezoelectric principles. Basically, all enzyme sensors work by immobilization of the enzyme system onto a transducer (Songa et al, 2009). This technique provides sensitive, miniaturized systems that can be used to detect unwanted microbial activity or the presence of a biologically active compound, such as glucose or a pesticide in food. Immunodiagnostics and enzyme biosensors are two of the leading technologies that have had the greatest impact on the food industry (Greiner and Konietzny, 2008).

7.6 Flow cytometry (FCM)
Specific detection of pathogenic strains can be achieved by Flow cytometry using immunofluorescence techniques, which allow microorganism detection at the single-cell

level. Although this technology can be used for food samples, it requires prior isolation of the target organism to generate antibodies (Comas-Riu, 2009). FCM finds wide application in milk and brewing quality control. The advantage of FCM is that it can also differentiate Viable Non-Culturable (VBNC) form of bacteria from healthy cultivable cells (Comas-Riu, 2009). This technology has the ability to detect microorganisms at relatively low concentrations in a short time, while multiple labeling allows the detection of different organisms or different stages in the same sample (Comas-Riu, 2009).

7.7 Polymerase chain reaction (PCR)

This technique involves the following steps: Isolation of DNA from the food, amplification of the target sequences, separation of the amplification products by agarose gel electrophoresis, estimation of their fragment size by comparison with a DNA molecular mass marker after staining with ethidium bromide and finally, a verification of the PCR results by specific cleavage of the amplification products and by restriction endonuclease or southern blot. Alternatively amplification products may be verified by direct sequencing or a second PCR (Greiner and Konietzny, 2008).

7.8 Pulsed-field gel electrophoresis (PFGE)

PFGE is a restriction-based typing method that is considered by many to be the "gold standard" molecular typing method for bacteria (Greiner and Konietzny, 2008). In this electrophoretic approach, DNA fragments are separated under conditions where there is incremental switch of the polarity of the electric field in the running apparatus. This technique allows for the resolution of DNA fragments up to 800 kb in size. When DNA is restricted with a restriction enzyme, PFGE provides a DNA "fingerprint" that reflects the DNA sequence of the entire bacterial genome. PFGE is a widely accepted method for comparing the genetic identity of bacteria (Greiner and Konietzny, 2008). PFGE typing has demonstrated a high level of reproducibility for food borne pathogens. A major advantage of this method is its universal nature making it useful in bacteria sub typing, however, its limitation is that it is time consuming.

7.9 Magnetic separation

One of the typical applications of this technique was reported by Mattingly (1984) & Safarik (1995). The author separated salmonella from food and faecal matter using myeloma protein and hybrid antibody (for O antigen), conjugated to a polycarbonate- coated metal bead. It has also been reported (Haik et al, 2008) that, Food sample like milk, yogurt, meat and vegetables can be tested. The challenge is in detecting E. coli is in the isolation of pathogenic strain from nonpathogenic strains. Immunomagnetic detection of listeria monosytogens has also been reported by Skjerve et al. (1990). While Johne et al (1989) using magnetic beads coated with polyclonal antibodies was able to detect and isolate the specific protein of s. aureus.

8. Near-infrared (NIR) spectroscopy

NIR has proven to be an effective analytical tool in the area of food quality control. The key advantages of NIR spectroscopy are (1) its relatively high speed of analysis, (2) the lack of a need to carry out complex sample preparation or processing, (3) low cost, and (4) suitability for on-line process monitoring and quality Control. The disadvantage of this method includes the requirement for large sample sets for subsequent multivariate analysis. (Michelini 2008; Cozzolino 2008). Recently, researchers at Zhejiang University in Hangzhou, China used Vis/NIR spectroscopy together with multivariate analyses to classify non-transgenic and transgenic tomato leaves (Xie et al, 2007).

8.1 X-ray
This is a relatively newer technology in food quality control. X -rays started making in-roads into the food industry in the early 1990s. The driving force behind this was the increasing number of foreign bodies which could not be identified by metal detectors. Other than contaminants like glass, bone, rubber, stone or plastic, some specific applications are also more challenging for metal detectors, such as fresh meat and poultry, or foil-wrapped products (Ansell 2008). X-ray inspection has considerable advantages in many food and beverage processing environments in that, it is easy to install, safe and simple to use, even without previous experience. It quickly and consistently identifies substandard products, reducing product recall, customer returns and complaints, therefore protecting manufacturers' brands and most importantly, preventing ill health.

8.2 Computer vision
Aguilera et al (2006) reported a computer system that consisted of four basic components: the illumination source, an image acquisition device, the processing hardware, and suitable software modules. Their study was focused on analyzing the relevance of computer vision techniques for the food industry, mainly in Latin America. The authors described how the use of these techniques in the food industry eliminates the subjectivity of human visual inspection, adding accuracy and consistency to the investigation. They also reported that, the technique can provide fast identification and measurement of selected objects, classification into categories, and color analysis of food surfaces with high flexibility. They opined that, since the method was non-contact and non-destructive, temporal changes in properties such as color and image texture can also be monitored and quantified.

9. Some recent reports on food quality control

9.1 Livestock
Nardone (2002) found that the meat and milk characteristics are more related with human health and with some factors affecting their quality. They used various techniques to control meat characteristics. The Molecular biology techniques was of great interest to the author as it gave insight to new product certification viz, species, breed, animal category (age, sex, etc). He also reported that milk quality may be influenced by some innovative technology among which is the Automatic Milking Systems (AMS). AMS increases milk yield and milking frequency from twice to three times or more per day requiring a minimum extra amount of labour, however, contradictory results are reported about the effects of AMS on milk quality. Several authors found that after the introduction of AMS milk quality decreased, particularly fat, proteins percentage while total bacterial plate count, SCC, freezing point and the amount of free fatty acids increased significantly. Nardone (2002). Aguilera et al, (2006) reported the usefulness of vision Q-Lab to assess the quality of food Samples. The system consists of a highly general hardware setting, able to support different

applications, and highly modular software, easily adapted to the measurement needs of diverse food products. The main results of this application, was to classify rice grains and lentils (Aguilera et al, 2006). Grain quality attributes are very important for all users and especially the milling and baking industries. An earlier report by Zayas et al (1996) showed the usefulness of machine vision to identify different varieties of wheat and to discriminate wheat from non-wheat components. In his own report, Katsumata (2007) showed that, visible light Photoluminescence (PL) peaking at around λ = 460 nm is characteristic of cereals, such as rice, wheat, barley, millet, flour, corn starch, peanut, under illumination of ultra-violet light at λ = 365 nm. They further reported that Peak intensity of PL and Distribution of PL intensity varies with variety and source of the specimens which was found to be fitted with a Gaussian curve. Visible light PL is suggested to be potentially useful technique for the non-destructive and quick evaluation of the cereals and other starchy products. Songa (2007) reported the use of amperometric nanobiosensor for determination of glyphosate and glufosinate residues in corn and soya bean samples. The author found that biosensor has the features of high sensitivity, fast response time (10 to 20 s) and long term stability at 4^0C (>1month). Detection limits was in the order of 10^{-10} to 10^{-11} M for standard solutions of herbicides and the spiked samples. The author found that herbicide analyses can be spiked on real samples of corn and soy beans, corroborating that the biosensor is sensitive enough to detect herbicides in these matrices.

9.2 Fruits, vegetables and nuts

Narendra and Hareesh (2010) observed that Computer vision has been widely used for the quality inspection and grading of fruits and vegetables. It offers the potential to automate manual grading practices and thus to standardize techniques and eliminate tedious inspection tasks. The capabilities of digital image analysis technology to generate precise descriptive data on pictorial information have contributed to its more widespread and increased use. Aranceta-Garza et al (2011) reported a PCR-SSCP method for the genetic differentiation of canned abalone and commercial gastropods in the Mexican retail market. Their study was aimed at creating molecular tools that can differentiate abalone (Haliotis spp), from other commercial fresh, frozen and canned gastropods based on18S rDNA and also identify specific abalone product at the species level using the lysine gene. The authors found that the methods were reliable and useful for rapid identification of Mexican abalone products and could distinguish abalone at the species level. The methods could genetically identify raw, frozen and canned products and the approach could be used to certify authenticity of Mexican commercial products or identify commercial fraud. Lehotay (2011) reported the qualitative analysis of pesticide residues in fruits and vegetables using fast, low-pressure gas chromatography – time of flight mass spectrometry (LP-GC/MS). The author demonstrated that, to increase the speed of analysis for GC-amenable residues in various foods and provide more advantages over the 40 traditional GC-MS approach, LP-GC/MS on a time-of-flight (ToF) instrument, which provides high sample throughput with <10 min analysis time should be applied. The method had already been validated to be acceptable quantitatively for nearly 150 pesticides, and in this study of qualitative performance, 90 samples in total of strawberry, tomato, potato, orange, and lettuce extracts were analyzed. The extracts were randomly spiked with different pesticides at different levels, both unknown to the analyst, in the different matrices. They compared automated software evaluation with human assessments in terms of false positive and negative results only to found that the result was not significantly different. Mustorp (2011) reported a robust ten-plex quantitative and sensitive ligation-dependent probe amplification method, the allergen- Multiplex, quantitative ligation-dependent probe amplification (MLPA) method, for specific detection of eight allergens: sesame, soy, hazelnut, peanut, lupine, gluten, mustard and celery. Ligated probes were amplified by PCR and amplicons were detected using capillary electrophoresis. Quantitative results were obtained by comparing signals with an internal positive control. The limit of detection varied from approx. 5 to 400 gene copies depending on the allergen. The method was tested using different foods spiked with mustard, celery, soy or lupine flour in the 1-0.001 % range. Depending on the allergen, sensitivities were similar or better than those obtained with PCR.

10. Conclusions

Without doubts, there is a need for continuous improvements in rapid diagnostic methods, analytical techniques as well as visionary and computational equipments required for food quality control of the future. For example, as it stands today, only about 41.5% of microbiology tests utilized rapid methods. The growth of diagnostic industry should result in increased rapid tests in the nearest future and this should result in improved performance. It is expected that, there will be significant economic benefits and the ability to practice proactive and risk prevention food safety programs. In fact various schools of thoughts have it that, by 2015, the companies should be able to utilize automation technology to screen incoming raw materials and in-process parameters with near real-time information; physical, chemical or biological, while utilizing these newer methods. With automation, time is saved and productivity increases. In this regard, the two main traditional and rapid methods: flow cytometry to provide the total microbial count rapidly and polymerase chain reaction (PCR) to detect microorganisms both quickly and specifically will continue to grow in the near future. Although some of these automated newer technologies are extremely rapid, there have been questions about their sensitivity. Some investigators agree that, the instruments can produce results in seconds, but they opine strongly that they are not sensitive enough. As instrument companies today are working to provide the best combination of speed and sensitivity, the challenge of the future still lies with their ability to produce one instrument combining accuracy, rapidity and sensitivity while ensuring that, the entire control process still ensures a minimal risk of contamination. In the nearest future, the mass spectrometry may be needed in food quality control. It is already proofing useful identifying unknown compounds, antibiotics and pesticides in raw materials and in detecting trace metals in foods. DNA-based assays, instruments, and software, all designed to work together including the emerging area of bioinformatics; sequencing, assay design, and chemical analysis are all capable of developing new ways for food quality control in the future. This chapter looked at the history, current and future of food quality control and also revealed a large number of approaches to enhancing food quality. It is therefore safe to conclude that, food quality control is an indispensable tool in the food industry. As enumerated above, the development of adequate, effective, rapid, and sensitive food quality control systems however, faces serious challenges driven by its capital intensive nature and sophisticated adulteration. While it may seem easier for the developed nations to match quality control with adulteration techniques, to make any meaningful progress in resource limited nations of the world, there is the need for collaborations between laboratories around the globe, just as it is necessary for regulatory agencies around the world to also collaborate both in sharing information and in technologies as well as capacity development. The existing legislations seem adequate, but implementation may be weak, there is need for international coordinated efforts to enforce the laws.

11. References

Adamson Melitta Weiss (2004): *Food in medieval times*, pp 64-67; Greenwood Publishing Group, 88 Post Road West, Westport, CT 06881.

Adu-Amankwa Pearl, (1999) Quality and Process Control in the Food Industry Food Research Institute, Published in The Ghana Engineer, http://practicalaction.org/practicalanswers/

Aguilera, Jose Miguel and David W. Stanley,(1999): *Microstructural Principles of Food Processing and Engineering*, Second Edition Springer, ISBN=0834212560.

Alexandrakis D, Downey G, Scannell AG. (2008),Detection and identification of bacteria in an isolated system with near-infrared spectroscopy and multivariate analysis. *J Agric Food Chem; 56: 3431-7.*

Alsberg. CL. (1970),*Progress in Federal Food control* In: Ravenel. MP.ed. A. Half century of Health. New york Times Pp. 211-220.

Ansell Tim: (2008) , X-ray a new force in food quality control, Al Hilal Publishing & Marketing Group.

Augustin Scalbert, Cristina Andres-Lacueva, Masanori Arita, Paul Kroon, Claudine Manach, Mireia Urpi-Sarda, and David Wishart: (2011): Databases on food phytochemicals and their health promoting effects *J. Agric. Food Chem., Just Accepted Manuscript Publication.*

Chaplain CV, (1970).History of state & municipal control of diseases In: Ravenel. MP, ed. A. Half century of Health, New york Arno Press and the New york Times: Pp 133-160

Codex Alimentarius Commission (1997), Report of the Twelfth Session of the Codex Committee on General Principles, ALINORM 97/33.

Comas-Riu Jaume, Núria Rius: (2009), Flow cytometry applications in the food industry; *J Ind Microbiol Biotechnol* 36: 999–1011.

Cozzolino D, Fassio A, Restaino E, Fernandez E, La Manna A. (2008). Verification of silage type using near-infrared spectroscopy combined with multivariate analysis. *J Agric Food Chem.;* 56: 79-83

FAO. (1998)*FAO Technical Assistance Programme: Food Quality and Safety. (ESN internal publication),Rome.*

FAO. (1999). *FAO* trade-related technical assistance and information. Rome. Food and Agriculture Organization of the United Nations. (2005), The State of Food Insecurity in the World.

Gravani, Robert B: (1986), *How to Prepare A Quality Assurance Plan, Food Warehousing.* Department of Health and Human Services, Public Health Service, U.S. Food and Drug Administration, Food Science Facts for the Sanitarian, Dairy and Food Sanitation.

Greiner Ralf and Ursula Konietzny http://www.worldfoodscience.org/cms/?pid=1003869 Modern Molecular Methods (PCR) in Food Control: GMO, Pathogens, Species Identification, Allergens.

Haik Yousef et al (2008); *MagneticTechniques for Rapid Detectionof Pathogens:* Reyad M. Zourob *et al.* (eds.), Principles of Bacterial Detection: Biosensors, Recognition Receptors and Microsystems,. I *Springer Science+Business Media, LLC 2008.*

Halász,Anna Radomir Lásztity, Tibor Abonyi, and Arpad Bata, (2009): *Decontamination of Mycotoxin-Containing Food* and Feed by Biodegradation Food Reviews International, volume 25, pp:284–298.

Handbook of organic food safety and quality, Edited by J Cooper, C Leifert, Newcastle University, UK and U Niggli, Research Institute of Organic Agriculture (FiBL), Switzerland, Woodhead Publishing Series in Food Science, Technology and Nutrition No. 148 n.d.
http://quality.com/details/print/807365/December/January2009.html
http://www.foodquality.com/mag/06012006_07012006/fq_06012006_SS1.htm

Hussain, M. A. (2010): Future insights: *Proteomics a power technique to ensure food quality and safety;* Proceedings China International Food Safety and Quality Conference, 10-11 Shanghai, PR China. p. 51.

Hutt, Peter Barton and Merrill Richard(1991): *public health and the healthy public:*a communitarian perspective on privatization and the FDA; *Food and drug law,* pp 6-14.

"Industrial Revolution,"(2004), Columbia Encyclopedia, Sixth Edition, Copyright (c).

Jane Byrne: (2008): *RFID temperature logger could enhance cold chain quality control,* News on Food and Beverage Processing and Packaging.

Jango-Cohen, Judith; (2005):*The History of Food;* Twenty-First Century Books, ISBN 0-8225-2484-8, Minneapolis, Minn.

Janni J, Weinstock BA, Hagen L, Wright S. (2008): Novel near-infrared sampling apparatus for single kernel analysis of oil content in maize. *Appl Spectrosc* 62: 423-6.

Johne et al.: (1989);Imunomagnetic separation of s. aureus , *J. Clin Microbiol,* 27(7): pp; 1631-1635.

Katsumata T. , T. Suzuki, H. Aizawa and E. Matashige (2007): Photoluminescence evaluation of cereals for a quality control application , *Journal of Food Engineering, Volume 78, Issue 2,* Pages 588-590.

Klaus G. Grunert (2005): Food quality and safety: consumer perception and demand, European Review of Agricultural Economics Vol. 32 (3) pp. 369–391.

Lehotay Steven J., Urairat Koesukwiwat, Henk van 5 der Kamp, Hans G.J. Mol, and Natchanun Leepipatpiboo (2011): Qualitative Aspects in the Analysis of Pesticide Residues in Fruits and Vegetables using Fast, Low-Pressure Gas Chromatography – Time-of-Flight Mass Spectrometry Low-pressure gas chromatography –mass spectrometry (LP-GC/MS *Journal of Agricultural and Food Chemistry,* just accepted manuscript.

Lucey John: (2006) "Management Should Serve as Role Models for Good Work Habits and Acceptable Hygienic Practices "Food Quality

Mattingly, J.A1. (984).: An enzyme immunoassay for the detection of all Salmonella using a ombination of a myeloma protein and a hybridoma antibody. *Journal of Immunological Methods* 73, 147-156.

Michelini E, Simoni P, Cevenini L, Mezzanotte L, Roda A. (2008): New trends in bioanalytical tools for the detection of genetically modified organisms: an update. *Anal Bioanal Chem* 392: 355-67.

Moros J, Llorca I, Cervera ML, Pastor A, Garrigues S, de la Guardia M. (2008): .Chemometric determination of arsenic and lead in untreated powdered red paprika by diffuse reflectance near-infrared spectroscopy. *Anal Chim Acta;* 613: 196-206

Nardone Alessandro (2002): Evolution of livestock production and quality of animal products, Proc. "39th Annual Meeting of the Brazilian Society of Animal Science" Brazil, 486-513.

Narendra V G and Hareesh K S (2010): Quality Inspection and Grading of Agricultural and Food Products by Computer Vision- A Review: *International Journal of Computer Applications* Volume 2 – No.1.

Natale Corrado Di et al (1997) ; Electronic-nose modelling and data analysis using a self-organizing map; *Meas. Sci. Technol.* 8 1236–1243.

Perrot. N. . Ioannou, I. Allais ͨ, C. CurtJ. Hossenloppand G. Trystram (2006): Fuzzy Concepts Applied to Food Control Quality Control Fuzzy Sets and Systems Volume 157, Issue 9, pp, 1145-1154.

Potter, Norman N. and Joseph H. Hotchkiss, (1995): *Food Science.* 5th Edition. New York: Chapman & Hall. pp. 90-112.

Quality Assurance / Control in Food Processing. Contained in: Food Fortification - Technology and Quality Control. (FAO Food and Nutrition Paper – 60)

R.V. Sudershan, Pratima Rao and Kalpagam Polasa.(2009): Food safety research in India: a review *As. J. Food Ag-Ind.* 2(03), 412-433.

Radomir Lasztity, Marta Petro-Turza, Tamas Foldesi (2004): H*istory of food quality standards, in Food Quality and Standards*, [Ed. Radomir Lasztity], in *Encyclopedia of Life Support Systems (EOLSS)*, Developed under the Auspices of the UNESCO, Eolss Publishers, Oxford ,UK.

Raju K. V. R.and Onishi Yoshihisa. (2002): *Report of the APO Seminar on Quality Control for Processed Food held in the Republic of China*, (02-AG-GE-SEM-02). This report was edited by. Raju K. V. R. http//foodquality.wfp.org/qualityprocedure/tabid/119/.aspxdefault

Roe R.S. (1956): *The food & Drugs Act- past,present & future* in Welch H marti-Ibunez. F..eds. The impact of the food & Drug Admin. on our society, New york : MD Publications Pp. 15-17.

Rust and Olson (1987): Coping With Recalls, Meat and Poultry Vol. No. 3, March.

Safarik.M, Safarikova,&M.J.Forsethe. (1995): Application of Magnetic Separation in applied Microbiology; *Journal of Applied Microbiology*, 78, 575-585.

"Short Narratives: the Early Modern Agricultural Revolution," World History at KMLA, (2002) ;http://www.zum.de/whkmla/apeur/narratives/NarrativesAgrRev.html

Skjerve E, L M Rørvik and O Olsvik. (1990): Detection of Listeria monocytogenes in foods by immunomagnetic separation. *Appl Environ Microbiol*, 56(11): 3478-3481.

Songa Everlyne A., Vernon S. Somerset, Tesfaye Waryo, Priscilla G. L. Baker, and Emmanuel I. Iwuoha. (2009): Amperometric nanobiosensor for determination of glyphosate and glufosinate residues in corn and soya bean samples; *Pure Appl. Chem.*, Vol. 81, No. 1, pp. 123–139.

Stina Lund Mustorp, Signe Marie Drømtorp, and Askild Lorentz Holck: Multiplex. (2011): Quantitative ligation-dependent probe amplification for determination of allergens in food, *J. Agric. Food Chem.*,Just Accepted Manuscript Publication .

Swetman Tony and Barrie Axtell, (2008) (updated): *Quality Control in Food Processing*, Technical Information Online-practical answers.mht

Thirupathi. V., Viswanathan .R. & Devadas. CT; (2006): *Science Tech Entrepreneur.*

Thomas Weschler, J. Stan Bailey, Jerry Zweigenbaum and Dipankar Ghosh. (2009): food quality magazine, December/January issue.

Vautz W.; D. Zimmermann; M. Hartmann; J. I. Baumbach and J. Nolte; J. Jung. (2006): *Ion mobility spectrometry for food quality and safety: Food Additives & Contaminants*: Part A: Chemistry, Analysis, Control, Exposure & Risk Assessment, Volume 23, Issue 11, Pages 1064 – 1073.

Wilm Karl Heinz. (2003): Chemical Contaminants, Our Food; Food Safety and Control System, www.ourfood.com www.ourfood.com

XIAO Jing and MA Zhongsu: Research on Food Security Risk Early Warning under Supply Chain Environment, research sponsored by Heilongjiang Province science and technology department (Project Grant No.GA06C101-02) n.d.

Xie L, Ying Y, Ying T.(2007): Quantification of chlorophyll content and classification of nontransgenic and transgenic tomato leaves using visible/near-infrared diffuse reflectance spectroscopy. *J Agric Food Chem*; 55: 4645-50.

Zayas, I.Y., Martin, C.R., Steele, J.L., Katsevich, A.(1996): "Wheat classification using image analysis and crush force parameters",Transactions of the *ASAE* 39 (6), pp. 2199-2204.

10

Quality Preservation and Cost Effectiveness in the Extraction of Nutraceutically-Relevant Fractions from Microbial and Vegetal Matrices

Marco Bravi[1], Agnese Cicci[1] and Giuseppe Torzillo[2]
[1]Chimica Materiali Ambiente, Sapienza Università di Roma, Roma,
[2]CNR Istituto per lo Studio degli Ecosistemi, Sesto Fiorentino,
Italy

1. Introduction

Terrestrial and microbial vegetal matrices are a major source of nutraceutically and pharmaceutically relevant chemical compounds of different nature. In several cases the consumption of the raw vegetal or microbial matrix has been part of established diet regimes and has provided the consumers with a host of once unknown dietary benefits. Nowadays, while the consumption of the raw matrix still provides the original functional value, the separation of bioactive-enriched fractions has enabled the production of nutraceuticals, while the addition of nutraceutical fractions to food previously lacking or partially possessing them has lead to the industrial production of functionalised or fortified food respectively.

The separability of nutraceutically relevant fractions depends on the combination of several different features of the carrier matrix and of the fraction to be separated, namely: size, aggregation state and physical (hardness) and chemical (composition) features of the embodying matrix, chemical nature, bonding and degree of dispersion of the fraction of interest in the embodying matrix. Physical, chemical and electrostatic interactions between the embodying matrix, the fractions to be separated and exogenous agents (equipment and process auxiliary substances, and the environment) affect the desired separation; the chemical nature of the solvent (if any), the specific energy applied by the physical agent (if any), the frequency and intensity of the mechanical (e.g., ultrasound) and electromagnetic (e.g., microwave) field (if any), the processing time and temperature and the presence of specific case-by-case unwanted substances (e.g., water, oxygen, metal ions) play a role in the final outcome of the recovery process of the desired fraction.

2. Sub- and supercritical fluid extraction

Sub- and supercritical fluid extraction are promising separation processes aimed at replacing traditional lengthy, laborious, low selectivity and/or low extraction yield, toxic chemical-using separation processes (Herrero et al., 2006).

The low viscosity and (relatively) high diffusivity inherent in the supercritical state confers these solvents better transport properties than liquids solvents have. Furthermore, the "adjustability" of viscosity, density and solvent power (Del Valle and Aguilera, 1999) by acting on fluid density through the change of pressure and/or its temperature make them amply tunable to the specific separation needs. Last, but not least, using solvents generally recognized as safe (GRAS) helps meet the requirements of food and pharma processes.

Carbon dioxide is the most commonly used because of its moderate critical temperature and pressure (31.1 °C and 7.39 MPa, respectively) and GRAS status. Supercritical CO_2 (Brunner, 2005): dissolves non-polar or slightly polar compounds; has a high solvent power for low molecular weight compounds, which decreases with increasing molecular weight; has high affinity with oxygenated organic compounds of medium molecular weight; has low solvent power for free fatty acids and their glycerides, even lower for pigments, and low solubility for water at temperatures below 100 °C; does not dissolve: proteins, polysaccharides, sugars and mineral salts; exhibits an increasing separating capability as pressure increases for compounds that are scarcely volatile, have a high molecular weight and/or are highly polar. Due to its low polarity, small amounts of "modifiers" (also called co-solvents), in the form of highly polar compounds such as ethanol or water, are often used to improve the separability of polar solutes.

Preparative systems for processing solid or liquid samples have different configurations; basically, they consist of solvent pump(s) deliverying the main solvent and any required modifyer throughout the system, an extraction cell (separation from a liquid) or column (solid), and one or more separators in which the extract is collected when the solvent is expanded.

Extraction from solids is usually carried out discontinuously (batchwise) in a single stage because solids handling in pressurised vessels is troublesome and separation factors are high. Fluid mixtures often suffer from low separation factors and multi-stage contacting (countercurrently for highest effectiveness) becomes a necessity. When separation factors approach unity and many theoretical stages are required for the separation preparative chromatographic systems may be set up on a process scale (Brunner, 2005).

Pressurised liquids (e. g. water and ethanol at an intermediate temperature between their boiling points and their critical temperatures and under the appropriate pressure to maintain them in the liquid state) mitigate the drawbacks of supercritical fluids (e.g. supercritical carbon dioxide): their scarce affinity for polar solutes and extensive capital costs. Subcritical water has been successfully used for the extraction of essential oils, nutraceuticals and bioactives, among which polyphenols (King and Srinivas, 2009).

Water is a highly polar solvent at room temperature and atmospheric pressure due to hydrogen-bonding which reflects in its high dielectric constant. While at room temperature water is unsuitable as a solvent for non-polar compounds, at increased temperature a lowered dielectric constant (from 80 at 25 °C to 27 at 250 °C and 50 bar, intermediate between those of methanol, 33, and ethanol, 24, at 25 °C), viscosity and surface tension and an increased diffusivity confer an increased solvent power for non polar substances. Thermally labile compounds may be degraded at elevated temperatures (Teo et al., 2010) but newly formed bioactives (antioxidants) have also been observed (Plaza et al., 2010).

By adjusting the prevailing temperature and under the required pressure to remain in the liquid state, water may be tuned to the purpose, which may be that of an extraction solvent and/or reaction medium. The main parameters that influence the selectivity and extraction efficiency of pressurised water extraction include temperature, pressure, extraction time, flow rates and concentration of modifiers/additives. The residence time of the solute (reactant) in the aqueous medium becomes a critical parameter in conducting extractions above the boiling point of water and for optimizing reaction conditions (King and Srinivas, 2009).

Quality Preservation and Cost Effectiveness in the Extraction of Nutraceutically-Relevant
Fractions from Microbial and Vegetal Matrices

189

2.1 Mathematical modelling of Pressurised Water Extraction (PWE)

The very core of PWE extraction may be divided in four sequential steps which take place in the extraction cell filled with sample materials: 1. desorption of solutes from the various active sites in the sample matrix; 2. diffusion of the extraction fluid into the matrix; 3. partitioning of solutes from the sample matrix into the extraction fluid and 4. chromatographic elution of the solutes out of the extraction cell to the collection vial. A two-step, single-site partition-based thermodynamic model is appropriate for PWE extraction (Kubátová et al., 2000 and 2002; Windal et al., 2000): 1. the compound is desorbed from its original binding sites in the sample matrix, under diffusion control but at a sufficiently fast rate to avoid being the overall limiting step; 2. the compound is eluted from the sample under thermodynamic partitioning control (K_D). The shape of an extraction curve would be defined by:

$$\frac{S_b}{S_o} = \frac{\left(1 - \frac{S_a}{S_o}\right)}{\frac{K_D \cdot m}{(V_b - V_a) \cdot \rho} + 1} + \frac{S_a}{S_o}$$

where S_a is the cumulative mass of the analyte extracted after volume V_a (ml), and S_b is the cumulative mass of the analyte extracted after volume V_b (data point b is next to data point a in experimental sequence). S_o is the initial total mass of analyte in the matrix. S_b/S_o and S_a/S_o are the cumulative fractions of the analyte extracted by the extraction fluid of the volume V_b and V_a, respectively. K_D is the distribution coefficient, ρ is the density of extraction fluid at given conditions (g/ml), and m is the mass of the extracted sample (g). The model depends on the extractant volume flowed, but not time.

The experimental device required for PWE is quite simple. Basically, the instrumentation consists of a water reservoir coupled to a high pressure pump to introduce the solvent into the system, a thermostating oven, where the extraction cell is placed and extraction takes place, and a restrictor or valve to maintain the pressure. Extracts are collected in a vial placed at the end of the extraction system. In addition, the system can be equipped with a coolant device for rapid cooling of the resultant extract (Herrero et al., 2006).

2.2 Predicting solubility via the hansen solubility parameter

Predicting solubility parameters capable of telling whether one substance can form a solution by dissolution in another can be done by several thermodynamical methods with a variable degree of theoretical base and reliability. Following the reasoning that dissolution is linked to likeness, i.e. similarity in bonding to itself, a solubility parameter can be defined for every substance as the square root of the cohesive energy density $\delta = (E/v)^{1/2}$ where v is the molar volume of the pure solvent, and E is its (measurable) energy of vaporization (Hildebrand and Scott, 1950). The total energy of vaporization of a liquid consists of multiple individual components, among which (atomic) dispersion forces, (molecular) permanent dipole–permanent dipole forces, and (molecular) hydrogen bonding (electron exchange) (Hansen, 1997). Dividing this by the molar volume gives the square of the total (or Hildebrand) solubility parameter as the sum of the squares of the Hansen D, P, and H components, that is:

$$\delta_T = E/V = E_D/V + E_P/V + E_H/V = \delta_D + \delta_P + \delta_H$$

Hansen parameters can be calculated from literature physical properties and solubility data, molecular structure, or group contribution methods (Hoy 1989; Hoftyzer and van Krevelen 1997; Stefanis and Panayitou 2008, Hansen 2007). These three parameters can be treated as coordinates for a point in the Hansen space; distance in the Hansen space predicts likeliness of reciprocal solubility of the substance pair. To calculate the distance (R_a) between Hansen parameters of substances S_1 and S_2 in Hansen space the following formula is used:

$$\left(R_a\right)^2 = 4 \cdot \left(\delta_{D_2} - \delta_{D_1}\right)^2 + \left(\delta_{P_2} - \delta_{P_1}\right)^2 + \left(\delta_{H_2} - \delta_{H_1}\right)^2$$

which defines the radius of a sphere centered in substance S_1 and identifying the domain of substances which possess equal or better solubility than substance S_2.

An interaction radius (R_o), a characteristic quantity of the substance to be dissolved, can be also defined. The interaction radius effectively defines a metrics for the solubilisation of the substance under concern. Together, R_a and R_o determine whether the pair is within (solubility) range. The ratio R_a/R_o is called the relative energy difference (RED) of the system; if RED < 1 the molecules are alike and will dissolve, if RED = 1 the system will partially dissolve and if RED > 1 the system will not dissolve.

Correlation methods are available for the prediction of the temperature dependence of the solute solubility parameter, such as the Jayasri and Yaseen method (1980). The Hansen solubility prediction method lends itself to a computational identification of the optimal solvent, which corresponds (in principle) to the mixture exhibiting the lowest RED value. Using this approach, minimum diameter Hansen spheres can be obtained by changing the composition and the temperature of the (subcritical) solvent phase and solute-solvent interaction can be predicted and optimised.

Srinivas et al. (2009) studied the solvent characteristics of subcritical fluid solvents at different temperatures interacting with various bioactive compounds from their natural sources; they tested the applicability of this predictive method on valuable nutraceutical solutes extracted from natural sources. The use of this method will be presented with reference to the extraction of flavonoids from grape pomace.

Flavonoids consist of anthocyanins, flavonols, procyanidins with antimicrobial, antiviral and antioxidant as well as food colorant properties. Many investigations (e.g. Cacace and Mazza, 2002) have been carried out on extraction of anthocyanins from berry substrates using subcritical water and water–cosolvent systems, whose results have shown that a maximum extraction yield is obtained in the 120 to 160 °C temperature range. A lower solvent–feed ratio requirement compared to traditional solvent extraction techniques is reported (King et al., 2003). Studies on the effect of temperature and particle size on subcritical water extraction of anthocyanins from grape pomace pointed to an optimal yield at 120 °C using a 150 micron-sized substrate (King et al., 2007). These studies also indicated the possibility of thermal degradation of the target solutes occurring at higher temperatures.

Srinivas et al. (2009) applied the Hansen solubility method to the extraction of malvidin-3,5-diglucoside with water–ethanol mixtures by investigating: 1. solubility in the pure solvents as a function of temperature and 2. solubility in mixtures of the two solvents as a function of concentration. They predicted a slightly higher miscibility with ethanol (radius of 7.74 MPa$^{1/2}$ vs 9.73 Mpa$^{1/2}$) at different temperatures (25 to 75 °C for ethanol; 100 to 200 °C for water): although both ethanol and water can dissolve the target anthocyanins at different

Quality Preservation and Cost Effectiveness in the Extraction of Nutraceutically-Relevant
Fractions from Microbial and Vegetal Matrices

191

conditions, ethanol would be a preferred solvent for the extraction of anthocyanins. Mixture calculations show a significant decrease in the interaction radius of the Hansen sphere with addition of 10% ethanol. Although the maximum solvent power is predicted at 80% ethanol, the improvement over 10% ethanol is minimal. The optimal solvent temperature using hydroethanolic mixtures is in the range 25 to 75 °C. Results are broadly coherent with data reported by Monrad et al. (2010) and plotted in Figure 1.

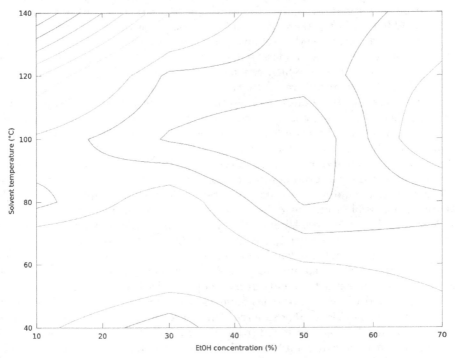

Fig. 1. Concentration of recovered Malvidin-3,5-O-diglucoside in hydroalcoholic extracts from subcritical extraction (Data from Monrad et al., 2010).

2.3 Multiple fluid process tuning for sub/supercritical fluid extraction

Multiple fluid processing involves the integration of two or more fluids held under pressure (above or below their critical temperature and pressure) applied as either mixtures or in a sequential manner for one or more unit processes (King and Srinivas, 2009). Their state may be variable (critical/subcritical/mixed). A large collection of solubility data is available in the literature (Gupta and Shim, 2007) and methods have been developed to predict it (such as the Hansen Solubility Parameter discussed above) to assist broad approaches in fraction separation and chemical transformation choices. Non polar gases can be tuned to the required solute capacity by adjusting their density (pressure) and to (ideally) match the required solubility parameter (polarity) by mixing them with polar modifier substances and possibly adjusting temperature and pressure. Polar substances, in their turn, may serve at one time as solvents and as reactants; water stands out in this respect because of its nil toxicity and cost and because it can potentially simplify process arrangement given that in most cases it is initially present in the vegetal matrix and its separation may be required for contact with non polar solvents (either organic or inorganic). Added substances may act as solubility enhancers (co-solvent: e.g. polar substances added to supercritical CO_2) or depressors (antisolvent: e.g. gases depending on their critical temperature); again, these added substances may also serve as reactants (e.g. hydrogen for hydrogenation reactions), which may be an atout despite the reduction in solvent power toward the other reactants. On the other hand, gas may be solubilised in subcritical liquid solvents, such as water. In addition to other properties (notably, for water, the dielectric constant) this may serve to tune pH by addition of CO_2, thereby obtaining a versatile medium with respect to acidic-based extraction chemistry and reactions (as in the extraction of flavonoids or in the pretreatment of lignocellulosic biomass) without the burden of a subsequent pH correction.

The change in solvent power which can be obtained by the change of pressure (in the supercritical state) or temperature (in the subcritical state) permits the implementation of multiple unit operations (fractionation of multiple compounds, chemical reaction) with only one running fluid. If pressure change to carry out reciprocal solute/solvent separation and fractionations are deployed for solute separation and energy consumption must be reduced, membrane separations (expecially nanofiltration and reverse osmosis) may be implemented to obtain bulk molecule size-based separation and avoid recompressing the solvent across the full pressure range. Performing a sequential arrangement of the operations may make important savings in equipment capital cost possible. Membrane coupling may be equally useful for compressible fluids (e.g. CO_2) and scarcely compressible fluids (e.g. water or hydroalcoholic solutions); these latter tend to yield dilute solutions that need concentration and membrane processing can obtain that cheaply and neatly.

Antioxidant-containing matrices (e.g. grape and olive waste) are a fruitful area in which to apply combinations of mixed critical and subcritical fluid and unit processing steps. Residence time of the extracted solute in the hot pressurized water must be minimized to prevent degradation of the anthocyanin moieties or their possible reaction with sugars to other products. Some evidence that side reactions in pressurized water could be generating antioxidant moieties has been gathered (King and Srinivas, 2009). An example of pipelined processing is then offered by sunflower oil triglyceride conversion to free fatty acids (in subcritical water) followed by enzymatic esterification of the free fatty acids to FAMES in supercritical CO_2 using lipase catalysis (Baig et al., 2008).

2.4 Cost-effectiveness analysis in supercritical fluid extraction

The extraction of valuable materials from solid substrates by means of SCFs has been carried out on a commercial scale for more than two decades. Large-scale processes are related to the food industry like the decaffeination of coffee beans and black tea leaves and the extraction of bitter flavours (α-acids) from hops. Smaller scale processes comprise the extraction and concentration of essential oils, oleoresins and other high-value flavouring compounds from herbs and spices, and the removal of pesticides from plant material. The extraction of edible oils would be a large-scale process, but, as for all commodity products, the value-added is not high, so the economy of the process is the main problem and must be considered separately for each case.

Oil from oleaginous seeds is traditionally produced by hexane extraction from ground seeds, with a possible thermal degradation of the oil and an incomplete hexane elimination. SC-CO_2 extraction of oil from seeds has been proposed in many cases (Coriander, fennel,

Quality Preservation and Cost Effectiveness in the Extraction of Nutraceutically-Relevant
Fractions from Microbial and Vegetal Matrices

193

grape, hyprose, sunflower etc--see Reverchon and De Marco, 2006) given that oilseed triglycerides solubilise well in SC-CO_2 at 40 °C and pressure higher than 280 bar. Main parameters of this process are particle size, pressure and residence time. After extraction, the extract solution is sent to a separator working at subcritical conditions so that the reduced solubility enables recovery of oil and elimination of gaseous CO_2 from oil. Alternatively, temperature variations may be used to recover the oil extracted in a process operated at a fixed pressure; energy consumption can be reduced if thermal integration is properly deployed throughout the plant (Reverchon and De Marco, 2006).

While several seed oils have been extracted this way up to the pilot scale, here we take sunflower oil as an example.

The main models of SFE from vegetal matrices are derived from classical models of mass transfer and extraction kinetics.

Several models were already published; they mostly assume plug flow of the fluid through the fixed bed of the solid matrix and mass transfer control located in the fluid phase (as done by Lee et al., 1986), in the solid phase (e.g. Reverchon, 1986), or in both (e.g. Sovová et al., 1994).

Most of them consider only one pseudocomponent, "the solute", characterised by one (averaged) solubility value obtained experimentally. The solid matrix of the seeds is porous, homogeneous, and with constant physical properties during extraction. The physical properties of the fluid are constant as well. Usually, pressure and temperature gradients that appear in the fixed bed are neglected. Perrut et al. (1997) considered one solid phase and one fluid phase, where external mass transfer is the controlling step. Axial dispersion in the bed is neglected and all pores are pre-filled by the solvent during the initial increase of pressure in the extractor until equilibrium is reached between the solid and liquid phase (initial condition). Oil concentration y* at the solid – SCO2 interface depends on an equilibrium relation, function of pressure and temperature y* = f(x, P, T):

$$\rho_f \cdot \varepsilon \cdot \left(\frac{\partial x_f}{\partial t} + v \cdot \frac{\partial x_f}{\partial z} \right) = j_f$$

with

$$v = \frac{Q/\rho_f}{\text{extractor section}} \cdot \frac{1}{\varepsilon}$$

j is the flux of solute that is exchanged between the solid and the fluid phase based on global mass transfer coefficient considered at equilibrium.

$$j_f = a_p\, k_f\, \rho_f\, (y^* - x_f)$$

Perrut et al. (1997) introduced the notion of "transition concentration" x_t, meaning that oil concentration in SCO_2 is equal to the thermodynamic solubility value y_o regardless of its concentration in the solid concentration as long as this latter is sufficiently rich (above threshold value x_t which is normally in untreated seeds); when the concentration in the solid diminishes oil concentration is determined with a partition coefficient K.

$$x < x_t \quad f(x) = K \cdot x$$
$$x \geq x_t \quad f(x) = y_o$$

Types of Foodstuff	Main adulterants
Milk and Milk products	Water, refined oil, separated (cream less) skim milk, solution, flours, animal fats, starch, dextrin.
Butter and Ice cream	Hydrogenated fats and animal fats, Artificial sweeteners,jelling agents, other fats, water nut flour, non permitted colours
Vegetable Oils and Fats	Coffee Powder Exhausted Coffee powder, starch, roasted dates, tamarind seeds, tea residues, other leaves with added color
Dry Beverages	Coffee Powder Exhausted Coffee powder, starch, roasted dates, tamarind seeds, tea residues, other leaves with added color
Wet Beverages	Spurious narcotics and other concentrated liquids, Artificial sweeteners (Saccharin), mineral acids, Colored invert sugar, high concentrated sugar solution, Cheap liquids, wine and beer water and juices.
Spices and Condiments	Whole turmeric Coating with lead chromate or coal tar dye starch or talc colored yellow with coal tar dye, Starch colored brown with coal tar dye, Coriander seed Other green colored seeds Powdered bran or sawdust colored green with dye, Starch colored red with coal tar dye, Argemone seeds, Artificial cumin seed like product, Dried papaya seeds and other plant gums.
Cereals and their products.	Wheat Stones, straws, low variety grains,Talc, chalk powder, tapioca flour, Maize flour, low variety rice and semolina.
Pulses and their products.	Yellow maize flour and Tapioca flour.
Miscellaneous Items	Seeds or nuts broken and colored, Pickle, jam, chutney and squash of cheap products. Also Sweets Artificial sweeteners and tapioca flour.

Table 1. List of common foodstuffs and their adulterants

Three parameters are adjusted to experimental extraction data (solubility, partition coefficient, transition concentration).

More accurate modelling can be obtained by adding: other dispersion axes (e.g., axial dispersion in the reactor volume and radial dispersion in the seed volume as done by Cocero and García, 2001); volume compartments; a shrinking core model; a dual-zone (intact cells and broken cells; Figure 2) continuous modelling inside the particle (Sovová, 2005). This latter model acknowledges that many cells have been broken by the pre-treatment but, inside the seed, cells are intact so that oil from the broken cells is "free" and can be directly extracted while oil in the intact cells must diffuse through broken cells. Correspondingly one equation must be written for the broken cells and one for the intact cells, where r is the fraction of broken cells in a crushed seed.

$$r \cdot \rho_s \cdot (1-\varepsilon) \cdot \frac{\partial x_1}{\partial t} = j_s - j_f$$

$$(1-r) \cdot \rho_s \cdot (1-\varepsilon) \cdot \frac{\partial x_2}{\partial t} = -j_s$$

The different fluxes between the broken cell and the fluid and between the intact and the broken cells are written (again resorting to the concept of transition concentration):

$$j_f = k_f \, a_p \, \rho_f \, (y^* - x_f) \; ; \; x_1 \neq x_t \text{ and } x_f < K \, x_t$$

$$j_s = k_s \, a_p \, \rho_s \, (x_2 - x_1)$$

The extraction of oil from oleaginous seeds might be one primary target of SFE if the added value of extracted vegetable oils were not very low. However, specialty oils or valuable components co-extracted with common oils could be reasonable target of SFE extraction (Brunner, 2005).

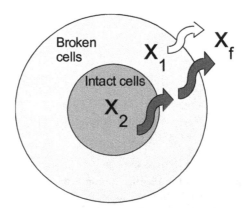

Fig. 2. Illustration of a dual-compartment modelling scheme for crushed seed extraction. (Reprinted from Boutin et al., 2011, with permission from Elsevier)

One drawback to widespread industrial scale adoption of supercritical fluid extraction use is the lack of realistic economical studies. In order to evaluate the feasibility of using SFE in the

Quality Preservation and Cost Effectiveness in the Extraction of Nutraceutically-Relevant
Fractions from Microbial and Vegetal Matrices

195

extraction of food products or components thereof, an estimate of the production cost must
be provided, generally by costing an approximate plant design providing the separation.
While the level of accuracy of the modelling (thermodynamics) and costing will affect the
reliability of the estimate, a rigorous approach is generally out of the scientific investigator's
reach and scope.

The assessment of the industrial economical feasibility of entire supercritical extraction
processes has been carried out by a few investigators. Citrus peel oil deterpenation (Diaz et
al., 2005), such as essential oil extraction from rosemary, fennel and anise (Pereira and
Meireles, 2007) and sunflower oil (Bravi et al., 2002). Then, given that yield and quality of
the product are simultaneous targets when investigating an extraction, an example of
recovery yield and product quality bridging will be given along the lines of Bravi et al.
(2003).

The solid substrate in most cases forms a fixed bed. The SCF flows through the fixed bed
and extracts the product components until the substrate is depleted. This extraction from
solids consists of two process steps, namely, the extraction, and the separation of the extract
from the solvent. During the extraction the supercritical fluid is fed and evenly distributed
at one end of the extractor where it flows, upward or downward, through a fixed bed of
solid particles of the vegetal/microalgal matrix and dissolves the extractable components.
The depletion boundary in the solid matrix will proceed in the direction of flow, as the
concentration of the extracted components in the solvent. The shape of the concentration
curve depends on the kinetic extraction properties of the solid material and the solvent
power of the SCF which, in turn, depend on operating conditions. For the solid as well as for
the solvent, the extraction is an unsteady process (Brunner, 1994).

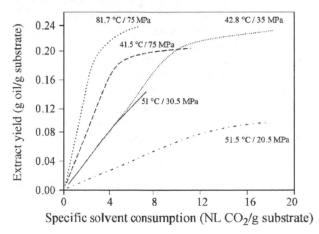

Fig. 3. Extraction yield vs mass of supercritical solvent flowed through the vegetal matrix
bed (With kind permission from Springer Science+Business Media: Gas extraction: an
introduction to fundamentals of supercritical fluids and the application to separation
processes, 1994).

A time profile of the extraction yield will be typically shaped as any one of the curves in Figure
3. The initial part of the curve is a straight line (constant extraction rate). At later times the
extraction rate decreases as the total amount of extractible substances in the substrate is

approached. Any extractible component in the solid follows the extraction course pattern. If mass transfer is fluid phase side-limited the extraction profile is represented by a straight line (the slope equalling the extraction rate); however, if mass transfer is split among the contacted phases the extraction has an exponential course. If the solvent enters the extractor free from the extractible compounds and there are no irreversible reactions of the extractible compounds with the substrate, the matrix can be totally depleted; otherwise the extraction curve approaches a non null asymptote (Brunner et al., 2005).

High-pressure processes are particularly energy intensive and economic feasibility depends on energy integration; the solvent cycle scheme plays a central role in this respect (Brunner, 1994). In supercritical fluid processes the solvent can be recirculated either in supercritical or in liquid state; correspondingly, the piece of equipment of choice would be a compressor or a pump. Pumps have a lower capital cost than compressors, but pump-based solvent cycles also require several heat exchangers and condensers and additional heat energy at low extraction pressures. Compressors have a higher capital cost than pumps, but compressor-based solvent cycles require only one heat exchanger and a limited thermal energy supply; at extraction pressures lower than 300 bar, the compressor-based system has higher electrical energy consumption and lower energy consumption compared to the pump cycle (Diaz et al., 2009).

Objective functions have been annualised cost, utility cost (Cygnarowicz and Seider, 1989), energy consumption (Diaz et al., 2000), net profit (Diaz et al, 2003), product cost (Bravi et al, 2002). The approaches range from DAE model resolution (Bravi et al., 2002), nonlinear programming (Cygnarowicz and Seider, 1989), mixed integer nonlinear programming (Diaz et al., 2000, 2003, 2005; Espinosa et al., 2005). Mixed integer programming permits the association of binary variables to design options (e.g. potential process units). Diaz et al. (2005) and Espinosa et al. (2005) performed the optimal design of process and solvent cycle by formulating and solving two nonlinear programming problems.

Rigorous mass transfer models are rare in such models: a dynamic optimization model for an extraction column, also including energy and momentum balances in the packed bed was developed and coded in gPROMS by Fernandes et al. (2007).

Alternative approaches include experimental data-based process optimization, which typically end up in nonlinear correlations among process variables (Létisse et al., 2007 adopted extraction temperature, solvent flow rate and operating time).

As an example here we discuss the optimisation approach followed by Bravi et al. (2002 and 2003). The optimisation requires 1. devising an industrially-feasible process layout and 2. identify optimal operating conditions and assess SFE-extracted sunflower oil economic economic acceptability in the food market. This latter aim requires the setup of a comprehensive mathematical model of the whole process (extraction section and recovery section).

The adopted process included multiple batch extractors in parallel, each containing a fixed bed of seeds, and multiple CO_2 recovery stages operating at decreasing pressures (Figure 4). The optimisation problem size was reduced by adopting: 1. Perrut's (1997) non-continuous and piecewise-linear solid-fluid equilibrium mathematical model of the extraction phase and extraction conditions (40 °C and 280 bar); 2. oil approximation as a single component; 3. mass transfer resistance occurring only in the solvent phase; 4. negligibility of in-extraction enthalpy variations; 5. requirement of constant extract flow rate.

A comprehensive model for the entire process describing extractors, expansion valve, separator, compressors, solvent recovery vessels, heat exchangers, pipe union joints and

Quality Preservation and Cost Effectiveness in the Extraction of Nutraceutically-Relevant
Fractions from Microbial and Vegetal Matrices

197

branches was set up. The BWR equation of state was chosen as a trade-off between accuracy and computational weight.

The performance criterion (objective function) adopted for the optimisation was the unit oil production cost, estimated by rigorously accounting the operating costs directly referenced by the process design (e.g., compression costs and duty requirements) and applying short-cut techniques for all the remaining operating costs (e.g., manpower) and for the equipment cost estimates (by the cost index method). Aim was finding the minimum unit oil production cost as a function of the time allotted for the extraction phase on each seed batch and of the prevailing pressure in the oil separator at constant: size of each extraction vessel, number of simultaneously flowed extraction vessels and circulating solvent flow rate.

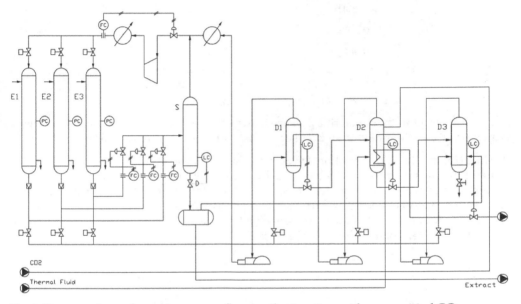

Fig. 4. Process scheme for contnuous sunflower oil extraction with supercritical CO_2 (Reprinted from Bravi et al., 2002, with permission from Elsevier).

A reduction of the required power expenditure for solvent recompression can be obtained by increasing the operating pressure of the oil separator but this reduces oil recovery. Cost minimisation led to optimal pressure in the separator of a single-extractor plant (100 bar). Furthermore, under the adopted equilibrium model, productivity collapses over time, when most of the seed bed has a very low oil concentration (Figure 5); from an economic standpoint product return decreases while the operating cost remains unchanged, thus leading to an increase of the average oil production cost. On the other hand, at the end of the extraction cycle the seed bed still contains an oil residue that cannot be exhausted by using re-circulated solvent but can be exhausted with hexane, yielding a lower-class product.

Reducing extraction time increases oil production rate at the expense of a lower recovery. In the example, extraction times of 20, 10, 5 minutes makes the best possible use of a plant with 3, 4, or 5 extractors. The authors found product cost to be minimum cost for a 4-extractor plant (0.67 Euro/kg)

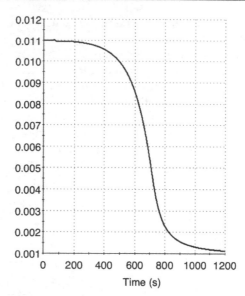

Fig. 5. Oil content in the supercritical phase leaving from a single extractor during a single extraction phase (Reprinted from Bravi et al., 2002, with permission from Elsevier)

Quality optimisation in extraction by SC-CO_2 can be obtained by suitably modelling the adopted quality parameters, which can be done relatively easily if these latter can be related to components whose concentration can be modelled by resorting to mass balances, mass transfer coefficients, and thermodynamic equilibria. Sunflower oil acidity modelling performed by Bravi et al. (2003) will be reported here as an example; another target may be the content of lipid-soluble alpha-tocopherol.

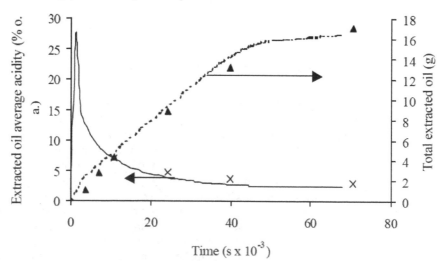

Fig. 6. Model prediction of instantaneous sunflower oil acidity and total extracted sunflower oil mass as a function of extraction time (From Bravi et al., 2003)

Quality Preservation and Cost Effectiveness in the Extraction of Nutraceutically-Relevant
Fractions from Microbial and Vegetal Matrices

199

The mathematical model of this process takes into account the mass and enthalpy balances of carbon dioxide and oil in the extractor, the expansion valve and the separator. Oil is considered as being made of two components: free oleic acid (as acidity is customarily reported), and a triacyglyceridic pseudo-component. Perrut's et al. (1997) mathematical model of the oil extraction was rewritten to account for two simultaneously extracted components. The parameter estimation was carried out by optimising the model fit on the data obtained by means of a suitable oil quality-oriented experimentation.

Batch extraction runs were carried out on ground sunflower seeds in Perrut's conditions at different CO_2 flow rates and the composition was recorded on the collected oil in four well-mixed batches per extraction run. The results shows an initial constant-rate extraction phase after which productivity declines; the combined time profile of the collected oil acidity as a function of the amount of oil extracted clearly shows that the acid component tends to concentrate in the first oil fraction (Figure 6).

From the point of view of process optimisation an improvement of the quality of oil can be obtained by two means: 1. by eliminating the first portion of the extracted oil, which gives the largest contribution to acidity and/or 2. by extending the oil extraction degree; in order to keep some more oil extracted in the initial stages of the operation, much more must also be extracted in the final stages to ensure that the overall acidity is within the limits. Both of these measures increase the oil production cost because the oil extraction rate is maximum at the beginning of the operation and then decreases continuously and asymptotically to zero. As suggested by Bravi et al. (2002) the only partially exhausted oil matrix can be treated in a conventional extraction plant using hexane; we add here that the first extracted oil, which features an excessive acidity but is hexane-free, could be de-acidified and sold as a different product.

When optimising SFE yield in a stream concentrated in some desired component, identifying yield and extraction kinetics as a function of temperature, solids particle size distribution and CO_2-to-solids mass ratio are a key step before modelling can be applied. However, industrially-relevant practices may reduce yield compared to lab vales.

As an example, based on Molero Gómez's et al. (1996) set of oil yield-relevant set of optimal extraction conditions from grape seeds (pressure 200 bar, temperature 40°C, seed fragment size 0.35 mm, seed moisture content up to 6.5% and processing time 2 h), Bravi et al. (2007) carried out a study aimed at investigate the yield and extraction kinetics of α-tocopherol-enriched grape seed oil as a function of temperature, solids particle size distribution and CO_2-to-solids mass ratio (hereinafter denoted as CSMR) and identifying the optimal extraction conditions defined as those ensuring a high yield in α-tocopherol-concentrated oil. Their experimental procedure did not include pre-soaking (elsewhere denoted as 'equilibration') of the seed matrix to better reflect the prospective process conditions in a forthcoming industrial use (From Bravi et al., 2003, with kind permission from AIDIC Servizi s.r.l.).

They confirmed that oil yield with CO_2 (14.4%) is slightly below that with hexane (15.4%) when the vegetal matrix is treated at a moderately low temperature and observed that temperature effects on yields must be analysed with care, as they may be the result of complex interactions between oil solubility in CO_2 and mass transfer coefficients, arguing that this may be due to a soaking degree which changes during the extraction itself. As Bravi et al. (2007) pointed out, this may lead to unexpected inverted yield vs temperature relationships.

The effect of fragment size is that of increasing the oil yield at any tested CSMR; this effect can be explained with the increase in the available surface area for mass transfer and with the reduction of the time required for the initial soaking of the vegetal matrix. From the practical point of view, the results of their work suggest that the grape seeds should be milled to a maximum size of 425 mm.

As far as α-tocopherol in the extract is concerned, its concentration in the oil extracted with SC-CO_2 increases with the extraction temperature; this result is coherent with the higher solubility of α-tocopherol in SC-CO_2 at 80°C than at 40°C measured by Chrastil (1982).

3. Optimisation of enzyme-assisted extraction

Whatever the solvent, purely solvent-based extraction (i.e., without the intervention of synergic agents) of bioactive compounds often suffers from low extraction yields, and requires long extraction times and leaves traces of the organic solvent used, generally toxic to some extent. When solvent extraction is carried out, resistance to solute migration to the bulk of the solvent may be controlled within the solid phase, within the liquid phase, or be split among the two. When the first, or even the last is true, non polar solvents may be able to overcome cellulosic barriers but may be unable to solubilise the desired compounds; in turn, polar solvents may be a suitable solvent for the desired compounds but may be unable to reach the solute location site inside the matrix. In both cases, reducing the mechanical hindrance to solute migration may make the operation significantly faster and reduce solvent requirements; as a side gain, it also reduces any degradation reaction that may affect the desired product during the extraction itself.

Enzymes, derived from bacteria, fungi, animal organs or vegetable/fruit extracts, have been used particularly for the treatment of plant material prior to conventional methods for extraction. Plant materials like vanilla, pepper, mace, mustard, fenugreek, rose, and citrus peel which have potential as a rich source of flavor have been studied for enzyme-assisted extraction of flavors. Similarly enzyme-assisted extraction of color has been studied in plant materials like marigold, safflower, grapes, paprika, tomato, alfalfa, and cherries (Sowbhagya and Chitra, 2010). Winemakers may make use of pectolytic enzymes to break down the middle lamella between the pulp cells and the pulp and skin cell walls releasing pigments, improving both juice yields and colour extraction (Ducruet et al., 1997). A list of some commercially relevant products recently obtained using enzyme-assisted extraction is reported by Puri et al. (2011).

Enzymes have been used to enhance extraction (e.g. flavonoids from plant material as by Kaur et al., 2010) while minimizing the use of solvents and heat and disrupt the pectin-cellulose complex (e.g. in citrus peel to enhance flavonoid production by Puri et al., 2011). A digestion step prior to extraction by solvents was found to be necessary to efficiently carry out the extraction from the raw material (Dheghan-Shoar et al., 2011).

Various enzymes such as cellulases, pectinases and hemicellulases are often required to disrupt the structural integrity of the plant cell wall; these enzymes hydrolyze cell wall components and degrade the pectin-cellulose complex in fruits, thereby increasing cell wall permeability, which results in higher extraction yields of bioactives.

The enzymatic reactions are usually conducted at low temperature (15 °C to 45 °C), the actual operating temperature being dictated by the trade-off between the two controlling phenomena (mass transfer enhancement by increased temperature which reduces required

Quality Preservation and Cost Effectiveness in the Extraction of Nutraceutically-Relevant
Fractions from Microbial and Vegetal Matrices

201

extraction time and thermal degradation of any thermolabile extracted compound). Above ~60 °C, heat alters the enzyme molecule irreversibly.

In order to carry perform cost-effective enzyme-assisted extractions the features and subtleties of enzyme catalysis must be considered and the appropriate enzyme or enzyme combination for the plant material selected must be identified.

Cost effectiveness and optimisation of enzyme-assisted extraction is obtained by identifying the optimal extraction conditions, aimed at maximising process profitability, which entails maximising the recovery rate of the target bioactive(s) while minimising their in-process degradation. Optimisation parameters, therefore, belong to two sets relevant to the enzymatic pretreatment and to the extraction phase respectively. The pretreatment should be optimised with regard to prevailing temperature and pH, pretreatment time, enzyme solution-to-solid ratio, solid particle size (distribution), enzyme load and enzyme composition; the extraction should be optimised with regard to prevailing temperature and pH, time, and solvent system deployed.

Synergism is due a special consideration when optimising enzyme-assisted extractions as it is for any enzyme-assisted hydrolysis. Considering the example of lycopene extraction from tomato waste, enzyme synergism may show up when they are acting simultaneously (e.g. Zuorro et al., 2011) or sequentially (Ruiz Teran et al., 2001). In the former case, enzyme preparations containing 50:50 pectinase and cellulase were found to have a significant synergistic effect (rate x18 w.r.t untreated material) with respect to simple enzymes (rate x3); in the latter, it was observed that cellulase (from *Trichoderma reesei*) and mixed enzyme cocktail (a mixture of arabinases, cellulases, hemicelullases, xylanases, and pectinases from *Aspergillus niger*) do not work efficiently when used together. Furthermore, results obtained using cellulase after the enzyme cocktail are similar to those observed when the cocktail is used alone. However, when cellulase was used first the extractive reaction proceeded with the highest efficiency (Table 1).

Enzymes used for pretreatment	Vanillin g/100 g
Water (control)	1.07 ± 0.08*
Cellulase + water	1.17 ± 0.06*
Cellulase + ethanol	2.70 ± 0.17*
Viscozyme + cellulase + water	1.17± 0.11
Viscozyme + cellulase + ethanol	2.66 ± 0.07
Cellulase + viscozyme + water	2.30 ± 0.10
Cellulase + viscozyme + ethanol	3.66 ± 0.04
Viscozyme + water	1.17 ± 0.05*
Viscozyme + ethanol	2.45 ± 0.21*

Table 1. Effect of enzyme treatment on vanillin extraction from vanilla beans. Viscozyme L. (Novozymes) is a mixture of arabinases, cellulases, hemicelullases, xylanases, and pectinases (Adapted from Ruiz Teran et al., 2001).

Enzyme-assisted extraction of bioactive compounds from plants has potential commercial and technical limitations: 1. the cost of enzymes (although this is going to decrease thanks to biofuel research); 2. inability of currently available enzyme preparations to hydrolyze some fractions of plant cell walls, limiting extraction yields of some compounds; 3. scale up to industrial scale may be troublesome because local process conditions in the equipment may be difficult to maintain at the larger scale.

Enzyme-assisted extraction optimization by traditional methods is time consuming and can ignore the interactions among various factors. The response surface method enables the evaluation of several process parameters simultaneously (along with their interactions up to the desired order) such as: time, temperature, pH, enzyme type and concentration. During cell wall degradation the polysaccharide–protein colloid can be degraded thereby creating an emulsion that interferes with the extraction. Therefore, non-aqueous systems are preferable for some materials because they minimize the formation of polysaccharide-protein colloid emulsions (Puri et al., 2011). Prior knowledge of the cell wall composition of the raw materials helps in the selection of an enzyme or enzymes useful for pretreatment; predictive methods would be useful to speed up the optimisation or keep up with product property changes, but nothing covering this gap has reached the open literature yet.

4. Optimisation of Microwave-Assisted Extraction (MAE)

Microwaves are electromagnetic waves in the frequency range 300 MHz to 300 GHz, that is between wavelengths of 1 cm and 1 m. Most small size microwave instruments operate at 2450 MHz and have an energy output between 600-700 W. At this frequency, the electric field changes the orientation of water molecules 2.45 x 109 times every second, while thermal agitation tends to restore the chaos inherent to the system. Thus creating an intense heat that can escalate as quickly as several degrees per second (depending on frequency and sample size, it has been estimated up to 100 °C/s at 4.9 GHz by Lew et al., 2002). The heating phenomenon is based on the interaction of the electrical field with the individual compounds of a material (possibly characterised by an inhomogeneous macrostructure). The transformation of electromagnetic energy in thermal energy occurs by two mechanisms: ionic conduction and dipole rotation. The ionic conduction generates heat due to the resistance of medium to ion flow. The migration of dissolved ions causes collisions between ions and molecules because the direction of the ions changes every time the electromagnetic field changes its orientation. Dipole rotation is related to the alternating movement of polar molecules, which try to line up with the electric field; thus only selective and targeted materials warm-up based on their dielectric constant.

The efficiency of the microwave heating depends on the dissipation factor of the material, $\tan \delta$, which measures the ability of the sample to absorb microwave energy and dissipate heat to the surrounding molecules as given by:

$$\varepsilon'' \tan \delta = \varepsilon'$$

where ε'' is the dielectric loss which indicates the efficiency of converting microwave energy into heat while ε' is the dielectric constant which measures the ability of the material to absorb microwave energy. The rate of conversion of electrical energy into thermal energy in the material is described by:

$$P = K \, f \, \varepsilon' \, E^2 \tan \delta$$

where P is the microwave power dissipation per unit volume, K is a constant, f is the applied frequency, ε' is the material's absolute dielectric constant, E is the electric field strength and tan ı is the dielectric loss tangent.

Quality Preservation and Cost Effectiveness in the Extraction of Nutraceutically-Relevant
Fractions from Microbial and Vegetal Matrices

203

Multiple collisions from this agitation of molecules generate energy dissipation and therefore a temperature increase whose value depends on the local intensity of energy dissipation, on the local specific heat, and on the local conductivity by which heat diffuses to or from neighbouring areas of the material when these latter are cooler or hotter.

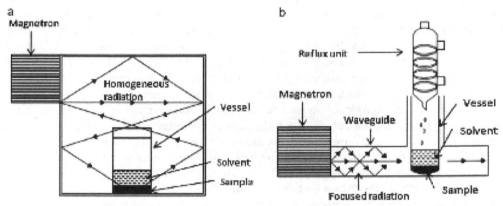

Fig. 7. (a) Closed type microwave system and (b) open type microwave system. (Source: Mandal et al., 2007)

Microwave-assisted extraction aims at supplying heating locally where the solvent or soaking medium is present, thus speeding up heating, with the aim of reducing bioactives degradation reactions and cause a discregation of the vegetal matrix structure. This latter aim requires that the basic structure of the extracted particles be stiff. Vegetal cells generally feature a stiff structure; microalgae lacking frustule or thick outer exopolysaccharide envelope may not benefit from microwave-assisted extraction (Pasquet et al., 2011), while wall-possessing microalgae benefit greatly from it (yields 3x-5x in a fraction of the time; Cravotto et al., 2008).

Microwave assisted extraction may be carried out in closed and open systems (Figure 7). Modifications of the basic scheme include operation under reduced pressure (for operation at reduced temperature and suction-enhanced migration of solutes); under nitrogen blanket (for increased protection against oxydation); without added solvent (resorting to constitutive/hydration water); with simultaneous microwave and ultrasound field (to facilitate the formation of cracks in the solid matrix, change the perceived polarity of the solvent and enhance the mass transfer) (Chan et al., in press).

Microwave-assisted extraction effectiveness strongly depends on: solvent choice, solvent to feed ratio, extraction time, microwave power, temperature, sample characteristic, effect of stirring.

Solvents which are suitable for conventional extraction techniques may not be suitable for microwave assisted extraction. Ethanol (used in the 40% to 100% concentration range) is a good microwave converter (good absorber as a dipole with a low specific heat) and is by itself suitable for extracting many active compounds from plants. However, modifiers can be added to solvents which are unsuitable per se to microwave capture in order to enhance their overall performance. Water was added as modifier to diethyl ether and ethanol or water can be added into poor microwave absorber such as hexane to enhance microwave heating efficiency (Alfaro et al., 2003; Ku et al., 2007). Room temperature ionic liquids interest lies in their negligible vapor pressure, wide thermal range in the liquid state, good thermal stability, tunable viscosity, miscibility with water and organic solvents, good solubility and extractability for various organic compounds (Du et al., 2007). Mixing carbonyl iron powders with the moist sample may increase absorption of microwave energy, particularly where solvent is limited (such as in SFMAE by Wang et al., 2006).

An optimum ratio of solvent to solid ratio ensures homogeneous and effective heating; above, heating may be insufficient and below mass transfer barrier may establish as the distribution of active compounds is concentrated in certain regions and their displacement out of cell matrix may be impaired (Mandal et al., 2010). The optimum ratio of solvent to solid ratio normally is in the interval 10--50 ml per g of solid matrix (Chan et al., in press).

Microwaves heat the sample locally and act as a driving force for damaging the plant matrix so that analytes can diffuse out and reach the solvent. Furthermore, the solvent viscosity and surface tension decrease improves mass transfer. Therefore, increasing the power will generally improve the extraction yield and result in shorter extraction time; however, the gain may be offset by thermal degradation of the solute, so that an optimal temperature (and microwave power) exist for each application (Chan et al, in press).

Process time (regardless of power) is from a few minutes to more than 1 h. Extending process time has been found to decrease the extraction yield due to thermal degradation, oxidation, or hydrolysis. The optimum microwave application mode should be adapted to the type of solid matrix and may be power-regulating (continuous medium power or pulsed high-power microwaves), for optimised degradation of the solid matrix; temperature-regulating (power is dependent variable), for optimised preservation of the bioactive extracts.

Prior to treatment the vegetal matrix is usually dried, powdered and sieved, avoiding too a small particle size which would be difficult to separate and would require equipment clean-up procedures later. Pretreating with water the vegetal matrix after drying may increase bioactives release thanks to the localised heating of the trapped soaking water which gives rise to cracks. The modified water content of the vegetal matrix also alters the balance between hydrolyzation (favoured) and oxidation of active compounds (Wang et al., 2006).

Stirring mitigates the negative effect of low solvent to feed ratio on extraction yield by improving mass transfer and thus avoiding the buildup of concentration gradients that impair the dissolution of bioactives bound to the sample matrix.

The determination of optimum MAE operating conditions is usually carried out through statistical optimization studies. A collection of optimised extraction conditions for a number of plant-sourced vegetal matrices is reported by Chan et al. (in press)..

A variation of MAE involves the deployment of microwaves without any added solvent. Historically, dry distillation was used by alchemists for sublimation and extraction. Solvent free microwave-assisted extraction (SFME) was conceived for laboratory scale applications in the extraction of essential oils from different kinds of aromatic plants. SFME is neither a modified microwave assisted extraction (MAE) which use organic solvents, nor a modified hydro-distillation which use a large quantity of water, both of which are more energy intensive due to the larger heat requirements to evaporate and condense the added solvents. Based on a relatively simple principle, this method involves placing plant material in a microwave reactor, without any added solvent or water. The heating undergone by

Quality Preservation and Cost Effectiveness in the Extraction of Nutraceutically-Relevant
Fractions from Microbial and Vegetal Matrices

205

constitutive water within the plant material expands the plant cells and leads to rupture of the glands and oleiferous receptacles, whereby essential oil is first freed and then evaporated together with the in situ water of the plant material. The vapours are continously condensed by a cooling system located outside the microwave oven. The water excess is refluxed to the extraction vessel in order to restore the in situ water to the plant material. A large number of different essential oils have been extracted by SFME (Lucchesi et al., 2004).

5. Optimisation of ultrasound-assisted extraction

Extraction enhancement by ultrasound is attributed to the propagation of ultrasound pressure waves, And to the resulting cavitation phenomena. The implosion of cavitation bubbles generates turbulence which accelerates diffusion and impingements and collisions which result in surface peeling, erosion, punching (Ugarte-Romero et al., 2006) and particle breakdown. This effect provides exposure of new surfaces further increasing mass transfer (Vilkhu et al., 2008).

Dry materials may swell, hydrate and increase their pore size under ultrasound treatment. Furthermore the particle size distribution of the vegetal matrix is shifted toward the smaller sizes, so that the cell surface directly exposed to extraction increases (Vinatoru, 2001).

Solvent selection is usually based on achieving high molecular affinity between the solvent and solute. When cavitation bubbles are generated in the bulk of the solvent phase by the ultrasound field, their hydrophobic surfaces increase the net hydrophobic character of the extraction medium (Vilkhu et al., 2008) so that its affinity toward non polar components is increased. Cavitation being initiated by drop of the total pressure below the saturation pressure of the solvent, ultrasound assistance in extraction with a given solvent will be affected by the physical properties of this latter (cavitation intensity decreases as vapour pressure and surface tension increase). When a supercritical solvent is used, cavitational events are impossible, since there is no liquid/gas phase boundary. However, other mechanisms are postulated, such as acoustic streaming and the presence of gas pockets in the solid causing cavitational collapse (Patist and Bates, 2008).

Ultrasound may also give rise to multiple simultaneous processes, such as extraction and (sono)chemical modification, whereby the food product may be modified by physical and chemical mechanisms. Cravotto et al. (2004), for instance, reported wax conversion to policosanol (common name for a mixture of C_{24}–C_{34} linear saturated fatty alcohols, a rich source of nutrients and pharmacologically active compounds) during rice bran extraction by using ultrasounds.

Although it is relatively easy to perform an ultrasound-assisted extraction at the laboratory scale, designing it for industrial scale is quite demanding. Some of the issues that need consideration when attempting to design an optimal ultrasound-assisted process at a significant scale have been reported by Vilkhu et al. (2008): 1. the nature of the tissue being extracted and the location of the components to be extracted with respect to tissue structures; 2. pretreatment of the tissue prior to extraction; 3. the nature of the component being extracted; 4. the effects of ultrasonics primarily involve superficial tissue disruption; 5. increasing surface mass transfer; 6. intra-particle diffusion; 7. loading of the extraction chamber with substrate; 8. increased yield of extracted components; 9. increased rate of extraction, particularly early in the extraction cycle enabling major reduction in extraction time and higher processing throughput.

Apparently, there is potential for ultrasonic cavitation to propagate free radicals (hydroxyl). Radical production should be quenched by the addition of small amounts of ethanol to cool cavitation bubbles and slow any radical-involving reactions (Vilkhu et al., 2008).

6. Optimisation for novel sources of bioactives: Microalgae production

The microalgae have in practice an interesting composition in regard to main components such as protein, polyunsaturated fatty acids (PUFA), pigments, and carbohydrates (Doucha 2009). The protein content is consistently high in micro algae. Some cyanobacteria (blue-green algae) are characterised by a high protein content (60-65%), not commonly found among higher plants. But, for full utilisation of the protein, special treatment of the microalgae is generally necessary. Moreover, microalgae are excellent producers of essential amino acids. However, until to date only three species are cultivated on industrial scale level: these are the cyanobacterium *Arthrospira*, the green algae *Chlorella* and *Dunaliella*. Their biomass is used for production of a rather limited range of products, most of them directed to the nutraceutical market. The success of these three species is due to the fact that they can be grown in a very selective medium (*Arthrospira* and *Dunaliella*), therefore contamination of parasites or competing organisms (microalgae, fungi, and others) is naturally prevented even in open reactors where it is possible to ensure a low cost of production for the biomass (Boussiba and Affalo, 2005), while *Chlorella* is endowed with a remarkably high growth rate sustained by organic acid addition in fermentor.

6.1 *Chlorella vulgaris*

Chlorella vulgaris cells contain β-1,3-glucan, polysaccharides and also a rich source of proteins, 8 essential amino acids, vitamins (B-complex, ascorbic acid), minerals (potassium, sodium, magnesium, iron, and calcium), β-carotene, chlorophyll, "CGF" (Chlorella growth factor), as well as other health-promoting substances (Hac´on-Lee et. al 2010). β-1,3-glucan is an active immunostimulator, a free- radical scavenger and a reducer of blood lipids (Ryll et al., 2003). However, various other health-promoting effects have been clarified (efficacy on gastric ulcers, wounds, and constipation; preventive action against atherosclerosis and hypercholesterolemia and antitumor action). *Chlorella vulgaris* biomass has colouring properties and has been tested with success as a pigment source for farmed products with functional activity (e.g. as antioxidants). The total annual production of *Chlorella* is estimated to be about 2000 t. The production process is based on the mixotrophic nature of the *Chlorella* strains and uses acetic acid as a carbon source. The production cost of biomass is not clear. However, based on claims that in those systems a high biomass concentration is achieved (more than 10 g/l) one may reach the conclusion that the production cost may be a figure close to those of open pond systems (10 -15 US$/Kg). However, the cost can raise to up to 30 US$/Kg, for example in the Central part of Europe where the adverse climatic conditions do not allow to grow the alga all the year around. Up to about 10 years ago most of the production took place in Taiwan and only 10-15% of the total production was carried out in green houses in Japan. The market for *Chlorella* products is limited to the Far East, mainly Japan. In some early works it was reported that *Chlorella* extracts may have affect the growth and production of lactic acid by lactic bacteria. The growth facilities are based on round concrete ponds mixed by a rotating arm which also provides CO_2 and acetic acid

Quality Preservation and Cost Effectiveness in the Extraction of Nutraceutically-Relevant
Fractions from Microbial and Vegetal Matrices

207

supply. Another development that should be mentioned is the attempt to set up new facilities for culturing *Chlorella* in Europe. One of them is located in Czech Republic (2 tons a year production capacity, Kopecky personal communication) which is based on inclined reactors that allow to maintain a fast flow rate of thin layer culture and as a result enables maintenance of high biomass concentrations and high volumetric productivity (Masojidek et al. 2010). The second facility is located in Germany, and uses tubular photobioreactors made with glass tubes arranged on a vertical fence and placed in a greenhouse. The total annual capacity is claimed to reach 150 tons per year (Pulz 2001). Although *Chlorella* market is still the largest one in term of gross revenue US$, one can expect that without developing new products and reducing the production cost *a sine qua non condition* for use of *Chlorella* biomass as feed additive in the animal feed market, it is difficult to expect an expansion of the production capacity.

6.2 *Arthrospira platensis*

At present *Arthrospira* (commercially indicated as *Spirulina*) represents the second most important commercial microalga in term of total market value US$ (after *Chlorella*), while in terms of total biomass produced, the *Arthrospira* market is twice or more of that occupied by *Chlorella* (Torzillo and Vonshak 2003). The major producers of *Arthrospira* are the DIC group of companies, Earthrise in California, USA, Hainan DIC Marketing in Hainan Island, China. On the whole these facilities produce about 1000 metric tons of *Arthrospira* annually (Belay et al. 2008; Sili et al. 2011). An other important *Arthrospira* producer is Cyanotech Corporation of Hawaii with an annual production of 300 tons. Other producers are located mainly in the Asia-Pacific region, particularly in China and India (Lee et. al 1997). The highest production capacity of *Arthrospira* biomass takes place in China. Recent estimates are that the total potential of the different sites of this country may exceed 2000 t. Production is carried out in raceway ponds of 2000-5000 m² in size and may contain between 400 and 1000 m³ of culture according to the dept adopted which can vary between 15 and 40 cm depending on season, desired algal density and, to a certain extent, the desired biochemical composition of the final product (Belay et. al 2008). The major share of the market for this organism is for health food involving crude biomass production. This has the advantages of simple processing (harvest and rudimentary handling) keeping production costs reasonably low, and of eluding the competition of the chemical industry which cannot match the wealth in nutritional bioactive components and attractiveness of natural products (Boussiba and Affalo 2005). *Arthrospira platensis* (Soletto et. al 2008, Harun et. al 2010) has commercialized as nutraceutical food, also a strong immune-stimulated molecule, Immulina®, can be extracted from it (Grzanna et. al 2006).

6.3 Dunaliella

This species is being grown as a source of beta-carotene. this carotenoid can accumulate in the cells grown under nutrient limitation and high sun light up to 12% of the dry weight (Ben-Amoz and Avron 1973). This pro-vitamin A product is widely used in the feed and food industry. Today, the product is available mainly in two forms: dried or extracted. Dried *Dunaliella*, in powder or pill form, is considered to be highest quality. The product is harvested by means of concentration and centrifugation, and thereafter dried by spray-drying. In this form, the product is mainly addressed to health food market for direct human consumption. The price is based on the beta-carotene content, and can reach about 2000 US$ per Kg of beta-carotene. The second kind of product is beta-carotene extracted into a vegetal oil. This product can be applied as a food colorant and a pro-vitamin additive for human consumption, for fish and poultry feed, or in the cosmetic industry as an additive to sunscreen products. From few reports it is estimated that the price, on the basis of beta-carotene content, varies in the range of US$ 500-600 per Kg of beta-carotene.

7. References

Alfaro M. J., Bélanger J. M. R., Padilla F. C., Jocelyn Paré J. R. (2003) Influence of solvent, matrix dielectric properties, and applied power on the liquid-phase microwave-assisted processes (MAP) extraction of ginger (*Zingiber officinale*). *Food Res Int* 36 (5): 499--504.

Baig M. N., Alenezi R., Leeke G.A., Santos R. C. D., Zetzl C., King J. W., Pioch D., Bowra S. (2008) Critical fluids as process environment for adding value and functionality to sunflower oil; a model system for biorefining, in: *Proceedings of the 11th Meeting on Supercritical Fluids*, Barcelona, Spain, May 4–7, 2008.

Belay A. (2008) Spirulina (*Arthrospira*): production and quality assurance. *In: Gershwin ME and Belay A (eds) Spirulina in human nutrition and health. pp1-25, CRC Press Taylor & Francis group, London, UK, pp.312.*

Ben-Amoz A., Avron M. (1983) On the factors which determine massive beta-carotene accumulation in the halotolerant alga Dunaliella bardawil. *Plant Physiol* 72: 593-597.

Boussiba S., Affalo C. (2005) An insight into the future of microalgal biotechnology. *Innovations in Food Tehnology (www.innovfoodtech.com).*

Boutin O., De Nadaïa A., Perez A. G., Ferrasse J.-H., Beltran M., Badens E. (in press) Experimental and modelling of supercritical oil extraction from rapeseeds and sunflower seeds, *Chem. Eng. Res. Des.*

Bravi M., Bubbico R., Manna F., Verdone N. (2002) Process optimisation in sunflower oil extraction by supercritical CO_2. *Chem. Eng. Sci.* 57: 2753 – 2764.

Brunner G. Gas extraction, in: An Introduction to Fundamentals of Supercritical Fluids and the Applications to Separation Processes, Springer, Berlin, 1994.

Cacace J. E., Mazza G. (2002) Extraction of anthocyanins and other phenolics from black currants with sulfured water. *J. Agric. Food Chem.* 50 (21): 5939--5946.

Carr A. G., Mammuccari R., Foster N. R. (2011) A review of subcritical water as a solvent and its utilisation for the processing of hydrophobic organic compounds. *Chem Eng J* 172: 1-17.

Chacón-Lee T. L. and González-Mariño G. E. (2010) Microalgae for "Healthy" Foods - Possibilities and Challenges. *Comp Rev Food Sci Food Saf, Vol. 9, (6): 655–675.*

Chana C.-H., Yusoffa R., Ngoha G.-C., Kung F. W.-L. (in press) Microwave-assisted extractions of active ingredients from plants – A review. *J, Chromatography A*

Cocero M. J., García J. (2001) Mathematical model of supercritical extraction applied to oil seed extraction by CO_2+ saturated alcohol--I. Desorption model. *J. supercrit. fluids* 20 (3): 229--243.

Cravotto G., Boffa L., Mantegna S., Perego P., Avogadro M., Cintas P. (2008) Improved extraction of vegetable oils under high-intensity ultrasound and/or microwaves. *Ultrason sonochem* 15 (5): 898--902.

Quality Preservation and Cost Effectiveness in the Extraction of Nutraceutically-Relevant
Fractions from Microbial and Vegetal Matrices

209

Cygnarowicz M. L., Seider W. D. (1989) Effect of retrograde solubility on the design optimization of supercritical extraction processes. *Ind. Eng. Chem. Res.* 28 (10): 1497--1503.

Dehghan-Shoar Z., Hardacre A. K., Meerdink G., Brennan C. S. (2011) Lycopene extraction from extruded products containing tomato skin. Int. J. Food Sci. Technol. 46: 365–371.

Diaz M. S., Espinosa S., Brignole E. A. (2003) Optimal solvent cycle design in supercritical fluid processes. *Lat Am Appl Res* 33 (2): 161--165.

Diaz S., Brignole E. A. (2009) Modeling and optimization of supercritical fluid processes. *J. Supercrit Fluids* 47: 611–618.

Diaz S., Espinosa S., Brignole E. A. (2005) Citrus peel oil deterpenation with supercritical fluids Optimal process and solvent cycle design. *J. Supercrit Fluids* 35: 49–61.

Diaz S., Gros H., Brignole E. A. (2000) Thermodynamic modeling, synthesis and optimization of extraction--dehydration processes. *Comp. Chem. Eng.* 24 (9-10): 2069--2080.

Doucha J. and Lívanský K. (2009) Outdoor open thin-layer microalgal photobioreactor: potential productivity.*J Appl Phycol 21:111–117.*

Du F.Y., Xiao X. H., Li G. K. (2007) Application of ionic liquids in the microwave-assisted extraction of trans-resveratrol from *Rhizma Polygoni Cuspidati. J Chromatography A* 1140 (1-2): 56--62.

Ducruet J. A., Dong Canal-Llauberes R. M., Glories Y. (1997) Influence des enzymes pectolytiques séléctionées pour l'oenologie sur la qualité et la composition des vins rouges, *Rev Franc Oenol* 155: 16–19.

Espinosa S., Diaz M. S., Brignole E. A. (2005) Process optimization for supercritical concentration of orange peel oil. *Lat Am Appl Res* 35 (4) 321--326.

Fernandes J., Ruivo R., Mota J. P. B., Simoes P. (2007) Non-isothermal dynamic model of a supercritical fluid extraction packed column. *J supercrit. fluids* 41 (1) 20--30.

Garcia-Salas P., Morales-Soto A., Segura-Carretero A. and Fernández-Gutiérrez A. (2010) Phenolic-Compound-Extraction Systems for Fruit and Vegetable Samples. *Molecules, 15: 8813-8826.*

Grzanna R., Polotsky A., Phan P.V., Pugh N., Pasco D., and Frondoza C.G.(2006) Immolina, a High–Molecular-Weight Polysaccharide Fraction of Spirulina, Enhances Chemokine Expression in Human Monocytic THP-1 Cells. *J.l of Alternative and Complementary Medicine. June 2006, 12(5): 429-435.*

Gupta R. B., Shim J., Solubility in Supercritical Carbon Dioxide, CRC Press, Boca Raton, FL, USA, 2007.

Hansen CM. 2007. Hansen solubility parameters: a user's handbook. 2nd ed. Boca Raton, Fla.: CRC Press.

Harrod M., Macher M.-B., Hogberg J., Moller P. (1997) Hydrogenation of lipids at supercritical conditions, in: Proceedings of the Fourth Italian Conference on Supercritical Fluids and their Application, Capri, Italy, September 7–10, 1997, pp. 319–326.

Harun R., Singh M ., Forde G.M. and Danquah M. K. (2010) Bioprocess engineering of microalgae to produce a variety of consumer products. *Renewable and Sustainable Energy Reviews 14:1037–1047.*

Herrero M., Cifuentes A., Ibáñez E. (2006) Sub- and supercritical fluid extraction of functional ingredients from different natural sources: Plants, food-by-products, algae and microalgae A review. *Food Chem* 98: 136–148.

Herrero M., Plaza M., Cifuentes A., Ibáñez E. (2010) Green processes for the extraction of bioactives from Rosemary: Chemical and functional characterization via ultra-performance liquid chromatography-tandem mass spectrometry and in-vitro assays. *J. Chromat A*, 1217: 2512–2520.

Hildebrand, J., Scott, R. L., *The Solubility of Nonelectrolytes*, 3rd Ed., Reinhold, New York, 1950.

Jain T., Jain V., Pandey R., Vyas A., Shukla S. S. (2009) Microwave assisted extraction for phytoconstituents – An overview. *Asian J. Research Chem.* 2 (1): 19-25.

Kaur, A., Singh, S., Singh, R. S., Schwarz, W. H., Puri, M. (2010) Hydrolysis of citrus peel naringin by recombinant α-L-rhamnosidase from Clostridium stercorarium. J. Chem. Technol. Biotechnol. 85: 1419–1422.

King J. W., Gabriel R. D., Wightman J. D. (2003) Subcritical water extraction of anthocyanins from fruit berry substrates. *Proceedings of the 6th Intl. Symposium on Supercritical Fluids – Tome 1*; April 28–30, 2003; Versailles, France.

King J. W., Howard L. R., Srinivas K., Ju Z. Y., Monrad J., Rice L. Super Green 2007. Pressurized liquid extraction and processing of natural products. *Proceedings of the 5th Intl. Symposium on Supercritical Fluids*; November 28–December 1, 2007; Seoul, South Korea.

King J. W., Srinivas K. (2009) Multiple unit processing using sub- and supercritical fluids. *J. Supercrit Fluids* 47: 598–610.

Kubátová A., Jansen B., Vaudoisot J. F., Hawthorne S. B. (2002) Thermodynamic and kinetic models for the extraction of essential oil from savory and polycyclic aromatic hydrocarbons from soil with hot (subcritical) water and supercritical CO_2. *J. Chromat A* 975: 175–188.

Lavecchia R, Zuorro A. (2008) Improved lycopene extraction from tomato peels using cell-wall degrading enzymes. Eur Food Res Technol 228:153–8.

Lee Y. K. (1997) Commercial production of microalgae in the Asia-Pacific rim. *J Appl Phycol* 9: 403-4511.

Létisse M., Rozières M., Hiol A., Sergent M., Comeau L. (2006) Enrichment of EPA and DHA from sardine by supercritical fluid extraction without organic modifier: I. Optimization of extraction conditions. *J. supercrit. fluids* 38 (1): 27--36.

Lew A., Krutzik P. O., Hart M. E., Chamberlin A. R. (2002) Increasing Rates of Reaction: Microwave-Assisted Organic Synthesis for Combinatorial Chemistry. *J. Comb. Chem.*, 4 (2): 95–105.

Lu Y., Yue X.-F., Zhang Z.-Q., Li X.-X., Wang K. (2007) Analysis of *Rodgersia aesculifolia* Batal. Rhizomes by Microwave-Assisted Solvent Extraction and GC–MS. *Chromatographia* 66 (5-6): 443-446.

Lucchesi M. E., Chemat F., Smadja J. (2004) Solvent-free microwave extraction of essential oil from aromatic herbs: comparison with conventional hydro-distillation. *Journal of Chromatography A* 1043: 323–327.

Quality Preservation and Cost Effectiveness in the Extraction of Nutraceutically-Relevant
Fractions from Microbial and Vegetal Matrices

211

Mandal V., Mandal S. C. (2010) Design and performance evaluation of a microwave based low carbon yielding extraction technique for naturally occurring bioactive triterpenoid: Oleanolic acid. *Biochem Eng J* 50 (1-2): 63--70.

Mandal V., Mohan Y., Hemalatha S. (2007) Microwave Assisted Extraction – An Innovative and Promising Extraction Tool for Medicinal Plant Research. Pharm Rev 1 (1, Jan-May): 7-18.

Masojidek J, Prasil O. (2010) The development of microalgal biotechnology in the Czech Republic. *J Ind Microbiol Biotechnol 37:1307–1317.*

Mercer P., Armenta E. E. (2011) Developments in oil extraction from microalgae. *Eur. J. Lipid Sci. Technol.* 24 (30-31): 539–547

Miller D. J., Hawthorne, S. B. (2000) Solubility of liquid organic flavor and fragance compounds in subcritical (hot/liquid) water from 298 to 473 K. *J Chem Eng Data* 45: 315-318.

Monrad J. K., Howard L. K., King J. W., Drinivas K., Mauromoustakos A. (2010) Subcritical Solvent Extraction of Anthocyanins from Dried Red Grape Pomace. *J. Agric. Food Chem.* 58: 2862–2868.

Pasquet V., Chérouvrier J. R., Farhat F., Thiéry V., Piot J. M., Bérard J. B., Kaas R., Serive B., Patrice T., Cadoret J. P., Picot L. (2011) Study on the microalgal pigments extraction process: Performance of microwave assisted extraction. *Proc Biochem* 46: 59–67.

Patist A., Bates D. Ultrasonic innovations in the food industry: From the laboratory to commercial production. *Inn Food Sci Emerg Technol* 9: 147 – 154.

Pereira C.G., Meireles M.A.A. (2007) Economic analysis of rosemary, fennel and anise essential oils obtained by supercritical fluid extraction. *Flavour Frag. J. 22 (5): 407-413.*

Plaza M., Amigo-Benavent M., del Castillo M. D., Ibáñez E., Herrero M. (2010) Facts about the formation of new antioxidants in natural samples after subcritical water extraction. *Food Res Int* 43: 2341-2348.

Plaza M., Herrero M., Cifuentes A., Ibáñez, E. (2009) Innovative Natural Functional Ingredients from Microalgae. *J. Agric. Food Chem.* 57: 7159–7170.

Pulz O. (2001) Photobioreactors: production systems for phototrophic microorganisms. *Appl Microbiol Biotechnol 57: 287–293.*

Puri M., Sharma D., Barrow C. J. (in press) Enzyme-assisted extraction of bioactives from plants. *Trends Biotech.*

Ryll J., Scheper T., Lotz M. (2003) Biotechnological production of β-1,3-Glucan in a technical pilot plant. Abstr. Eur. Workshop Microalgal Biotechnol., Germany, p. 56, 2003).Sili C., Torzillo G., Vonshak A. (2010) *Arthrospira.* In: Ecology of cyanobacteria. Whitton BA, Pott M (eds), Kluwer Academia Publishers. Dordrect/ London/ Boston (in press).

Sovová H. (2005) Mathematical model for supercritical fluid extraction of natural products and extraction curve evaluation. *J. supercrit. fluids* 33 (1): 35--52.

Sowbhagya H. B., Chitra V. N. (2010) Enzyme-Assisted Extraction of Flavorings and Colorants from Plant Materials. Crit Rev Food Sci Nutr 50 (2): 146-161.

Sowbhagya H. B., Chitra V. N. (2010) Enzyme-Assisted Extraction of Flavorings and Colorants from Plant Materials. Crit Rev Food Sci Nutr 50:146–161.

Srinivas K., King J. W., Monrad J. K., Howard L. R., Hansen C. M. (2009) Optimization of Subcritical Fluid Extraction of Bioactive Compounds Using Hansen Solubility Parameters. J Food Sci 74 (6): 342--354.

Teo C. C., Tan S. N., Yong J. W. H., Hew C. S., Ong E.S. (2010) Pressurized hot water extraction (PHWE). *J. Chromatography A.* 1217 (16): 2484--2494.

Torzillo G., Pushparaj B., Masojidek J. and A. Vonshak A. (2003) Biological constraints in algal biotechnology. *Biotechnol. Bioprocess Eng. 8: 339–348.*

Torzillo G., Vonshak A. (2003) Biotechnology for algal mass cultivation. In: Recent advances ion marine biotechnology (Fingerman M., Nagabhushanam R. eds) Volume 9 Biomaterials and Bioprocessing, Science Publishers, Inc. Enfield (NH), USA, pp. 45-77.

Ugarte-Romero E., Feng H., Martin E., Cadwallader R., Robinson J. (2006) Inactivation of *Echerichia coli* in apple cider with power ultrasound. *J Food Sci* 71: 102–109.

Vilkhu K., Mawson R., Simons L., Bates, D. (2008) Applications and opportunities for ultrasound assisted extraction in the food industry – A review. *Inn Food Sci Emerg Technol* 9: 161–169.

Vinatoru M. (2001) An overview of the ultrasonically assisted extraction of bioactive principles from herbs. *Ultrasonics Sonochemistry* 8: 303–313.

Wang Z., Ding L., Li T., Zhou X., Wang L., Zhang H., Liu L., Li Y., Liu Z., Wang, H., Zeng H., He H. (2006) Improved solvent-free microwave extraction of essential oil from dried *Cuminum cyminum* L. and *Zanthoxylum bungeanum* Maxim. *J. Chromatography A* 1102 (1-2): 11--17.

Windal I., Miller D. J., De Pauw E., Hawthorne S. B. (2000) Supercritical Fluid Extraction and Accelerated Solvent Extraction of Dioxins from High- and Low-Carbon Fly Ash. Anal. Chem. 72: 3916-3921.

Zuorro A., Fidaleo M., Lavecchia R. (2011) Enzyme-assisted extraction of lycopene from tomato processing waste. *Enz. Microb. Technol.* 49 (6-7): 567-573.

11

Employment of the Quality Function Deployment (QFD) Method in the Development of Food Products

Caroline Liboreiro Paiva and Ana Luisa Daibert Pinto
University Federal of Minas Gerais
Brazil

1. Introduction

Currently, in a more intensive way, companies have been forced to adapt to the new competitive market. The technological industry changes' occurring since the 80's brought implications for the international competition, specially the demarcation of new areas of global competition. That happens due to the acceleration of technological changes added by the shortening of the life cycle of products and processes, besides the increasing of the products differentiation. In fact, what is observed is that these factors have not only led companies to restructure their production systems and their types of management, but above all, to guarantee the capacity to deliver to a market, products even more sophisticated. The ability to realize alternatives to compete in the market, to develop strategies and to invest in appropriate training that is what has ensured the survival and profitability of organizations in this new structure.

In terms of food industry competition, what is observed is that the integrity of the product has become the main focus. This means that product excellence, in the food industry, goes beyond simply offering goods with basic attributes. These attributes have only become a precondition for the company to keep playing the competitive game. In fact, nowadays, the products must not only satisfy, but above all, surprise their customers. What is realized is that these consumers have accumulated experience with several products and become sensitive to small differences in many ways. This means that innovations in products and processes increase the excellence standard of product, making the process of development an essential factor for the competition of enterprises.

Certainly, the best projects require staff competence, efficiency in the work, on the exchange of information between functions and on the understanding of the market needs into technical language. They also require efficiency in problem solving and in the use of its resources. In this sense the Quality Function Deployment (QFD) method has proven to be efficient in order to translate in a more effectively way the needs and expectations of consumers, to promote greater interaction between the teams involved in the project, to accelerate the solution of problems and to reduce the development time.

The QFD was invented in the late 1960's in Japan. Within the context of TQC (Total Quality Control), the Japanese model of quality management system was responsible to cause a revolution in the production system of that country. All that was possible, due to the emphasis on the product quality considering the point of view of the customer. QFD is the unfolding, step by step, of functions or operations that make up the product quality. The methodology seeks to solve the problems inherent to the product's development process in their early stages, in a way that the critical points that determine the quality of the product and the manufacturing process are established in the phase of their design and controlled during the development stages. The methodology also ensures the achievement of quality because it works with a focus on consumer needs. More specifically, it translates the consumer's requirements into technical language and then ensures their satisfaction along the process of product's development.

The quality matrix is the tool used to organize the consumer's needs into technical information. The matrix goal is to define the pattern, quantitative or qualitative, of each attribute of quality of the final product. The other matrices are due to the quality matrix and aim to detail the project so that all the factors that contribute to the achievement of the final product are designed, as characteristics of the intermediates products, parameters of the manufacturing process, raw materials and inputs.

In addition, the QFD method assists the management of product development process because it coordinates the flow of information and organizes activities in terms of functions. It promotes the functional integration and rapid resolution of problems.

With all that, the purpose of this chapter is to describe the potential use of the QFD method into product development in food companies. The study initially intends to contextualize the management of product development in the food industry and show the QFD method as a tool capable of directing, in a practice way, how to plan and conduct the activities of the process of product development. So the steps for the application of QFD in the development of a food product will be detailed. In addition, support tools within the marketing research and sensory analysis will be suggested.

2. Product differentiation: A strategy adopted by the food industry

The food industry never has launched so many new products as it has in recent years. Due to factors such as technological development, increasing of competitiveness in the sector due to the growth of the competition such in and out of the countries, and greater consumer demand which incorporated new values to its preferences, the shelves of supermarkets receive daily new products (Athayde, 1999).

The focus on markets niche is one of the strongest trends today in the food sector. There is a search for products that provide pleasure to be consumed, such as the sophisticated products with high added value, or looking for fun products aimed at children. Likewise, products that refer to a particular region of the world, or of exotic flavors are searched by another portion of the market of processed foods.

Allied to all that, a strong feature of the new releases is the convenience in food consumption. This requirement is related to changes on consumers' lifestyle. The growing participation of women in the labor market, added by the increasing mobility of consumers, reduced the demand for ingredients to prepare meals at home and increased the offer of practical foods that can be consumed at any time, and of ready to go foods or pre-prepared.

Industry has also been required to apply new technologies in the development of food and beverages, specially the search for new ingredients. The changes in consumption habits it is driven by the concern for the health, aesthetics and environment. It demands food products of low-calorie, healthier and natural and environmentally friendly. A strong trend is the launch of products, which besides the presence of the sensory and nutritional quality, also present health benefits, so-called functional foods.

It is also important to emphasize the growing importance of equipment suppliers. Companies specializing in process engineering, who believe the research as a basis for technological innovation, have an important role in the development of food products (Earle, 1997). New technologies are able to provide new concepts of product, new alternatives for use, and being difficult to be imitated by competitors.

Finally, in addition to the significant number of new products available on the market in recent years, it is worth noting the great contribution of the sector of packaging for the food market, making it possible that these strategies of differentiation, segmentation and consumer convenience can be realized.

3. Stages of product development process

The process of product development, outlined in a model, consists of a sequence of activities ordered in time or a set of tasks that aim to facilitate the management of the process as a whole.

There isn't a standard development model that fits all circumstances and conditions experienced in a company. However, if the company adapts your way of management to a model more suited to its environment, probably the company will get better performance in their innovation processes. Anyway, the consensus is that development must be conducted so that the product reaches the market as quickly as possible, providing to the product the quality expected by customers and having costs optimized.

Students of product development management have different ways of representing the necessary steps to this process. Picture 1 seeks to represent the basic steps; steps that will assist the planning, the development of the product itself and the release of the same. Of course the product will have a greater chance of market success with this process if there is efficient management. For Clark & Wheelwright (1993), this means that the company should have skills to quickly identify opportunities, which often leads them to introduce new products ahead of their competitors. The best projects require also the team's competence, work efficiency, the exchange of information between functions and translation of the markets needs in technical language. They also required efficiency in problem solving and in the use of resources.

In Picture 1, the process of product development is represented by *stage-gates*. The *stages* are the various stages of development and the *gates*, decision points that precede each stage, opening or closing the door to continue the project (Cooper, 2001). These *gates* serve as critical steps for assessing the projects. The results of these evaluations are reflected in the decision to continue the project, drop it, stop it or resume it on another occasion.

Before joining the project into the development phase, the organization must seek the means that will ensure that the product will reach customers needs. Several market research – research of needs and desires of the consumers, competitive analysis and concept testing - will help to define more precisely the concept of the product. The first step in this direction is to translate the information inside and outside of the company in technical language, until you define the product's concept. For this, extract the data through market research, group discussions, customer complaints and tacit knowledge of employees. The various functions involved in the process are then in charge of mapping information and developing the work.

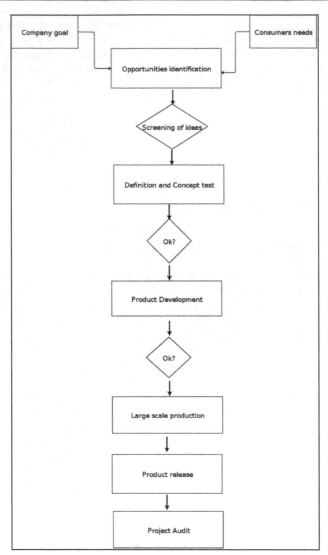

Fig. 1. Stages of the process of product development

In the stage of identifying opportunities, the company must seek ways to generate ideas for the new products. This can be achieved by internal efforts, through research in the departments of R&D, through contests to stimulate ideas for new products, or through the Customer Services, in meetings, using *brainstorming* techniques, or by stimulating a business culture that valorize the opinions and ideas of employees. On the other hand, the ideas for new products can come from external sources such as quantitative or qualitative research with target consumers. Other sources of ideas can come from university research publications or specialized organs, experience and knowledge of sales staff, contact with suppliers and also reverse lookup on products of competitors.

In this step it should be also made a prior assessment of the market for each idea, considering its size, segments and potential. It should also be evaluated the feasibility of manufacturing the product, the ability to be accepted by the market as well as their vulnerability towards competitors products or substitutes.

The ideas should then go through a team that will evaluate and select the ideas by checking out promising, profitable or those that must be rejected. Every idea that is nominated as possible to be developed will then go to the stage of definition and testing of its concept. The concept of a product can be defined as the expected benefits to meet the needs and expectations of consumers. The concept definition phase must determine the target audience, what are the main benefits that the product will present and a more appropriate occasion to consume it.

After the definition of the product's conception is convenient to test it. The test of the concept is a marketing research technique used to assess the market potential of the concept. Provides estimates of intent to purchase and sales volume. Define "who" would use the product, in which "circumstances" and how "often".

Finally, we must make the financial analysis of the project. The size of the market, the expected market share, the price analysis, along with technical cost estimates of equipment and for product launch are the inputs needed to make such an analysis. Once the project is defined, the only thing needed is that top management approves it, so the development of the product can be started.

Only then the product will go into the product development stage itself. However, it is necessary to first make the process and product planning. Regarding the product is necessary to define the product requirements, such as: the ingredients that will be needed, the most suitable additives, the quantity/volume that will be marketed. It is still important to define the requirements of legislation, such as: what will be the standard of identity and quality of the product, if the planned additives are allowed by the competent organizations and what the limit of application, and also the labeling requirements.

In relation to the manufacture of the product, it is necessary to first specify the parameters of the process, which involves the study of manufactured technology and the parameters of quality and of process that need to be controlled in the manufacturing line.

The development process then proceeds to the phase of preparation of the prototypes, usually in an industrial kitchen for the definition of the formulation and of the sensory products parameters. Soon after it should be performed sensory tests in one or more prototypes, if possible with a sample of the target market, in order to verify the acceptance of the product.

Thus, the development of the product passes to the manufacturing phase of the pilot which consists in the manufacture of the product on a small industry scale, in order to define the quality parameters of intermediate products and process parameters that will be monitorized. Likewise, tests should be made of pilot products: sensory tests, again if possible with a sample of the target market, in order to verify if the product remains viable. Only then the company will plan the production on an industrial scale.

Soon afterwards the company can produce on an industrial scale, to launch the new product. In the launch phase is necessary to determine: a release date, geographical location and potential consumers in the target market. It is necessary to establish an advertising plan, that would include promotional and dissemination strategies.

Finally, the last step is to evaluate the project developed. The company has an opportunity to implement its system of product development through learning gained during the implementation of each individual project. The project audit aims to verify the strong and weak points and to define strategies for improving the performance of future projects. It is believed that only a deep understanding of the causes of problems and circumstances in which they occur, will allow the company to improve the performance of development activities, by improving the procedures, processes, management skills, methods, making the company able to develop a faster process, more efficient in the use of resources and in the development of products of higher quality.

4. The QFD Method

The QFD Method, *Quality Function Deployment*, originated in Japan in the late 60's, as a result of the study of the professors Akao and Mizuno (Mizuno, 1969). On this occasion, the movement for the Total Quality in that country had already achieved very significant results. The ideas of quality emerged after World War II starting with the Statistical Process Control (SPC) and evolved in the late 60's, to a much broader approach, in which it was already understood at the system level, and not only in technical terms or isolated functions, but also in management terms, and thus practiced throughout the whole organization. To get an idea, in 1968, the Quality Control (QC) in Japan had already reached the point where virtually all firms made usage of the QC in some way (Mizuno, 1969).

However, there was a gap in establishing the quality into a level of development of products. There were questions about what points should be considered in the design phase of projects that could operationalize the quality planning of both products and processes. There were also difficulties in ensuring that the planned quality was actually executed in the phase of serial production (Mizuno, 1969).

Then arises from these needs, the initial concepts of *Quality Deployment*, and in 1972, after conducting some researches, the ideas become practically implemented. In 1978, it was published the book "*Quality Function Deployment*" which gave a new impetus to the dissemination of QFD, causing it to be quickly implemented in several companies in the country.

Currently, QFD inspires a strong interest in the world, generating ever-new applications, practitioners and researchers each year. This method is in use in several countries in the world such as South Africa, Germany, Australia, Brazil, China, Spain, United States, Italy, India, Japan, Mexico, United Kingdom, Sweden and others, not only in product development, but also in developing manufacturing processes, software, services, etc. (Akao & Mazur, 2003; Chan & Wu, 2002).

In the U.S., QFD has become known in 1983 after conducting a seminar on the subject in Chicago. It was initially introduced in the *3M Corporation*. Currently, the use of QFD in the U.S. is in almost all industry sectors, particularly in the automotive, electronics, software and services industry. It is also used by the space industry.

In 1996 a survey was conducted through a collaboration of Tamagawa University and the University of Michigan on the applications of QFD in the U.S. and Japan. It was selected 400 companies from each country. 146 Japanese companies (37%) and 147 American (37.6%) responded to the survey. According to the results, 31.5% of Japanese companies and 68.5% of Americans use the QFD (Akao & Mazur, 2003).

In Europe, QFD is also well known and many application cases have been reported. In other parts of the world, one can mention the innovative applications of QFD in Australia in the area of strategic planning and development of new business or improving existing business (Melo Filho & Cheng, 2007).

In emerging countries such as Brazil, QFD was introduced in 1989 and the concern now is how to make the method more effective, better understood and applied (Akao & Ohfuji, 1989). In China, the Quality Bureau from the State Bureau of Technical Supervision, a national agency of The People's Republic of China, has invited Professor Akao to give QFD seminars in Peking and Shanghai since 1994. India has shown a strong interest in the application of QFD, specially in software industry and in manufacturing industries such as trucks, automobiles, and farm tractors (Akao & Mazur, 2003).

The true in general is that the QFD method has ensured the achievement of project quality because it relays in one point that is the most cited by scholars of the subject as essential to the success of the product: a focus on customer needs. In addition, assists in managing the development process because it coordinates the flow of information and organizes activities in a function level. Thus promoting cross-functional integration and quick problem solving.

4.1 Method's approach

The QFD method, as it was originally designed by the professors Akao and Mizuno, includes the deployment of information, called the Quality Deployment (QD) and displays of work, addressed as Quality Function Deployment narrowly defined or restricted (QFDr).

In the first approach, QFD works detailing the necessary information to the innovation process. For that, are used tables, matrices, and the conceptual model, called the basic units of the QD. On the tables the data are organized, which in turn will be linked into the matrices. The interaction between the matrices is shown in the conceptual model (Akao, 1996).

The beginning of the process of extracting information in the QFD always starts from a table, so it is considered as the elementary unit of the method. It has the main purpose of deploying the information, always starting from a more general level to a more concrete. Using data from market research or internal information of the company, the work team uses the tables to detail the information, which are then arranged so that they are grouped according to their level of abstraction. Thus, the characteristics, requirements or functions that aren't so explicit, they become more clear for the working group.

The use of matrices in QFD aims to translate succinctly the relationship between two tables. It is a way of storing information and at the same time, to visualize the degree of interaction between each element of a table in relation to all the other elements of the other.

The conceptual model is the structure within the QFD that allows the visualization of the path taken to deploy the information until they get the technical standard processes. According to the sequence of matrices, it is able to verify a relation of cause and effect between the characteristics of the final product, its components, their functions, costs, raw materials and intermediate processes for their manufacture. Thus, it has been stored in a visible and detailed way, all product design and process.

The second approach of the method refers to the deployment of the work (QFDr). The technical and management procedures are established to ensure that all functions involved in the activities have their tasks previously established. The QFDr aims to specify who will do the job and how it will be done. Thus, from this deployment of the work it can be generated a set of documents, such flowchart of product development and a plan to manage the product development activities. The first determines the functional areas involved in each stage of development and the procedures for carrying out the work. The second specifies the schedule for each activity within the project.

4.2 Elaboration of the quality matrix of the final product

The Quality Matrix of the finished product is the first matrix that should be developed within the QFD method. In it are contained all information relating to the finished product. This topic displays an example of developing step-by-step from the quality matrix of a functional ready to bake dough for pies (Pinto & Paiva, 2010). While the development of other matrices of raw materials, intermediate products and processes are not treated in this chapter, the understanding of this first matrix will benefit the reader to understand how the matrices are made in the context of QFD.

4.2.1 Listing of primitive data

The primitive data are informations written in colloquial language, which can be collected through interviews or questionnaires with consumers, through discussion with focus groups or can be extracted from consumer complaints. They may also be got from opinions of company employees and in the news world. When the consumer does not directly express their needs, the imagination of scenes, or occasions of consumption, facilitates the description of the item required.

To meet the needs of the target market related to the dough for pies, there was a market research through semi-structured interviews with a sample of thirty possible consumers of the product. In the interview it was assessed the characteristics that the interviewers hoped to find in the ready dough for pies through the deployment of the scene in the manner, place and circumstances under which they would like to consume the product.

With the primitive information obtained, it was listed the greatest possible number of consumer desires. An example of this conversion is when an interviewee said that "the dough for pie should be used for both pies – sweet and salty," and the translation of primitive data for a required item was that "the dough for pie has to have a neutral flavor."

4.2.2 Establishment of the required qualities

At this stage you just have to format the primitive language, obtained from the market or from the deployment of scenes, observing certain rules. It is important that the terms of the customer requirements are simple, summarized in a single sentence, without explanation and did not have double meaning, making sure that the desired quality is clear. Whenever possible, you should be careful to avoid expressions in the form of denial, employing for this, affirmatives expressions.

Then the customer requirement qualities must be arranged in a table, the table of required qualities. This table is assembled from right to left. From the more concreted level to the more abstract. Generally, for food products, the markets requirements are grouped in terms of looks or appearance, flavor, texture, ease of preparation.

For the elaboration of the table it should be observe the following script: sentences with the same content should be eliminated to avoid repetition. The sentences should be arranged so that they can be viewed in only one frame (tertiary level, Table 1). It must then be joined in groups of four or five sentences with similar content and add expressions of customer requirements that represents the groups formed (secondary level, Table 1). With the phrases of similar content from the previous procedure must be formed other groups and add expressions of customer requirements to represent the groups formed (primary level, Table 1).

Primary Level	Secondary Level	Tertiary Level
Looks nice	Nice texture	Being soft
		Being crunchy
		Being a dough that dissolves easily in the mouth
	Nice color	Have a color next to cream/beige
	Appealing aspect	Have an uniform size
		Have an uniform thickness
Being tasty	Pleasant aroma	Have an appetizing aroma
	Pleasant flavor	Have a neutral taste
Satisfaction of the preparation	Being fully	Being fully
Being safe	Being safe	Being safe
Being healthy	Being healthy	Being functional
		Have a padronized caloric value

Table 1. Customer requirements to the functional dough for pies.

In the example of the functional dough for pies, it was constructed a table of deployment of the required qualities mainly from the joining of different sensory aspects of the product (Table 1).

4.2.3 Establishment of the quality characteristics
From the customer requirement of the tertiary level, must be extracted the technical characteristics of the finished product. At this point you have to convert the world of market into the technological world, drawing as much as possible, technical characteristics that will be easy to be measured. To do this, you should use the following reasoning: "How the required quality could be assessed in the final product?"

Then the table of quality characteristics should be built the same way as the table of the required qualities was. It should be built in groups thinking in the objectives of measurement or types of analysis to be carried out in the final product. For example, in the case of food products, in physico-chemical, microbiological and sensory analysis (Table 2).

Primary Level	Secondary Level	Tertiary Level
Physico-chemical characteristics	Physico-chemical characteristics (cold dough)	Thickness of the dough for pie
		Diameter of the dough for pie
		Ash content
		Moisture of the dough for pie
		Baking time
		Fiber content
		Carbohydrate content
		Protein content
		Fat content
Sensory characteristics	Visual (baked dough)	Color
		Integrity
	Taste (baked dough)	Aroma
		Neutral flavor
		Soft texture
		Crispness
		"Hollow" texture
Microbiological characteristics	Microbiological characteristics (cold dough)	Coliforms at 45°C
		Salmonella sp/25g
		B. cereus
		Estafilococcus coagulase positive

Table 2. Table of quality characteristics of the dough for pies

4.2.4 Establishing the correlations in the quality matrix

In the central part of the matrix it's necessary to make the correlation of each required quality with each characteristic quality. To this must be observed the following rules:
1. Judge each relationship independently.
2. Assign symbols for each correlation which correspond to numeric values. The meanings can be:
    ◎ or 9: Strong correlation;
    ○ or 6: There is a correlation;
    △ or 3: Possible correlation
3. For each required quality should be at least one strong correlation.
4. The symbols can not be concentrated in one place only.

5.   There should not be an item excessively marked with symbols.

With this information it was possible to build the quality matrix of a functional ready to bake dough for pies (Picture 2). There was done a correlation between these required qualities by the market and the quality characteristics of the finished product, assigning values 3, 6 or 9.

4.2.5 Establishment of the planned quality

The first column of the planned quality is the degree of importance. This must be established by the survey with consumers of the target market. When the survey is done, it is necessary to launch the averages values obtained in the matrix.

In the example of the functional dough for pies, the degree of importance of the required qualities has been established through research with thirty-two prospective buyers, where the interviewed indicates how important each characteristic was on a scale from 1 (unimportant) to 5 (very important). The medians for each attribute were also launched into the matrix.

After the development of the prototypes in industrial kitchens or in a pilot plant, it should be performed a search for sensory analysis with a sample of the target market. For this, should perform an affective sensory test with samples of one or more developed prototypes and a competitor's product, if any.

In the survey of the sensory analysis can be used items of the second level of the table of deployment of required qualities. Based on the type of scale used in the sensory test, it should be launched into the matrix the averages or medians of the results of sensory analysis.

In the case of the functional dough for pies, for the sensory analysis of the products developed, it was used the test of acceptance by the hedonic scale, varying gradually from 1 to 9, based on attributes like or dislike. Fifty tasters commented on all the attributes initially listed as important to the market. The medians of the attributes evaluated in the sensory analysis of the product developed were included in the matrix (Picutre 2). To compare the performance of the products developed for each required quality, we used nonparametric statistical test of Mann & Whitney (Siegel & Castellan, 2006) in order to distinguish the preferred.

Then you must define, through consensus among the development team, the column called planned quality, taking into account the degree of importance and the performance of the company and competitors. See the example in Picture 2.

The rate of improvement is established through the ratio between each value of planned quality by the performance of the product in the sensory analysis. Then it should be to establish which required qualities are considered strong, medium or weak selling points, i.e., which attributes will be attractive to the market, which items are attractive to the consumers in an average way and which attributes are obvious or mandatory to the product. In the column of the selling point, items considered to be attractive to the market receiving the note 1.5, intermediate items, the value 1.2 and those considered obvious, is assigned the value 1.0.

To calculate the absolute weight, multiply the degree of importance by the rate of improvement and also by the selling point. The relative weight of each required quality is the corresponding percentage of the absolute weight.

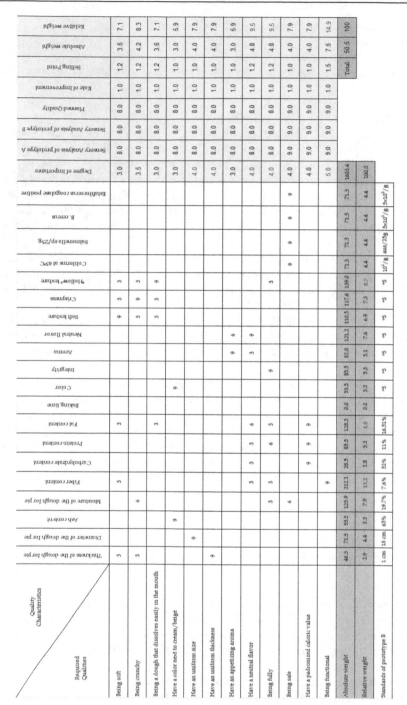

Fig. 2. Quality Matrix of a functional ready to bake dough for pies

4.2.6 Establishment of the designed quality

Now, it should be calculated the absolute weight of the quality characteristics at the bottom of the quality matrix. To do this, multiply the relative weight of each required quality by the numerical values of correlations, and add up these products vertically. The relative weight of each quality characteristic is the corresponding percentage of the absolute weight. For example, in Picutre 2, the absolute weight of 46.3 of the quality characteristic "dough thickness" was calculated by multiplying (7.1 x3) + (8.3 x 3) + (7.9 x9).

In the case of the functional dough for pies, were established physico-chemical, microbiological and sensory specifications for each quality characteristic of the finished product. The physico-chemical specifications were obtained by laboratory tests. The microbiological standards have been established according to the Brazilian law and the sensorial by a trained sensorial team.

5. Overview of the application of QFD in food products

QFD has been used in the food industry since 1987, i.e., its use is recent (Costa et al., 2001). But only after the 90's is that the articles were published showing the benefits of using the method in food products (Benner et al., 2003).

According to Souza Filho and Nantes (2004), the literature has dealt with much more organizational benefits and potential for improving the technical quality than the implementation process of QFD. Few articles describe how QFD has been used in real products and discuss their own experiences. Thus there isn't a deeper theoretical in the application of this method. To Charteris (1993) and Govers (1996), the strategic importance of QFD to contribute to the competitive advantage of firms may explain the reluctance of companies to share such important information on QFD.

Importantly, too, that one factor, perhaps one of the main, is the difficulty of finding scientific study of the application of QFD into the food industry, is due to the reliance on specific sensory attributes of each product. QFD was originally developed for the development of boats, automobiles and automobiles pieces. The technical characteristics related to this type of product are characteristics that have defined shapes and dimensions, specifications and parameters that can be measured accurately. However, the area of food has a different characteristic; the food may have different perceptions of consumer to consumer. Sensory perceptions are intrinsic to every human being and despite the food being technical specifications, mainly physico-chemical and microbiological parameters, the sensory perceptions are very difficult to specify. Allied to this, food ingredients have slightly larger deviations than pieces of heavy industry and those may change due to interactions between them or due to the process applied (Favaretto, 2007). Thus, the technicians of the food industry wishing to use the QFD tool must idealize the necessary changes in the method so that it becomes applicable in the development of a food product. It is important that simplifications are made to the product, its ingredients and their interactions so that the matrices can be used without any difficulty.

Cheng & Melo Filho (2010), in a survey on the contour and depth of application of QFD in the 500 largest Brazilian companies, came to the conclusion that is recent the introduction of the method in those companies and that is still a long way to go and a need for greater understanding and support of top management in the implementation of this method by companies.

Although the use of QFD is recent, some companies, however, begin to realize the advantages of their use and are already getting great results with the implementation of the method. Table 3 presents some applications of QFD in food product development in the last 12 years.

Authors	Applications of QFD	Results of the use of QFD
Antoni (1999)	Turkey dry fermented sausage	Easy understanding of the real needs to be addressed in the product. Interpretation of the first Quality Matrix as the voice of the customer divided into two parts: final consumer requirements and demands of the point of sale, unfolding them into the final product characteristics. Correlation between the raw material characteristics and the final product characteristics.
Viaene & Januszewska (1999)	Chocolate couverture	Approximation of the areas of Marketing and of Food Science and Technology, reduction of the final cost and increase of the success potential in launching the new product due to the participation of the consumers belonging to this target segment into the process.
Tumulero et al. (2000)	Salty crackers	Product improvement, sales increase and expansion of market share.
Marcos (2001); Marcos & Jorge (2002)	Table Tomato	Higher interaction of functional areas involved in the development process and establishment of sales strategies based on the analysis of the consumer market. Reduction of losses.
Chaves (2002)	Yogurt	Identification of the most important aspects from the standpoint of quality. Translation of the needs and desires of customers and structural changes to meet the expectations of a specific market.
Magalhães (2002)	Packaged pasteurized milk	Increase on the sales of the product
Stewart-Knox & Mitchell (2003)	Food products with reduced fat	Market identification, consumer awareness and participation of suppliers.

Table 3. Applications of QFD in food product development

Authors	Applications of QFD	Results of the use of QFD
Cortés & Da Silva (2005)	Yogurt	Translation of customer requirements and restructuring of the product according to market expectations. Commitment of all members of the company.
Gonçalves & Silva (2005)	Production of meals: integration of QFD x APPCC	Identification of the potential dangers of contamination in the generic process to produce meals. Verification of the phases of meals production that require greater attention from staff. The results of the items prioritized in the matrix served as an aid in the selection and training process of employees.
Waisarayutt & Tutiyapak (2006)	Instant rice noodles	Based on customer requirements, determination of the most important technical specifications of the product and the important parameters of the process. Development of control plans in accordance with the process parameters.
Delgado & Pedrozo (2007)	Micro food company in Peru, "Delicias del Sur".	Identification of customer needs and competitors' actions. Development of new food products based on differentiation through the incorporation of raw materials and inputs from local resources of functional character. Strengthening relationships in the areas of the organization.
Favaretto (2007)	Soft drink	Decrease in the time of product development, quality assurance of product and service offered to the customers. Culture change in the organization.
Ferreira & Miyaoka (2007)	Sweet Milk based on soy	Identification of real needs and desires of the audience. Serving an attractive niche market, who are lactose intolerant, vegetarians and people seeking a healthier diet. Improved quality and cost reduction.

Table 3. (continued): Applications of QFD in food product development

Authors	Applications of QFD	Results of the use of QFD
Matsunaga (2007)	Breaded chicken	Identification of the attributes most valued by consumers. Higher interaction between research and product development area and marketing area.
Miguel et al., (2007)	Consumer profile of pineapple "Pérola"	Identification of critical quality attributes at the purchase time, establishment of the degree of importance of each attribute required by the market and interpretation of sensory analysis. Identification of points to be improved within the supply chain, minimizing losses and maintaining and improving the quality of the final product.
Anzanello et al., (2009)	Christmas turkey	Identification of the process parameters, product and resources to be prioritized in the adoption of improvements. Identification of resources seen as bottlenecks to meet the demands prioritized in QFD. Global view while driving improvements in products.
Cheng & Melo Filho (2010)	Noble products with embedded (ready dishes like ham lasagna)	Anticipating the possible changes of habits and attitudes, adding value to the product.
Garcia (2010)	Fluid milk	Disclosure of the characteristics to be worked in order to improve product quality.
Kawai (2010)	Table tomato	Identification of the required characteristics by consumers and the establishment of a competitor that has the largest market share. Reduce losses and increase the income of the producer. Improved communication between sectors.
Pinto & Paiva (2010)	Functional dough ready for pies	Identification of the characteristics that consumers attach as of greater importance and prioritization those into the product development.

Table 3. (continued): Applications of QFD in food product development

Authors	Applications of QFD	Results of the use of QFD
Rodrigues (2010)	Improving food services at the Universidade Estadual de Campinas (Unicamp)	Identification of the quality characteristics most important and which should be tailored to the improvement of food service establishments of Unicamp. Improved workflow and execution of solutions more quickly and affordably.
Vatthanakul et al., (2010)	Product of gold kiwifruit leather	Defining the importance of the various sensory characteristics for the design of new products.
Zarei et al., (2011)	Canning industry	Identification of enablers of Lean Manufacturing viable to be implemented in practice in order to increase the lean production of food supply chain.

Table 3. (continued): Applications of QFD in food product development

Based on the work presented, it is emphasized that QFD enables the company to develop products that meet the diverse and growing demands of its customers, serving different market niches, in a short time, and is characterized by the efficiency in storage and transmission of information during the multifunctional product development activity. In addition, the use of the method QFD in several companies have pointed out others advantages as: resolving problems, reducing the development time and using of prototypes in a more objective way. The QFD is effective in order to direct, in a practice way, how to plan and conduct activities necessary to the process of the product development.

6. Sensory analysis and support tools to market research

The application of QFD in the food industry is complex and the literature does not allow establishing its full application in the development of these products. For this reason, adaptations that take into account the characteristics of the product "food" are necessary for the successful implementation of QFD in this industry. The adjustments should pursue a better integration of aspects of market research with the sensory evaluation of food, aiming to reveal the perception of consumers in a jointful way, in relation to the general market attributes (shape, size, ease of use, etc.) and to the sensory attributes that make them decide to accept and purchase of food. The integration of these aspects should also consider the current model of food consumption in the region and the market segment defined for the product.

In developing a new product is essential to optimize parameters such as shape, color, appearance, odor, flavor, texture, consistency and the interaction of different components, in order to achieve a full balance that translates into an excellent quality and have good acceptability. Thus, according to Monteleone and colleagues (1997), studying the relationships between the sensory attributes and the acceptance of a product can be very useful to formulate or improve a product, as well as to evaluate potential market opportunities. Understanding what drives the acceptance of a product is very important in light of a competitive market like nowadays. Meeting the most valued attributes makes more oriented the product development, benefiting not only the area of research and product development, as well as marketing, reducing distances between these two areas. In this sense, the methods of sensory evaluation of foods and evaluation of consumer desires can be an aid when applied the method QFD into the food product development.

In sensory analysis the differences between products (discriminatory tests), the intensity of a sensory attribute of quality (descriptive tests), or the degree of acceptance, preference or rejection for a product (affective tests) are measured by human senses. However, it is necessary to consider that the sensory perceptions can not be measured directly. Therefore to assess the individual stimulation received in the sensory evaluation is used scales that allow the quantification of them, as the specific objective of the evaluation (Bech et al. 1994; Chaves & Sproesser, 1996; Ferreira et al. 2000).

In acceptance tests, the hedonic scale is the most used and widespread. It is a structured scale, with nine sentences, balanced, considered easy to use and to understand (says the pleasant and unpleasant states in the body). The evaluation of the hedonic scale is converted into numerical scores, and statistically analyzed to determine the difference in the degree of preference between samples (ABNT, 1998, IFT, 1981; Land & Shepherd, 1988). Optionally you can use the hybrid hedonic scale, a continuous scale of 10cm, and with the advantages to meet the assumptions of the models of analysis of variance (ANOVA) as the normality of the residuals and homoscedasticity, and increase the discrimination between the samples.

The traditional methodologies for analyzing the affective tests data, like the ANOVA, used to compare more than two averages in the study, and the averages test, to determine the significance to a particular level of confidence, have shown limitations and shortcomings. For each product evaluated is obtained the average of the consumer's group, assuming therefore that all respondents have the same behavior, ignoring their individuality. Thus, there may be occurring loss of important information about different market segments (Meilgaard et al. 2006; Polignano et al., 1999, Reis & Minim, 2006).

In order to analyze the affective data, taking into account the individual response of each consumer and not just the average of the consumer group that evaluated the products, it was developed a technique called Preference Mapping, which has been widely used by scientists of the area of sensory analysis (Behrens et al., 1999). This technique is often employed to identify groups of consumers who respond uniformly and that differ from other groups by age, sex, attitude, need, eating habits and/or responses to the product's attributes. This gives the opportunity to interpretate the different areas of action of the market (Polignano, 2000, Westad et al., 2004).

As it can explain better the consumer preference for each attribute, the Preference Mapping is an interesting tool to collect the real needs of customers and turn them into project qualities. It can be very useful during the development of food products, particularly in the construction phase of the Designed Quality of the Quality Matrix (Polignano, 2000).

Cluster analysis is a set of statistical techniques whose objective is to seek a classification according to the natural relations that the sample shows, forming groups of objects (individuals, companies, cities or other experimental unit) according to the similarity in relation to some predetermined criteria, thus reducing the dimensionality of the data. According to Hair et al. (1998), the groupings (or clusters) resultants should have a high internal homogeneity (within groups) and high external heterogeneity (between groups).

Cluster analysis is useful in developing new products to meet consumer profiles. This can identify the profile of each group (age, marital status, psychological characteristics, etc.) that will define whether there are different demands (market segmentation). The profile is defined by the characteristics that make up the cluster, based on the concept of similarity.

In addition to these approaches, there is the publication of articles dealing with the use of Conjoint Analysis in the food industry. The Conjoint Analysis is a technique that allow to study the combined effect exerted by two or more independent variables on a dependent variable (Carneiro et al., 2003). Based on a decomposition analysis, the Conjoint Analysis determines the contribution of the studied factor levels expressed in samples or combinations of the consumer response (acceptance, preference or purchase intent). It is an analysis technique that can be used to identify the attributes/levels that most influence the selection, purchase and acceptance of products, after the sensory evaluation of them.

The Conjoint Analysis has been widely applied in the development of new products in all industrial sectors, in particular in the establishment of concepts, competitive analysis, selection of target market segment, the definition of price and advertising strategy (Drummond, 1998).

Another tool that has been used is the Repertory Grid. Kelly presented it in 1955, which advocates the theory that people act as scientists evaluating the world around them, creating hypotheses and establishing descriptors of what it is seen. The Repertory Grid method is a term used to describe a set of techniques related to the Kelly's theory (1955), which can be used to investigate the individual definitions in the perception of characteristics of his surroundings. This method is very flexible and allows it to be applied according to the researcher's interest (Mac Fie & Thomson, 1994).

A descriptor is defined as the way in which two items are similar and, in some way, different from a third party. In the sensory field, the samples are arranged in triads. In each set, two samples are kept together and one away from the taster and it describes how the samples are alike and how different. In addition to identifying the differences, the tasters should also describe the extremes of each descriptor raised, in order to build a scale so that samples can be quantified. The data collected are analyzed by GPA (Generalized Procrustes Analysis), which allow to be established a configuration with the main and common descriptors to the tasters. This is a multivariate analysis, that establish the consensus map of the data (Mac Fie & Thomson, 1994).

Another method that helps to identify the product perception by the consumer is called Focus Group. Focus Group can be defined as a planned session to obtain individual perceptions of such a product or service, in a peaceful environment, through the moderation of groups formed by six to nine people (Macfie & Thomson, 1994) or eight to twelve people (Fuller , 2011). It is a qualitative technique of group discussion, which allows interaction between people. The moderator leads the discussion to the topics of interest, listening to people, without interfering directly. This technique allows to raise the participants' perceptions about the subject matter. According to Fuller (2011), the main function of the Focus Group is to determine consumer reaction to the objects of study.

The Kano method has as it's main goal to evaluate the influence of components of products in consumer satisfaction (Sauerwein et al., 1996). This method aims to rank the attributes in four characteristics groups according to the degree of care and satisfaction: indifferent (characteristics that do not affect consumer satisfaction), expected (mandatory characteristics), proportional (characteristics that customer satisfaction is proportional to the degree of care) and attractive (characteristics that customer satisfaction does not diminish if not offered, but increases if met) (Fonseca, 2002).

Using this model it is possible to improve the process of understanding the requirements from a classification that can help prioritize development resources. The identification of mandatory requirements for certain attributes and opportunities of future innovation, which customers have not even being classified as needs now, allowing the development and the use of criteria in a more efficient way for resource allocation. The process of translating requirements into product characteristics or services should also benefit from the information obtained using the Kano model (Guimarães, 2003).

Kano helps in the process of product development, prioritizing the characteristics that really have an impact on customer satisfaction. Beyond that, this may be very well combined with QFD, identifying the relative importance of the needs raised by consumers, contributing to the product development more focused and better targeted to the audience. The classification of characteristics can also be a guideline to define the attributes to be worked out for different market segments, creating product differentiation in that market.

7. Conclusion

The effective development of products has become the competitive advantage for many companies, especially in the food industry. For that, as shown in this chapter, projects must achieve the best levels of quality, of efficiency and speed in the elaboration of products. Certainly this requires an organizational effort of the entire company. What the firm plans, i.e., its development strategy, and how the company makes its planning – its development management - will determine the expressiveness of the product in the market. In this sense, it can be affirmed that the competitive advantage of these organizations is based on the capacity of its technical staff, in the procedures and organizational structure, in the strategies established to guide the process, in the methods used, in the way that the top management interacts with the process and yet, in the organization of the team according to the level of complexity of the project.

The QFD method as pointed out in this work is not a mechanism that addresses all of the aspects above, important for structuring the system of product development. However, by offering a way to treat the necessary information to the process and to plan the activities, the method has boosted the development system in several companies.

Ultimately, it is worth noting that, specifically for the food industry, some aspects can enhance the use of the method. Among them those can be mention:

Existence of a support infrastructure to the process, with sufficiently equipped laboratories and trained personnel to conduct the analysis.

The intrinsic quality of food products can be evaluated by physico-chemical, microbiological and sensory analyses. Thus, the values of the quality characteristics during construction of the matrices are determined by such methods. Various techniques of sensory analysis assist in analyzing and gathering market information and then building a quality matrix, as reported in this chapter.

Likewise, the pilot plants and experimental kitchens within companies, contribute greatly to the development of prototypes and conduction of activities in a more quickly way.

Knowledge and application of various statistical techniques in the process of product development.

Such techniques facilitate the identification of opportunities and strategic positioning of products, the identification of factors that affect customer preference and the market segmentation, supporting the use of the QFD method.

Assurance system of the consolidated quality

The consolidated quality management benefits directly the development process, as it aims to optimize the exchange of information, the commitment of the people involved in the activities, the specification of process parameters and the standardization of this procedures, as well, developing products with little variability. Certainly, these factors create a safer environment for the conduct of projects, facilitating, in particular, the specification of product technical parameters, of raw materials, packaging and of the manufacturing process.

8. References

ABNT (1998). Associação Brasileira de Normas Técnicas. *NBR 14141*: escalas utilizadas em análise sensorial de alimentos e bebidas. Rio de Janeiro, 1998.

Akao, Y. & Mazur, G. (2003).The leading edge in QFD: past, present and future. *International Journal of Quality & Reliability Management*. Vol. 20, No.1, pp. 20-35, ISSN 0265-671X.

Akao, Y. & Ohfuji, T. (1989). Recent aspects of Quality Function Deployment in service industries in Japan. *Proceedings of the International Conference on Quality Control*, ISBN 0-915299-41-0 3, Rio de Janeiro, 1989.

Antoni, I. (1999). *Desenvolvimento de um Embutido Fermentado de Carne de Peru pelo Método do Desdobramento da Qualidade*. 136 p. Dissertação (Mestrado em Tecnologia de Alimentos). Faculdade de Engenharia de Alimentos, Universidade Estadual de Campinas (Unicamp), Campinas.

Anzanello, M.; Lemos, F. & Encheveste, M. (2009). Aprimorando Produtos Orientados ao Consumidor Utilizando Desdobramento da Função Qualidade (QFD) e Previsão de Demanda. *Produto & Produção*, Vol. 10, No. 2, jun., 2009, pp. 01 – 27, ISSN 1516-3660.

Athayde, A. (1999). Indústrias agregam conveniências aos novos produtos. *Engenharia de Alimentos*, São Paulo, Vol. 24, 1999, pp. 39-41.

Bech, A.; Engelund, E.; Juhl, J.; Kristensen, K. & Poulsen, C. (1994). *QFood – optimal design of food products*. MAPP working Paper no.19. MAPP Centre, Aarhus. March, 1994; pp. 2-12; ISSN 09072101.

Behrens, J.; Silva, M. & Wakeling, I. (1999). Avaliação da aceitação de vinhos brancos varietais brasileiros através de testes sensoriais afetivos e técnica multivariada de mapa de preferência interno. *Revista da Sociedade Brasileira de Ciência e Tecnologia de Alimentos*, Campinas, Vol. 19, No. 2, May, 1999, ISSN 0101-2061.

Benner, M.; Linnemann, A.; Jongen, W. & Folstar, P. (2003). Quality function deployment (QFD) – can it be used to develop food products?, *Food Quality and Preference*, Netherlands, Vol. 14, pp. 327-339, ISSN 0950-3293.

Carneiro, J.; Silva, C.; Minim, V.; Regazzi, A.;Deliza, R. & Suda, I. (2003). Princípios básicos da Conjoint Analysis em estudos do consumidor. *Revista da Sociedade Brasileira de Ciência e Tecnologia de Alimentos*, Campinas, Vol. 37, No. supl, 2003, pp. 107-114, ISSN 0101-2061.

Chan, L. & Wu, M. (2002). Quality function deployment: a literature review. *European Journal of Operational Research, Vol.* 143, 2002, pp. 463-497, ISSN 0377-2217.

Charteris, W. (1993). Quality function deployment: a quality engineering technology for the food industry. *Journal of the Society of Dairy Technology*, Vol.46, No 1, 1993, February, 1993, pp. 12–21, ISSN 0037-9840.

Chaves, J. & Sproesser, R. (1996) *Práticas de laboratório de análise sensorial de alimentos e bebidas*. Publisher: UFV, Viçosa, Brazil.

Chaves, O. (2002). *Aplicação do método de desdobramento da função qualidade na industrialização do leite de consumo em Minas Gerais*. 86 p. Dissertação (Mestrado em Economia Rural). Departamento de Economia Rural, Universidade Federal de Viçosa, Viçosa.

Cheng, L. & Melo Filho, L. (2010). *QFD: desdobramento da função qualidade na gestão de desenvolvimento de produtos*. (2 ed). *Publisher:* Blucher, ISBN 9788521205418, São Paulo.

Clark, K. & Wheelwright. S. (1993). *Managing new product and process development*. Free Press, ISBN 0-02-905517-2, New York.

Cooper, R. (2001). *Winning at new products. Accelerating the process from idea to lanch*. (3ªed), Publisher: Basic Books. Cambridge, Massachusetts. ISBN 0738204633.

Cortés, D. & Da Silva, C. (2005). Revisão: Desdobramento da Função Qualidade – QFD – conceitos e aplicações na indústria de alimentos. *Brazilian Journal Food Technology*, Vol. 8, No. 3, 2005, pp. 200-209, ISSN 1517-7645.

Costa, A.; Dekker, M. & Jongen, W. (2001). Quality function deployment in the food industry: a review. *Trends in Food Science and Technology*, Vol.11, No. 9–10, 2001, pp. 306–314, ISSN 0924-2244.

Delgado, G. & Pedrozo, E. (2007). Inovação de Produtos Alimentícios: Alimentos funcionais a partir de produtos locais. *Proceedings of IV Convibra: Brazilian Administration Virtual Congress*, São Paulo/SP. Brazil, dez., 2007.

Drumond, F. (1998). *Ténicas Estatísticas para o Planejamento do Produto*. Fundação Christiano Ottoni, Belo Horizonte.

Earle, M. (1997). Changes in the food product development process. *Trends in Food Science & Technology*, Vol. 8, 1997, pp. 19-24, ISSN 0924-2244.

Favaretto, R. (2007). *Modelo de aplicação de QFD no desenvolvimento de Bebidas*. 96 p. Dissertação (Mestrado em Profissional em Engenharia Mecânica) – Faculdade de Engenharia Mecânica, Universidade Estadual de Campinas (Unicamp), Campinas.

Ferreira, G. & Miyaoka, A. (2007). *Estratégias de desenvolvimento do doce de leite à base de soja*. 30 p. Monografia (Graduação no Curso de Engenharia de Produção). Engenharia de Produção da Universidade Federal de Viçosa, Viçosa

Ferreira, V.; Almeida, T.; Pettinelli, M.; Silva, M.; Chaves, J. & Barbosa, E. (2000). *Análise sensorial: testes discriminativos e afetivos*. Sociedade Brasileira de Ciência e Tecnologia de Alimentos, Campinas.

Fonseca, M. (2002). *Uma abordagem para a redução de custos no desenvolvimento de produtos alimentícios*. 82 p. Tese (Mestrado em Engenharia de Produção). Faculdade de Engenharia de Produção, Universidade Federal do Rio de Janeiro, Rio de Janeiro.

Fuller, G. (2011). *New food product development: from concept to marketplace*. (3ed.), CRC Press, ISBN 13:978-1-4398-1865-7, Florida.

Garcia, A. (2010). *Uso do método DFQ (Desdobramento da Função Qualidade) para melhoria da qualidade do leite fluido*. 181 p. Tese (Doutorado em Tecnologia de Alimentos). Faculdade de Engenharia de Alimentos, Universidade Estadual de Campinas, Campinas.

Gonçalves, T. & Silva, C. (2005). Proposta de utilização do quality function deployment (QFD) no sistema de análise de pontos críticos de controle (APPCC) na produção de refeições. *Proceedings of XII SIMPEP*, Bauru, SP, Brazil, Nov., 2005.

Govers, C. (1996). What and how about quality function deployment (QFD). *International Journal of Production Economics*, New York, Vol.46-47, 1996, pp. 575-585, ISSN 0925-5273.

Guimarães, L. (2003). QFD - Analisando seus aspectos culturais organizacionais. *Revista Qualidade*, Vol., No. 128, Jan., 2003, pp.56-66.

Hair J.; Anderson, R.; Tatham, R. & Black, W (1998). Cluster analysis. In: *Multivariate data analysis*. Hair, J, Black, B, Babin, B & Anderson, R (5.ed.), pp. 469-518, Prentice Hall, Upper Saddle River.

IFT. (1981). Sensory Evaluation Division. Guidelines for the preparation and review of paper reporting sensory evaluation date. *Journal of Food Technology*, Vol.35, No.4, 1981, pp.16-17, ISSN 0022-1163.

Kawai, S. (2010). *Desenvolvimento de Tomate de Mesa, com o uso do método QFD (Quality Function Deployment), comercializado em um supermercado.* 217 p. Tese (Doutorado em Tecnologia Pós-Colheita). Faculdade de Engenharia Agrícola. Universidade Estadual de Campinas, Campinas.

Macfie, H. & Thomson, D. (1994). *Measurement of food preferences.* (1ed.), Springer, ISBN 9780834216792, New York.

Magalhães, G. (2002). *Incorporação da Qualidade Desejada pelos Consumidores ao Leite Pasteurizado Utilizando o Desdobramento da Função Qualidade.* 77 p. Dissertação (Mestrado em Ciência e Tecnologia de Alimentos). Programa de Pós-Graduação em Ciência e Tecnologia de Alimentos. Universidade Federal de Viçosa, Viçosa.

Marcos, S. & Jorge, J. (2002). Desenvolvimento de tomate de mesa, com o uso do método QFD (Desdobramento da Função Qualidade), comercializado em um supermercado. *Horticultura Brasileira*, Brasília, v. 20, n. 3, p. 490-496, setembro 2002.

Marcos, S. (2001) *Desenvolvimento de tomate de mesa, com o uso do método QFD (Quality Function Deployment), comercializado em um supermercado.* 199 f. Tese (Doutorado em Ciência de Alimentos). Faculdade de Ciência de Alimentos. Universidade Estadual de Campinas, Campinas.

Matsunaga, P. (2007). *Identificação de atributos sensoriais de pedaços empanados de frango mais valorizados pelo consumidor.* 121p. Dissertação (Mestre em Alimentos e Nutrição). Faculdade de Engenharia de Alimentos. Universidade Estadual de Campinas, Campinas.

Meilgaard, M.; Civille, G. & Carr, B. (2006). *Sensory Evaluation Techniques*, (4. ed.), FL: CRC Press, ISBN 0849338395, Boca Raton.

Melo Filho, L. & Cheng, L. (2007). QFD na garantia da qualidade do produto durante seu desenvolvimento: caso em uma empresa de materiais. *Produção*, Vol. 17, No.3,Dec., 2007, . p. 604-624, ISSN 0103-6513.

Miguel, A.; Spoto, M.; Abrahão, C. & Silva P. (2007) Aplicação do método QFD na avaliação do perfil do consumidor de abacaxi "Pérola". *Ciência e Agrotecnologia*, Vol. 31, No. 2, 2007, pp. 563-569, ISSN 1413-7054.

Mizuno, S. (1969). Company-wide quality control activities in Japan. *Reports of Statistical Application Research*, Vol.16, No.3, 1969, pp.68-77.

Monteleone, E.; Carlucci, A.; Caporale, G. & Pagliarini, E. (1997). Use of slope analysis to characterize preference for virgin olive oil. *Italian Journal of Food Service*, Vol. 9, No. 2, 1997, pp. 133-140, ISSN 1120-1770.

Pinto, A. & Paiva, C. (2010). Developing a functional ready to bake dough for pies using the Quality Function Deployment (QFD) method. *Revista da Sociedade Brasileira de*

Ciência e Tecnologia de Alimentos, Campinas, Vol.30, No.(Supl.1), mai., 2010, pp 36-43, ISSN 0101-2061.

Polignano, L. (2000). *Desenvolvimento de produtos alimentícios: implentação da ferramenta mapa de refrência e estudo da sua articulação com a matriz da qualidade*. 268 p. Dissertação (Mestre em Engenharia de Produção). Escola de Engenharia. Universidade Feredal de Minas Gerias, Minas Gerais.

Polignano, L.; Cheng, L. & Drumond, F. (1999). Utilização dos mapas de preferência como técnicas auxiliares do QFD durante o desenvolvimento de produtos alimentícios. *Proceedings of I Brazilian Conference on Management of Product Development*, Belo Horizonte, Brazil, 1999.

Reis, C. & Minim, V. (2006) *Testes de aceitação*. In: Análise sensorial: estudos com consumidores., Minim V., pp. 67-83, Publisher UFV, ISBN 85-7269-282-7, Viçosa, Brazil.

Rodrigues, N. (2010). *Aplicação da matriz da qualidade do QFD – Desdobramento Da Função Qualidade – para avaliar serviços de alimentação do campus da Unicamp*. 190 p. Tese (Doutorado em Tecnologia Pós-Colheita). Faculdade de Engenharia Agrícola. Universidade Estadual de Campinas, Campinas.

Sauerwein, E.; Bailom, F.; Matzler, K. & Hinterhuber, H. (1996). The Kano Model: how to delight your consumers. *Preprints of the IX International Working Seminar on Production Economics*, Innsbruck/Igls/Austria, February, 1996.

Siegel, S. & Castellan Jr, N. (2006). *Estatística não-paramétrica para ciências do comportamento*. (2 ed), Artmed, ISBN 85-363-0729-3, Porto Alegre.

Souza Filho, M. & Nantes, J. (2004). O QFD e a análise sensorial no desenvolvimento de produtos. *Proceedings of Production Engineering Symposium*, Bauru, Nov., 2004.

Stewart-Knox, B. & Mitchell, P. (2003). What separates the winners from the losers in new food product development? *Trends in Food Science and Technology*, Vol. 14, No.1–2, 2003, pp. 58–64, ISSN 0924-2244.

Tumelero, N.; Ribeiro, J. & Danilevicz, A. (2000). O QFD como ferramenta de priorização para o planejamento da qualidade. *Proceedings of Brazilian Conference on Management of Product Development*, São Carlos, Aug., 2000.

Vatthanakul, S.; Jangchud, A.; Jangchud, K.; Therdthai, N. & Wilkinson, B. (2010). Gold kiwifruit leather product development using quality function deployment approach. *Food Quality and Preference*, Vol.21, No.3, 2010, pp. 339–345, ISSN 0950-3293.

Viaene, J. & Januszewska, R. (1999). Quality function deployment in the chocolate industry. *Food Quality & Preference*, Vol. 10, pp. 377–385. ISSN, 0950-3293.

Waisarayutt, C. & Tutiyap, O. (2006). Application of Quality Function Deployment in Instant Rice Noodle Product Development. *The Kasetsart Journal: Natural Sciences*. Vol.40 (Suppl.) 162-171 p.

Westad, F.; Hersleth, M. & Lea, P. (2004) Strategies for consumer segmentation with applications on preference data. *Food Quality and Preference*, Vol. 15, No. 7-8 , 2004, pp.681-687, ISSN 0950-3293.

Zarei, M.; Fakhrzad M. & Paghaleh, M. J. (2011). Food supply chain leanness using a developed QFD model. *Journal of Food Engineering*, Vol. 102, pp. 25–33, ISSN 0260-8774.

Permissions

The contributors of this book come from diverse backgrounds, making this book a truly international effort. This book will bring forth new frontiers with its revolutionizing research information and detailed analysis of the nascent developments around the world.

We would like to thank Benjamin Valdez, Michael Schorr and Roumen Zlatev, for lending their expertise to make the book truly unique. They have played a crucial role in the development of this book. Without their invaluable contribution this book wouldn't have been possible. They have made vital efforts to compile up to date information on the varied aspects of this subject to make this book a valuable addition to the collection of many professionals and students.

This book was conceptualized with the vision of imparting up-to-date information and advanced data in this field. To ensure the same, a matchless editorial board was set up. Every individual on the board went through rigorous rounds of assessment to prove their worth. After which they invested a large part of their time researching and compiling the most relevant data for our readers. Conferences and sessions were held from time to time between the editorial board and the contributing authors to present the data in the most comprehensible form. The editorial team has worked tirelessly to provide valuable and valid information to help people across the globe.

Every chapter published in this book has been scrutinized by our experts. Their significance has been extensively debated. The topics covered herein carry significant findings which will fuel the growth of the discipline. They may even be implemented as practical applications or may be referred to as a beginning point for another development. Chapters in this book were first published by InTech; hereby published with permission under the Creative Commons Attribution License or equivalent.

The editorial board has been involved in producing this book since its inception. They have spent rigorous hours researching and exploring the diverse topics which have resulted in the successful publishing of this book. They have passed on their knowledge of decades through this book. To expedite this challenging task, the publisher supported the team at every step. A small team of assistant editors was also appointed to further simplify the editing procedure and attain best results for the readers.

Our editorial team has been hand-picked from every corner of the world. Their multi-ethnicity adds dynamic inputs to the discussions which result in innovative outcomes. These outcomes are then further discussed with the researchers and contributors who give their valuable feedback and opinion regarding the same. The feedback is then collaborated with the researches and they are edited in a comprehensive manner to aid the understanding of the subject.

Apart from the editorial board, the designing team has also invested a significant amount of their time in understanding the subject and creating the most relevant covers. They scrutinized every image to scout for the most suitable representation of the subject and create an appropriate cover for the book.

The publishing team has been involved in this book since its early stages. They were actively engaged in every process, be it collecting the data, connecting with the contributors or procuring relevant information. The team has been an ardent support to the editorial, designing and production team. Their endless efforts to recruit the best for this project, has resulted in the accomplishment of this book. They are a veteran in the field of academics and their pool of knowledge is as vast as their experience in printing. Their expertise and guidance has proved useful at every step. Their uncompromising quality standards have made this book an exceptional effort. Their encouragement from time to time has been an inspiration for everyone.

The publisher and the editorial board hope that this book will prove to be a valuable piece of knowledge for researchers, students, practitioners and scholars across the globe.

List of Contributors

Joana Tulha, Joana Carvalho, Rui Armada, Fábio Faria-Oliveira, Cândida Lucas, Célia Pais, Judite Almeida and Célia Ferreira
University of Minho/CBMA (Centre of Molecular and Environmental Biology), Portugal

Saulat Jahan
Research and Information Unit, Primary Health Care Administration, Qassim, Ministry of Health, Kingdom of Saudi Arabia

Emeje Martins Ochubiojo
National Institute for Pharmaceutical Research and Development, Nigeria

Asha Rodrigues
Physical and Materials Chemistry Division, National Chemical Laboratory, India

José C.E. Serrano, Anna Cassanyé and Manuel Portero-Otin
Department of Experimental Medicine, University of Lleida, Spain

Rommy N. Zúñiga
Center for Research and Development CIEN Austral, Puerto Montt, Chile

Elizabeth Troncoso
Pontificia Universidad Católica de Chile, Santiago, Chile

Ine M. Salazar-Vega, Maira R. Segura-Campos, Luis A. Chel-Guerrero and David A. Betancur-Ancona
Facultad de Ingeniería Química, Campus de Ciencias Exactas e Ingenierías, Universidad Autónoma de Yucatán, Yucatán, México

Heidi Riedel, Nay Min Min Thaw Saw, Divine N. Akumo, Onur Kütük and Iryna Smetanska
Technical University Berlin, Department of Food Technology and Food Chemistry, Methods of Food Biotechnology, Germany

Ihegwuagu Nnemeka Edith
Agricultural Research Council of Nigeria, Nigeria

Emeje Martins Ochubiojo
National Institute for Pharmaceutical Research and Development, Nigeria

Marco Bravi and Agnese Cicci
Chimica Materiali Ambiente, Sapienza Università di Roma, Roma, Italy

Giuseppe Torzillo
CNR Istituto per lo Studio degli Ecosistemi, Sesto Fiorentino, Italy

Caroline Liboreiro Paiva and Ana Luisa Daibert Pinto
University Federal of Minas Gerais, Brazil